The Geometry of Kerr Black Holes

The Geometry of Kerr Black Holes

Barrett O'Neill

A K Peters
Wellesley, Massachusetts

Editorial, Sales, and Customer Service Office

A K Peters, Ltd.
289 Linden Street
Wellesley, MA 02181

Library of Congress Cataloging-in-Publication Data

O'Neill, Barrett.
    The geometry of Kerr black holes / Barrett O'Neill.
        p. cm.
    Includes bibliographical references and index.
    ISBN 1-56881-019-9
    1. Kerr black holes--mathematics. 2. Geometry. I. Title.
QB843.B55054   1995b
523.8'875--dc20   93-21643
                                                                  CIP

Typesetting: Eigentype Compositors
Printing and binding: Hamilton Printing Company

Printed in the United States of America
99 98 97 96 95   10 9 8 7 6 5 4 3 2 1

For HJE

Contents

Preface

This book is an account of the global geometry of Kerr spacetime, the relativistic model of the gravitational field of a rotating central mass. Actually, the Kerr exact solution is a family of spacetimes depending on parameters M (mass) and a (angular momentum per unit mass). Schwarzschild spacetime, where there is no rotation, is the limiting case $a = 0$, and comparisons with the Schwarzschild case recur throughout the book.

Only three of the five chapters deal exclusively with Kerr spacetime. Chapter 1 supplies general background material, and Chapter 5 is an exposition of Petrov types and null congruences, but with the Kerr case used throughout as the main example. Its purpose also is to give some perspective on the place of Kerr spacetime within all spacetimes. Chapter 2 establishes the basic geometry of the Kerr metric; Chapter 3 constructs maximal analytic extensions and examines their global structure; and Chapter 4 (the longest) is a detailed study of Kerr geodesics. The introduction to each chapter gives a fuller summary of its contents.

As the title of the book indicates, it is Kerr geometry that is studied—curvature, geodesics, isometries, totally geodesic submanifolds, topological structure, and so on. Of course, the physical interpretations of these geometrical invariants are emphasized, but there is no attempt to "do physics." Here Kerr spacetime is viewed only as the arena in which physics and related disciplines operate—from observational astronomy to quantum gravity theory.

However, as the title may suggest, we are interested not just in the exterior of a rotating star but in the entire extent of maximally extended (uncharged) Kerr black holes. The two main Kerr subfamilies, slowly rotating ($a^2 <$ M^2) and rapidly rotating ($a^2 >$ M^2), are separated by the extreme case ($a^2 =$ M^2). The fast case is the least interesting, both physically and geometrically, and is largely subsumed in the other two. The extreme case, though exceptional, deserves attention at least as a stepping stone to the more intricate slowly rotating case $a^2 <$ M^2. Here the maximal Kerr black hole is a vast, symmetrical spacetime, whose construction pattern is suggested in Figure 3.10 of Chapter 3.

Mathematicians interested in Lorentz geometry will find many of its novelties— particularly its differences from positive-definite Riemannian geometry—illustrated by Kerr spacetime, for example, (1) causal character, the separation of vectors, geodesics, etc., into spacelike, timelike, or null types; (2) the construction of maximal analytic extensions of metrics initially known only locally; and (3) the existence of "singularities," which the theorems of Penrose and Hawking show to be natural, not pathological, in Lorentz geometry.

To make the book accessible to students and researchers with varied interests, prerequisites have been kept to a minimum. In particular, this book is definitely not a sequel to my book *Semi-Riemannian Geometry* (Academic Press, 1983).

A major obstacle in the vast literature on Kerr spacetime is the great variety of notational systems or formalisms that are used. In spite of temptations, there are only three in this book: elementary tensor calculus, differential forms, and (in Chapter 5) the Newman–Penrose formalism. The specifically Kerr notation is also quite conventional.

To my best knowledge, many results in the book are new, in particular, the determination of the isometry groups (Sections 3.7 and 3.8 in Chapter 3) and topological structure (Section 3.9 in Chapter 3), and much of the analysis of geodesic orbits (Sections 4.7 through 4.10 in Chapter 4). The latter topic requires some comment. The ultimate goal is to classify all Kerr geodesics and to describe their global trajectories in the maximally extended spacetime—not just in the equatorial plane or Kerr exterior. However, when the number of cases becomes oppressive, I have chosen to specialize to *timelike* geodesics, taking the view that null geodesics are in effect a simpler special case. On the other hand, the (unphysical) spacelike geodesics would be more difficult, as evidenced by their behavior in various submanifolds.

The book is organized as follows. Chapters are divided into sections, with the sixth section of Chapter 3, for example, numbered as 3.6. Within a given section a single sequence of numbers designates collectively the theorems, lemmas, remarks, etc. (but not the figures). Thus Corollary 3.6.4 is the fourth such item in

Section 3.6, not the fourth corollary. References to the bibliography are given in the form "author, date." A few special topics are reviewed in appendixes at the back of the book, where there is also an index of commonly used notations.

The basic computations for the book were all done by hand, but difficult cases were checked using the computer programs *Mathematica* (Wolfram Research, Inc., Champaign, Illinois) or *Maple* (Waterloo Maple Software, Waterloo, Ontario, Canada). Differential form computations were all checked on *Maple*. *Mathematica* was used extensively for numerical computation and the preparation of figures.

Many authors I have learned from are named in the bibliography, but a list of other people who have helped me might be almost as long—my students in several courses and my teachers, recent and past, formal and informal. To all I give my thanks.

Introduction

Understanding our solar system has been the most important single impetus to the development of modern science. Newton's comprehensive description in *Principia Mathematica*—founded on his gravitational law and laws of motion—absorbed the work of Kepler, Copernicus, Galileo, and many earlier theorists. Two centuries later the profound influence of Newton's ideas, in science and beyond, made it difficult to accept the growing evidence that a flaw existed in the Newtonian model of the solar system. The flaw was small, but it persisted. After the most careful estimates of the perturbations produced by other planets, the predicted position of the perihelion of Mercury was still falling behind its observed position at the rate of about 43 seconds of arc per century.

1915. Where Newton had used the motion of the moon around the earth as a guiding example in his work, Einstein used this deviation of Mercury. The central ideas of general relativity were already in place; at issue was the exact form of the curvature term G in the Einstein equation $G = kT$. A precise relativistic model of the sun's gravitational field was not needed; Einstein used a simple polynomial approximation. Late in this year he suceeded, and the 43 second lag was eliminated.

1916. A few weeks later, Einstein, working in Berlin, received a paper from Karl Schwarzschild, an astronomer who, though no longer young, was serving in the German army in Russia. Hospitalized by an illness that soon proved mortal,

Schwarzschild had time to discover the desired precise relativistic model, and *Schwarzschild spacetime* replaced the Newtonian model as the best description of the gravitational field of an isolated, spherically symmetric star. But only a few theorists were familiar with relativity, and significant experimental tests were not possible in earth-bound laboratories.

1920s. With the end of the World War, further astronomical tests of general relativity were begun, notably by an expedition led by the British physicist Arthur Eddington. The goal was to compare two observations of a star, one near the sun, the other far from it, to see if gravity could in fact "bend" light as Einstein had predicted. The successful result led to enormous popular and scientific interest in relativity, and the cooperative development of astronomy and relativity in this decade was explosive.

A crucial area of study was stellar evolution. The gravitational collapse of a normal star such as our sun is prevented by nuclear burning. As its fuel is used up, the star must contract, and many end as white dwarfs. These are about the size of the earth but with masses comparable to that of the sun—thus with densities of thousands of tons per cubic inch.

1931. The first relativistic model of the interior of a white dwarf, by the astrophysicist S. Chandrasekhar, produced a simple curve relating its mass and radius. Surprisingly, the larger the mass the smaller the radius. In fact, if the mass is more than about 1.2 solar masses the radius is so small that the star cannot stabilize: Further collapse is inevitable.

1934. W. Baade and F. Zwicky predicted that this collapse strips the atoms of their electrons, packing the nucleii together as a *neutron star*. These are only 10 to 15 miles in diameter and consequently have enormous densities. While general relativity is useful in the study of white dwarfs, for the superdense neutron stars it is a necessity—Newtonian physics no longer applies.

1939. The first theoretical appearance of black holes occurs in a paper by Robert Openheimer and H. Snyder: "When all thermonuclear sources of energy are exhausted, a sufficiently heavy star will collapse. Unless [something can somehow] reduce the star's mass to the order of that of the sun, this contraction will continue indefinitely" ... past white dwarfs, past neutron stars, to an object cut off from communication with the rest of the universe.

Such discoveries redirected attention to the Schwarzschild model of the exterior of a star. Until then it had generally been assumed that the model becomes singular at the Schwarzschild radius $r_* = 2M$. This seemed of no great significance since in the case of our sun, for example, whose radius is about 700,000 km, the Schwarzschild model's presumed singularity is buried in at $r_* = 3$ km. But for black holes, whose bulk has effectively vanished, it was gradually realized that radius $r_* = 2M$ is not a singularity but rather a "horizon" from which nothing, not even light, can emerge.

1954. *The Reports of the State University of Kazan* (a city 300 miles east of Moscow in the then Soviet Union) contained this year a classification of spacetimes given by the young physicist A. Z. Petrov. The families of radially ingoing and outgoing light rays in the Schwarzschild model show that it has what is now called *Petrov type D*. Petrov's classification was slow in making its way into the mainstream, but was crucial to the next major development.

All stars rotate. For a very slowly turning star like the sun this is not very important, but when a star collapses, conservation of angular momentum implies that its rate of rotation increases. Thus, for example, a neutron star can be expected to rotate at a fantastic rate. It is believed that *pulsars* are rapidly rotating neutron stars.

But the theoretical star that produces Schwarzschild spacetime does not rotate. In 1915 this static model had been found in only a few weeks, so it must have been considered fairly easy to set it spinning. However, years passed without success.

1963. The British-educated New Zealand physicist Roy Kerr, working at the University of Texas, adopted a shrewd strategy. Bearing in mind that Schwarz-schild spacetime has Petrov type D, he did not aim directly at the elusive rotating model, but instead examined an algebraically simple class of type D metric tensors. The long-sought metric appeared.

Kerr's minimal one-and-one-half-page description of his discovery (1963) was followed two years later by elaborate detailed calculations.

1967–68. R. H. Boyer and R. W. Lindquist (1967) made Kerr spacetime more accessible by introducing the elegant coordinate system that now bears their names. In the same paper they found maximally extended Kerr spacetimes and investigated their geodesics. However, a full analysis of Kerr geodesics became possible only with the discovery of a fourth geodesic first-integral by Brandon Carter, a student of Stephen Hawking at Cambridge University. Carter's paper (1968) remains the best brief exposition of the global properties of Kerr spacetime.

1968–75. From the 1960s theoretical and astronomical evidence mounted for the existence of black holes, including very massive ones formed by the collapse of great clusters of stars at the center of galaxies. Speculatively, the high densities following the big bang may have formed *primordial* holes, including tiny ones. But even for single collapsing stars their wide variety of physical properties might be expected to produce a diversity of black holes. However, a series of results prin-cipally by Werner Israel, Brandon Carter, Stephen Hawking, and David Robinson leads to the conclusion that a collapsing star loses individual characteristics, set-tling down to a final state uniquely determined by mass and rate of rotation—thus leaving the Kerr model as the prime black hole of nature.

BACKGROUND

This chapter gives a concise exposition of most of the material needed for the study of Kerr black holes in the chapters that follow. The topics discussed include manifold theory, tensor calculus, differential geometry, general relativity, and differential forms. The amount of detail varies considerably from topic to topic. Roughly speaking, brevity is possible when the topic is covered in a variety of readily available sources. The section on manifold theory, for example, does little more than record basic definitions and fix notation. By contrast, the special topic of extensions of analytic manifolds requires more attention because it is usually dealt with informally but is crucial to the construction of Kerr spacetime.

Tensor calculus (Sections 1.2 and 1.3) is a generally accepted common language for differential geometry and relativistic physics, with the fundamentals expressed invariantly as well as in coordinate terms. The Cartan calculus of differential forms (Section 1.8 and Appendix B) is at least close to general acceptance, and it is the most efficient way to compute the curvature of Kerr spacetime—and manifolds in general. Perhaps less widely known is the Newman–Penrose formalism (Chapter 5) which is particularly well-suited to analysis of the relation between the curvature and geodesics of spacetimes.

1.1   Manifolds

Roughly speaking, a manifold is a topological space whose local equivalence to Euclidean space $\mathbf{R}^n$ permits calculus to be globally established on it. Accordingly, we assume a familiarity with the basic calculus of $\mathbf{R}^n$. If $\phi = (f_1, \ldots, f_n)$ is a mapping from an open set $\mathcal{U}$ of $\mathbf{R}^m$ into $\mathbf{R}^n$ we say that ϕ is *smooth* (or *infinitely differentiable* or C^∞) if each of the real-valued functions f_j $(1 \le j \le n)$ has continuous partial derivatives of all orders.

MANIFOLDS AND COORDINATE SYSTEMS

A *coordinate system* of dimension n in a topological space S is a homeomorphism ξ from an open set $\mathcal{U}$ of S onto an open set $\xi(\mathcal{U})$ of $\mathbf{R}^n$. The open set $\mathcal{U}$ is a *coordinate neighborhood* of ξ, and the real-valued functions $x^1, \ldots, x^n$ on $\mathcal{U}$ such that $\xi = (x^1, \ldots, x^n)$ are called its *coordinate functions*.

A *smooth manifold M* of *dimension n* is a Hausdorff space furnished with a collection $\mathcal{A}$ (called a *smooth atlas*) of n-dimensional coordinate systems such that

1. The coordinate neighborhoods for all $\xi \in \mathcal{A}$ cover all of M, and
2. Any two $\xi, \eta \in \mathcal{A}$ *overlap smoothly*; that is, the composite maps $\xi \circ \eta^{-1}$ and $\eta \circ \xi^{-1}$ are smooth in the usual Euclidean sense (as above).

Notable manifolds are *Euclidean n-space* $\mathbf{R}^n$ and the *n-sphere* S^n. For $\mathbf{R}^n$ the identity map, alone, constitutes an atlas, while S^n can be covered by two stereographic coordinate systems. An open set in a manifold is itself a manifold in a natural way. The cartesian product $M \times N$ of manifolds M and N becomes a manifold using product coordinate systems $\xi \times \eta: \mathcal{U} \times \mathcal{V} \to \mathbf{R}^m \times \mathbf{R}^n$. If $\xi = (x^1, \ldots, x^m)$ and $\eta = (y^1, \ldots, y^n)$, we often write merely $\xi \times \eta = (x^1, \ldots, x^m, y^1, \ldots, y^n)$, ignoring the projection maps of $M \times N$ onto M and N.

By convention the atlas $\mathcal{A}$ defining a manifold M is enlarged by adding to it all coordinate systems that overlap smoothly with those initially in $\mathcal{A}$. For example, $\mathbf{R}^2$, which was defined by its usual cartesian coordinate system, now admits polar coordinates, etc.

SMOOTH MAPPINGS

A map $\psi: M \to N$ of manifolds is *smooth* provided that all its expressions $\zeta \circ \psi \circ \xi^{-1} = (f_1, \ldots, f_n)$ in terms of coordinate systems are smooth in the Euclidean sense. In short, the ζ coordinates of $\psi(p)$ depend smoothly on the ξ

coordinates of p. In particular, smooth real-valued functions $f: M \to \mathbf{R}^1$ are well defined on M. The set $\mathfrak{F}(M)$ of all such functions f on M forms a commutative ring under the usual definitions of addition and multiplication of functions.

A *diffeomorphism* $\psi: M \to N$ is a smooth map that has an inverse ψ^{-1} that is also smooth; in this case, M and N are said to be *diffeomorphic*. Diffeomorphic manifolds are considered to be the same from the viewpoint of manifold theory.

TANGENT VECTORS

The crucial definition in manifold theory is that of tangent vector. The idea is to axiomatize the directional derivative property of vectors, defining a *tangent vector v to M at $p \in M$* to be an $\mathbf{R}$-linear map $v: \mathfrak{F}(M) \to \mathbf{R}$ with the Leibnizian property

$$v[fg] = v[f]g(p) + f(p)v[g] \quad \text{for all} \quad f, g \in \mathfrak{F}(M).$$

The set of all tangent vectors to a manifold M at the point p is denoted by $T_p(M)$. This set becomes a vector space, called the *tangent space* to M at p, under the natural definitions of (1) scalar multiplication by a real number and (2) addition of functions. The developments that follow support the picture of a tangent vector v at p as an arrow starting at p and of $T_p(M)$ as the linear approximation of M near p. A good example is the usual Euclidean tangent plane to a surface M in $\mathbf{R}^3$.

If ξ is a coordinate system in a manifold M (that is, in the atlas of M), then at any point p in the coordinate neighborhood of ξ there are n tangent vectors $\partial_1|_p, \ldots, \partial_n|_p$ defined by $(\partial_j|_p)[f] = (\partial f/\partial x^j)(p)$. The partial derivative here is the ordinary Euclidean i^{th} partial derivative, evaluated at $\xi(p)$, of the *coordinate expression* $f \circ \xi^{-1}$ of f, where $f = (f \circ \xi^{-1})(x^1, \ldots, x^n)$. We picture these *coordinate vectors* $\partial_j|_p$ as tangent to the coordinate curves of ξ.

The following useful result implies that, at any point, the coordinate vectors form a basis for the tangent space at that point.

Theorem 1.1.1 (Basis Theorem). *If $x^1, \ldots, x^n$ is a coordinate system on a neighborhood $\mathcal{U}$ of a point p, then $v = \Sigma_j v[x^i]\partial_j|_p$ for all $v \in T_p(m)$.*

(Unless otherwise specified, indices run from 1 to $n = \dim M$.)

The union of all the tangent spaces $T_p(M)$ of a manifold M becomes a manifold, called the *tangent bundle TM of M*, as follows. The *projection* $\pi: TM \to M$ sends each $v \in T_p(M) \subset TM$ to p. If $x^1, \ldots, x^n$ is a coordinate system on $\mathcal{U} \subset \mathcal{M}$, define $\dot{x}^i \colon \pi^{-1}(\mathcal{U}) \to \mathbf{R}$ by $\dot{x}^i(v) = v[x^i]$ for all v. Then $(x^1, \ldots, x^n, \dot{x}^1, \ldots, \dot{x}^n) \colon \pi^{-1}(\mathcal{U}) \to \mathbf{R}^{2n}$ is a coordinate system in TM.

CURVES

In general, a *curve* in a manifold M is a smooth mapping α from an open (possibly infinite) interval $I \subset \mathbf{R}$ into M. For each $s \in I$ the curve has a *tangent vector* $\alpha'(s) = (d\alpha/ds)(s)$; this is the tangent vector to M at $\alpha(s)$ such that

$$\alpha'(s)[f] = \frac{d(f \circ \alpha)}{ds}(s)$$

for all $f \in \mathfrak{F}(M)$.

By convention, unless the issue is explicitly raised, all relevant objects (e.g., curves, real-valued functions) are assumed to be smooth. Consistent with any invariant definition of smoothness is the following generic definition: An object X is *smooth* provided all coordinate expressions for X consist of Euclidean-smooth functions.

DIFFERENTIAL MAPS

Given a mapping $\psi \colon M \to N$, for each point $p \in M$ there is a linear transformation $d\psi|_p \colon T_p(M) \to T_{\psi(p)}(N)$ called the *differential map* of ψ at p and characterized by the identity

$$d\psi(v)[g] = v[g \circ \psi] \quad \text{for all } v \in T_p(M) \text{ and } g \in \mathfrak{F}(M).$$

The differential maps for all $p \in M$ can be regarded as a single map $d\psi \colon TM \to TN$ of tangent spaces, so we omit the subscript p on $d\psi$.

Another characterization is this: $d\psi$ *preserves tangents*, that is, if α is a curve in M, then $d\psi$ carries each vector $\alpha'(s)$ to the tangent vector $(\psi \circ \alpha)'(s)$ of the image curve $\psi \circ \alpha$ in N.

VECTOR FIELDS

A *vector field* V on a manifold M is a function that assigns to each point p of M a tangent vector V_p to M at p. Vector fields are intuitive objects, familiar as representations of force fields or the velocities of a steady-state fluid flow. The set

$\mathfrak{X}(M)$ of all (smooth) vector fields on M is a module over $\mathfrak{F}(M)$, that is, the expected algebraic rules hold for the operations of pointwise addition $X + Y$ and multiplication fX by a function $f \in \mathfrak{F}(M)$.

The definition of tangent vector lets us interpret a vector field V as the differential operator $f \to Vf$, where the latter is the real-valued function such that $(Vf)(p) = V_p[f] \in \mathbf{R}$ for all $p \in M$. That V is smooth means, in invariant terms, that f smooth implies Vf smooth. This interpretation makes V a *derivation* on the ring $\mathfrak{F}(M)$.

In terms of a coordinate system, the Basis Theorem gives $V = \Sigma V[x^i]\partial_i$. This is the fundamental link between the invariant definition of V and its tensor description (Section 1.2) as an n-tuple of functions $V^i = Vx^i$.

There is a remarkable operation that combines two vector fields V, W into a new vector field $[V, W]$, called the *Lie bracket* of V and W, characterized by

$$[V, W]f = V(Wf) - W(Vf) \quad \text{for all} \quad f \in \mathfrak{F}(M).$$

The bracket operation is bilinear over the real numbers, skew-symmetric, and satisfies the *Jacobi identity*:

$$[U, [V, W]] + [V, [W, U] + [W, [U, V]] = 0.$$

In elementary mathematics the bracket operation is neglected, perhaps because, for coordinate vector fields, $[\partial_i, \partial_j] = 0$. (*Proof:* independence of order of partial differentiation.)

INTEGRAL CURVES

A vector field V on M has another interpretation: as a differential equation whose solutions are curves α, called *integral curves*, that at each point have the tangent specified by V. Explicitly,

$$\alpha'(s) = V_{\alpha(s)} \quad \text{for all} \quad s \in I.$$

In terms of cooordinates,

$$\alpha'(s) = \Sigma \alpha'(s)[x^i]\partial_i = \Sigma d(x^i \circ \alpha)/ds \ \partial_i, \quad \text{and}$$
$$V = \Sigma V^i \partial_i, \quad \text{where} \quad V^i = Vx^i.$$

Then the vector equation above becomes a linear system of ordinary differential equations

$$\frac{d(x^i \circ \alpha)}{ds} = V^i(x^1, \ldots, x^n) \quad \text{for} \quad i = 1, 2 \ldots, n.$$

Local existence and uniqueness theorems apply (since we assume V smooth), and it follows that for each point p of M there is a unique integral curve α_p, that has largest possible domain and starts at p: $\alpha_p(0) = p$. Integral curves α, β of V can meet only if one is a reparametrization of the other; explicitly, $\beta(s) = \alpha(s + c)$ for some constant c.

A vector field X is *complete* provided each of its maximal integral curves is defined on the entire real line. Suppose X is complete; then for each $s \in \mathbf{R}$ let $\psi_s: M \to M$ be the (smooth) mapping that lets each point $p \in M$ "flow" to the point $\alpha_p(s) \in M$. Evidently, $\psi_0 = id$, and $\psi_{s+t} = \psi_s \circ \psi_t$ for all s, t. Hence each ψ_s is a diffeomorphism, since it has smooth inverse ψ_{-s}. The collection $\{\psi_s : s \in \mathbf{R}\}$ is called the *flow* of X.

If X is not complete, its flow is defined only locally. Explicitly, for each point $p \in M$ there is a neighborhood $\mathcal{U}$ of p and a number $\varepsilon > 0$ such that for every $q \in \mathcal{U}$, α_q is defined at least on $(-\varepsilon, \varepsilon)$. Thus $\psi_s: \mathcal{U} \to M$ is defined and has properties as above whenever s and t are sufficiently small.

ONE-FORMS

One-forms on a manifold are the objects dual to vector fields. At each point p of M let $T_p(M)^*$ be the dual space of the tangent space $T_p(M)$. Thus elements of $T_p(M)^*$, are linear maps $T_p(M) \to \mathbf{R}$, called *covectors*. A *one-form* θ on M is a covector field, that is, θ assigns to each point p of M a covector $\theta_p \in T_p(M)^*$. Like $\mathfrak{X}(M)$, the set $\mathfrak{X}^*(M)$ of all (smooth) one-forms on M is a module over $\mathfrak{F}(M)$ under pointwise addition $\theta + \omega$ and multiplication by a function, $f\theta$.

A vector field $V \in \mathfrak{X}(M)$ and a one-form $\theta \in \mathfrak{X}^*(M)$ combine naturally to give the function $\theta V = \theta(V) \in \mathfrak{X}(M)$ such that $(\theta V)(p) = \theta_p(V_p)$ for all $p \in M$.

Each function $f \in \mathfrak{F}(M)$ gives rise to a one-form, namely, its *differential df*, given by $(df)(v) = v[f]$ for all tangent vectors v. This elegant (and practical) definition clarifies the mysterious "dx" of elementary calculus.

The dual to the Basis Theorem is this: On the domain of a coordinate system, every one-form can be expressed in terms of the *coordinate one-forms $dx^1, \ldots, dx^n$* as $\theta = \Sigma\theta(\partial_i)dx^i$. (*Proof:* Apply both sides to ∂_j .)

In particular, this yields the famous formula $df = \Sigma\partial f/\partial x^i dx^i$, since $(df)(\partial_i) = \partial_i(f) = \partial f/\partial x^i$.

SMOOTH FIBER BUNDLES

The notion of fiber bundle is important in understanding the structure of Kerr spacetime. Here are some general definitions. A *(left) action* of a Lie group G on a manifold F is a smooth map $(g, q) \to gq$ from $G \times F$ to F such that

$g_1(g_2(q)) = (g_1 g_2(q)$ for all $g_1, g_2 \in G$ and $q \in F$, and, if e is the identity element of G, then $eq = q$ for all q. Thus each $g \in G$ determines a map $\lambda_g : F \to F$ sending q to gq and λ_g is a diffeomorphism, since it has inverse $\lambda_{g^{-1}}$. The action is *effective* if $\lambda_g = id$ implies $g = e$. For Lie group theory, see for example, Helgason (1978) or Warner (1983).

In Kerr applications, G will be the circle group S^1 acting effectively on the 2-sphere S^2 by rotation around the z-axis. Explicitly, for each "angle" $\alpha \in \mathbf{R}$ the rotation R_α of $S^2 = \{v \in \mathbf{R}^3 : |v| = 1\}$ is given by

$$R_\alpha(x, y, z) = (x \cos \alpha - y \sin \alpha, x \sin \alpha + y \cos \alpha, z).$$

A fiber bundle is a generalization of product manifold $B \times F$ in which only projection π on B remains, and the fibers $b \times F$ can be globally twisted (e.g., projection of a Möbius band onto its central axis B). The bundle remains a local product, and its global twisting is modulated by the action of a Lie group on F.

By definition, a *smooth fiber bundle* consists of a smooth map $\pi : E \to B$, an action (as above) of a Lie group G on a manifold F, and a collection Φ of mappings ϕ such that

1. Local product condition: Each $\phi \in \Phi$ is a diffeomorphism $\mathcal{U} \times F \to \pi^{-1}(\mathcal{U})$ such that $\pi(\phi(b, q)) = b$ for all $q \in F$ and all $b \in \mathcal{U}$, an open set of B, and
2. Overlap condition: if $\phi, \psi \in \Phi$ have intersecting domains $\mathcal{U} \cap \mathcal{V}$ in B there is a smooth map $\gamma : \mathcal{U} \cap \mathcal{V} \to G$ such that

$$\psi(b, q) = \phi(b, \lambda_{\gamma(b)}(q)) \qquad \text{for all } b \in \mathcal{U} \cap \mathcal{V} \text{ and } q \in F.$$

Terminology: π is the *projection*, E the *total manifold*, B the *base*, F the *fiber*, G the *structural group*, Φ the set of *bundle coordinate systems* ϕ.

Such a bundle is a (real) *vector bundle* if the fiber is $\mathbf{R}^n$ acted on as usual by the general linear group. An example is the tangent bundle TM of a manifold M, where each coordinate system $x^1, \ldots, x^n$ in M produces a bundle coordinate system $\phi(b, q) = \Sigma q_i(\partial_i|_b)$ for $\pi : TM \to M$.

CIRCULAR COORDINATES

Once the notion of manifold is established, there is no harm defining a new manifold M by using generalized coordinate systems η in M that map not just into $\mathbf{R}^n$ but into any n-dimensional manifold Q. (Then following η by ordinary coordinates in Q gives ordinary coordinates in M.)

For example, the usual polar coordinates r, ϑ in the plane $\mathbf{R}^2$ actually form a coordinate system only if restricted to a subset such as $\mathbf{R}^2$–(ray from 0 to ∞). Consider ϑ, in the obvious way, as a map into S^1. Then taking $Q = \mathbf{R}^1 \times S^1$

makes (r, ϑ) a generalized coordinate system on the larger domain $\mathbf{R}^2 - \{0\}$. This minor generalization will prove quite useful.

SMOOTH vs ANALYTIC

In the definition of smooth atlas $\mathcal{A}$, suppose that all overlap functions $\xi \circ \eta^{-1}$ and $\eta \circ \xi^{-1}$ are analytic in the usual Euclidean sense, that is, their coordinate expressions are given by convergent power series. Then $\mathcal{A}$ is an *analytic atlas* and makes M an *analytic manifold*. For example, $\mathbf{R}^n$ and S^n as defined earlier are analytic.

Since analytic functions are smooth, every analytic manifold M is also a smooth manifold. Thus there is no needed to redefine the machinery of smooth manifold theory in the analytic case. Objects on an analytic manifold M are *analytic* if their coordinate expressions—in terms of analytic coordinate systems—are given by convergent power series.

Analyticity imposes the following powerful rigidity property: if two analytic mappings $M \to N$, with M connected, agree on some one open set $\mathcal{U}$ in M then they are globally identical. Equivalently, an analytic mapping defined $\mathcal{U} \subset M$ has at most one analytic extension to all of M. Easy examples show that this property fails for mappings that are merely smooth.

1.2   Tensors

There are a number of equivalent ways to describe tensors; the following is the simplest coordinate-free definition of tensor fields on a manifold M.

THE INVARIANT APPROACH

Let $\mathfrak{X} = \mathfrak{X}(M)$ be the vector fields on M, and $\mathfrak{X}^* = \mathfrak{X}^*(M)$ the one-forms. These are modules over the ring $\mathfrak{F} = \mathfrak{F}(M)$ of real-valued functions on M.

Definition 1.2.1 *Let r and s be nonnegative integers, not both zero. A (smooth) tensor field of type (r, s) on M is an $\mathfrak{F}$-multilinear function*

$$A: \underbrace{\mathfrak{X}^* \times \ldots \times \mathfrak{X}^*}_{r} \times \underbrace{\mathfrak{X} \times \ldots \times \mathfrak{X}}_{s} \to \mathfrak{F}.$$

A tensor field of type (0,0) is a function $f \in \mathfrak{F}$.

Thus, for every choice of one-forms $\theta^1, \ldots, \theta^r$ and vector fields $X_1, \ldots, X_s$ an (r, s) tensor field A produces a real-valued function $A(\theta^1, \ldots, \theta^r, X_1, \ldots, X_s)$, and A is linear over $\mathfrak{F}$ in each slot (i.e., in each variable separately). In particular, functions f can be factored out of each slot; for example,

$$A(\theta^1, \ldots, f\theta^i, \ldots, \theta^r, X_1, \ldots, X_s) = f A(\theta^1 \ldots, \theta^i, \ldots, \theta^r, X_1, \ldots, X_s).$$

Let $\mathfrak{T}^r_s = \mathfrak{T}^r_s(M)$ be the set of all type (r, s) tensor fields on M. The usual definitions of addition and of multiplication by a function make $\mathfrak{T}^r_s$ a module over $\mathfrak{T}$.

It is easy to verify that $\mathfrak{T}^0_1 = \mathfrak{X}^*$, and $\mathfrak{T}^1_0$ can be identified with the vector fields $\mathfrak{X}$ by interpreting $X \in \mathfrak{X}$ as the function $\mathfrak{X}^* \to \mathfrak{F}$ such that $X(\theta) = \theta(X)$ for all θ. (This is a close relative of the reflexivity isomorphism $V \approx V^{**}$ for vector spaces.)

Another useful interpretation can be described by a special case. Let $A \colon \mathfrak{X} \times \mathfrak{X} \to \mathfrak{X}$ be an $\mathfrak{F}$-bilinear function; then the definition $A(\theta, X, Y) = \theta(A(X, Y))$ makes $A \approx A$ a tensor of type $(1,2)$. The next section includes several examples of geometrically important tensor fields that occur naturally in this form. As a nonexample, consider the bracket operation $[,] \colon \mathfrak{X} \times \mathfrak{X} \to \mathfrak{X}$. It is bilinear over the real numbers $\mathbf{R}$, but not over functions since $[fX, Y] = f[X, Y] - YfX$. Thus the bracket operation is not a tensor.

Only tensors of the same type are added, but there is an easy way to multiply tensors of arbitrary types. For example, if A has type $(1,1)$ and B has type $(1,2)$, their *tensor product* $A \otimes B$ is the $(2,3)$ tensor given by

$$(A \otimes B)(\theta, \omega, X, Y, Z) = A(\theta, X)B(\omega, Y, Z)$$

for all one-forms θ, ω, and vector fields X, Y, Z.

The tensor product is $\mathfrak{F}$-bilinear and associative but is generally not commutative (e.g., $X \otimes Y \neq Y \otimes X$). However, commutativity does holds for tensors A and B if A is *covariant*, that is, has type $(0, s)$, and B is *contravariant*, that is, has type $(r, 0)$.

THE CLASSICAL APPROACH

By considering the coordinate expressions for tensors (as defined above) we recover their classical index description.

Definition 1.2.2 *Let* $x^1, \ldots, x^n$ *be a coordinate system on* $\mathcal{U} \subset M$. *The* components *of a tensor* $A \in \mathfrak{T}^r_s = \mathfrak{T}^r_s(M)$ *are the real-valued functions*

$$A^{i_1, \ldots, i_r}{}_{j_1, \ldots, j_s} = A(dx^{i_1}, \ldots, dx^{i_r}, \partial_{j_1}, \ldots, \partial_{j_s}) \quad on \quad \mathcal{U},$$

where all indices run from 1 to $n = \dim M$.

Here the superscripts are *contravariant indices*, the subscripts *covariant indices*. The operations on tensor fields described invariantly above are easily expressed in coordinate terms. For example, if a (1,1) tensor A is applied to a one-form $\theta = \Sigma \theta_i dx^i$ and vector field $X = \Sigma X^j \partial_j$, the result is

$$A(\theta, X) = \Sigma \theta_i X^i A(dx^i, \partial_j) = \Sigma A^i{}_j \theta_i X^j.$$

Convention. We use the following weak form of the Einstein summation convention: The summation sign is not omitted, but summation is understood over all repeated indices. Frequently (but not always) the repeated index occurs exactly twice—once up, once down. Indices typically run from 1 to $n = \dim M$.

The components of a sum of (r, s) tensors are the sums of the components. The components of a tensor product are the products of their components in this sense: if A is a (1,1) tensor and B is (1,2), then the (2,3) tensor $A \otimes B$ has components

$$(A \otimes B)^{ij}{}_{k\ell m} = A^i{}_k B^j{}_{\ell m}.$$

It is best to *define* tensors invariantly, because with a definition in terms of components we are under obligation to check independence of coordinate system. Nevertheless, the classical index approach has advantages of explicitness and flexibility, together with the built-in error detection feature *index balance*, which asserts that the same unsummed superscripts and subscripts must appear on each side of a tensor equation.

FRAME FIELDS

There is a mild but valuable generalization of the classical approach to tensors in which the coordinate vector fields $\partial_1, \ldots, \partial_n$ of a coordinate system are generalized to an arbitrary *frame field* $\{E_1, \ldots, E_n\}$. The latter consists consists of n vector fields E_i that supply a basis for the tangent space $T_p(M)$ at each point p of their common domain $\mathcal{U} \subset M$. Coordinate differentials $dx^1, \ldots, dx^n$ are now replaced by the *dual one-forms* $\omega^1, \ldots, \omega^n$ of the frame field, these characterized by $\omega^i(E_j) = \delta^i{}_j$. The one-forms $\{\omega^i\}$ constitute the *coframe field* dual to $\{E_i\}$. Then for any vector field X we have the *duality formula* $X = \Sigma \omega^i(X) E_i$.

The components of a tensor field on M relative to a frame field are defined in exactly the same way as for a coordinate system. For example, suppose A is a tensor of type (1,2). Described invariantly it is an $\mathfrak{F}(M)$-multilinear function $\mathfrak{X}^*(M) \times \mathfrak{X}(M) \times \mathfrak{X}(M) \to \mathfrak{X}(M)$. In terms of a coordinate system, A has components $A^i{}_{jk} = A(dx^i, \partial_j, \partial_k)$. In terms of a frame field, the components are $A^i{}_{jk} = A(\omega^i, E_j, E_k)$. Addition of tensors and tensor products are described in terms of these components exactly as in the coordinate case.

One advantage of this generalization is that frame fields can often be defined on larger domains than can coordinate systems. For instance, it is easy to find a single globally defined frame field on a torus T, but four coordinate systems are required to cover all of T.

Change of coordinates is now replaced by change of frame fields. Given a frame field $\{E_j\}$ with dual forms $\{\omega^i\}$, let $\{F_j\}$ be a new frame field with dual forms $\{\theta^i\}$ on the same domain $\mathcal{U}$. There are unique $n \times n$ matrix-valued functions a and b on $\mathcal{U}$ such that

$$F_j = \Sigma a_j{}^m E_m, \quad \text{and} \quad \theta^i = \Sigma b^i{}_q \omega^q.$$

By duality, $\Sigma b^i{}_m a_j{}^m = \delta^i{}_j$. In the special case of coordinate change, the matrices a and b are just the (inverse) Jacobian matrices $(\partial y^i / \partial x^j)$ and $(\partial x^i / \partial y^j)$.

To see how tensor components alter under change of frame field, it suffices to consider a (1,1) tensor A with components $A^i{}_j$ relative to $\{E_i\}$. Then its components relative to the new frame field $\{F_j\}$ are

$$A(\theta^i, F_j) = A(\Sigma b^i{}_q \omega^q, \Sigma a_j{}^m E_m) = \Sigma b^i{}_q a_j{}^m A^q{}_m.$$

Conversely, any component assignment that obeys this *tensor transformation rule* determines a unique tensor. In the case above, for example, A can be recovered from $\{A^q{}_m\}$ by defining $A(\theta, V) = \Sigma A^q{}_m \theta_q V^m$. The transformation rule shows that this definition is independent of the choice of frame field.

CONTRACTION

Associated with a linear operator T (on a finite-dimensional vector space) is its *trace*, a number given by $\Sigma A^m{}_m$ for every matrix $(A^i{}_j)$ of T. There is an direct generalization to tensors, described in frame field terms as follows. Choose one contravariant index and one covariant index of the components of A; then ignoring the other indices, take the trace as above. (This is certain to be an invariant operation since we know the trace is.) For example, suppose A has type $(2,3)$, and pick the first contravariant index (superscript) and the second covariant index (subscript). The resulting tensor $C_2^1 A$, called a *contraction* of A, has components $\Sigma A^{mi}{}_{jmk}$. Thus, in general, contraction reduces type from (r, s) to $(r - 1, s - 1)$.

TENSORS AT A POINT

Let p be a point of M. In Definition 1.2.1 replace $\mathfrak{F}$ by $\mathbf{R}$, and the $\mathfrak{F}$-modules $\mathfrak{X}$ and $\mathfrak{X}^*$ by the real vector spaces $T_p(M)$ and $T_p(M)^*$. The modified definition describes an (r, s) *tensor at the point* $p \in M$.

A tensor field A on M determines at each $p \in M$ a tensor A_p at p called its *value* at p. For example, suppose A has type $(2,0)$. Given any $v, w \in T_p(M)$, choose vector fields V, W on M such that $V_p = v$ and $W_p = w$. Now $A(V, W)$ is a function on M. Define $A_p(v, w)$ to be the value at p of $A(V, W)$. Then

1. $A_p(v, w)$ is independent of the choices of V and W, hence A_p is well-defined, and

2. $A_p \colon T_p(M) \times T_p(M) \to \mathbf{R}$ is $\mathbf{R}$-bilinear, hence is a tensor at p.

Evidently we can further replace $T_p(M)$ by any finite-dimensional real vector space V, and $T_p(M)^*$ by the dual space V^* of V, and define (r, s) tensors as before. For example, in these terms, an *inner product* g on V is a symmetric, positive definite $(0,2)$ tensor on V. We call g a *scalar product* if positive definite is weakened to *nondegenerate* (that is, $g(v, w) = 0$ for all $w \in V$ implies $v = 0$).

For a treatment of tensors emphasizing applications to relativity, see Dodson, Poston (1991).

1.3   Differential Geometry

METRIC TENSORS

The foundation of mainstream differential geometry is a smooth manifold M furnished with a *metric tensor* g. Explicitly, g is a smooth covariant $(0,2)$ tensor that assigns to each point p of M a scalar product g_p on the tangent space $T_p(M)$. Then M is called a *semi-Riemannian manifold*. For vector fields V and W on M, the real-valued function $g(V, W)$ will usually be denoted by $<V, W>$, with similar notation for individual tangent vectors.

If the metric tensor is positive definite (i.e., all the scalar products g_p are inner products), then g is a *Riemannian metric* making M a *Riemannian manifold*. But for relativity theory an indefinite metric is needed: If each g_p is a Lorentz scalar product (see Section 1.5), then g is a *Lorentz metric* making M a *Lorentz manifold*.

On the domain $\mathcal{U}$ of a coordinate system $x^1, \ldots, x^n$ the components of the metric tensor are $g_{ij} = g(\partial_i, \partial_j) = \, <\partial_i, \partial_j>$. Hence $g = \Sigma g_{ij} dx^i \otimes dx^j$. However, the metric tensor itself is often replaced by its *line-element*, $q = ds^2$ which gives at each point the associated quadratic form of g_p. Thus, the value of ds^2 on a vector field V is $g(V, V) = \, <V, V>$.

The metric tensor can be reconstructed from its line-element by polarization. Though a line-element is not a tensor, it can be expressed in terms of a coordinate system as $ds^2 = \Sigma g_{ij} dx^i dx^j$, where the juxtaposition of differentials is just ordinary multiplication of functions. To verify this, consider a vector field $V = \Sigma V^i \partial_i$. Then

$$(ds^2)(V) = \, <V, V> \, = \, <\Sigma V^i \partial_i, \Sigma V^j \partial_j> \, = \Sigma V^i V^j <\partial_i, \partial_j>$$
$$= \Sigma g_{ij} V^i V^j, = \Sigma g_{ij} dx^i(V) dx^j(V) = (\, \Sigma g_{ij} dx^i dx^j)(V).$$

EXAMPLES

For integers $0 \le \nu \le n$, let $\mathbf{R}_\nu^n$ be the manifold $\mathbf{R}^n$ furnished with line-element $\Sigma \varepsilon_i (dx^i)^2$, where $\varepsilon_i = \pm 1$ is -1 for $i \le \nu$ and $+1$ for $i > \nu$. $\mathbf{R}_\nu^n$ is called *semi-Euclidean n-space of index ν*. When $n = 0$, this is ordinary Euclidean n-space (a Riemannian manifold), and when $\nu = 1$ and $n \ge 2$ it is *Minkowski n-space* (a Lorentz manifold).

Given two semi-Riemannian manifolds M and N, their product manifold has a natural metric tensor for which $<v, w>$ is the sum of the scalar products of the projections of v and w on M and on N. Any open subset $\mathcal{U}$ of a semi-Riemannian manifold M becomes itself semi-Riemannian by using the metric of M on $\mathcal{U}$. The richest source of semi-Riemannian manifolds is as submanifolds of semi-Euclidean spaces (see Section 1.7).

ISOMETRIES

A mapping $\psi \colon M \to N$ of semi-Riemannian manifolds is an *isometry* if it is a diffeomorphism and *preserves scalar products* in this sense:

$$<d\psi(v), d\psi(w)> \, = \, <v, w> \quad \text{for all tangent vectors } v, w \text{ to } M.$$

Then M and N are said to be *isometric* and are regarded as geometrically the same. For example, any two open intervals in $\mathbf{R}^1$ are diffeomorphic, but they are isometric if and only if they have the same length.

Here is a coordinate criterion for isometry: Suppose $ds_N^2 = \Sigma \bar{g}_{ij} dy^i dy^j$ on $\mathcal{V} \subset N$ and $ds_M^2 = \Sigma g_{ij} dx^i dx^j$ on M. A mapping $\phi \colon \mathcal{U} \to \mathcal{V}$ given in coordinates by $y^i = y^i(x^1, \ldots, x^n)$ is an isometry on $\mathcal{U}$ provided ds_N^2 pulls back under ϕ to ds_M^2. Explicitly, when $dy^i = \Sigma \partial y^i / \partial x^m dx^m$ is substituted into ds_N^2 and $y^i = y^i(x^1, \ldots, x^n)$ is substituted into $\bar{g}_{ij}$, the result is exactly ds_N^2.

In later applications to Kerr spacetime this clumsy-looking criterion will be quite simple.

COVARIANT DERIVATIVES

Given vector fields V, W on a semi-Riemannian manifold M, we want to define a new vector field $\nabla_V W$ whose value at each point p is the vector rate of change of W in the direction V_p. On semi-Euclidean space $\mathbf{R}_\nu^n$ a natural choice for ∇ is

$$\nabla_V (\Sigma W^i \partial_i) = \Sigma V[W^i] \partial_i,$$

where $\partial_1, \ldots, \partial_n$ are the coordinate vector fields of the natural cartesian coordinate system of R_ν^n (the particular index ν is irrelevant here). An operation with the same basic properties as this ∇ can be uniquely obtained in general.

Theorem 1.3.1 (Fundamental Theorem of Semi-Riemannian Geometry). *On a semi-Riemannian manifold M there is a unique function $\nabla \colon \mathfrak{X} \times \mathfrak{X} \to \mathfrak{X}$ such that*

(1) $\nabla_V W$ *is $\mathfrak{F}$-linear in V,*
(2) $\nabla_V W$ *is $\mathbf{R}$-linear in W,*
(3) $\nabla_V(fW) = V[f]W + f\nabla_V W$ *for all $f \in \mathfrak{F}$.*
(4) $[V, W] = \nabla_V W - \nabla_W V$,
(5) $X\langle V, W \rangle = \langle \nabla_X V, W \rangle + \langle V, \nabla_X W \rangle$.

∇ is called the *Levi-Civita connection* of M, and for particular vector fields, $\nabla_V W$ is the *covariant derivative* of W with respect to V.

Because of property (1), $\nabla_v W$ is well-defined for individual tangent vectors v, as $(\nabla_V W)|_p$, where V is any vector field such that $V_p = v$.

In terms of a coordinate system on a neighborhood $\mathcal{U} \in M$, ∇ is described by *Christoffel symbols*, the real-valued functions $\Gamma^k{}_{ij}$ on $\mathcal{U}$ such that $\nabla_{\partial_i}(\partial_j) = \Sigma \Gamma^m{}_{ij}\partial_m$ for all i, j. Then

$$\Gamma^k{}_{ij} = \tfrac{1}{2}\Sigma g^{k\ell}[\partial g_{\ell j}/\partial x^j + \partial g_{\ell j}/\partial x^i - \partial g_{ij}/\partial x^\ell].$$

The covariant derivative ∇ is adapted to a curve α in M as follows. If Y is a vector field on α, then (if $\alpha' \neq 0$) the *covariant derivative of Y along* α is defined to be $Y' = \nabla_{\alpha'} Y$. In terms of coordinates

$$Y' = \sum_k \left\{ \frac{dY^k}{ds} + \sum_{ij} \Gamma^k{}_{ij} \frac{dY^i}{ds}\frac{dx^j}{ds} \right\}\partial_k.$$

Here the coordinate function $x^i \circ \alpha$ of the curve is abbreviated to just x^i (and the formula is valid for arbitrary α).

If $Y' = 0$, the vector field Y is said to be *parallel*, and for parameter values s_1, s_2 we say that $Y(s_2)$ is obtained from $Y(s_1)$ by parallel translation along α.

The tangent α' of α is a vector field on α, and the covariant derivative of α' is the *acceleration* $\alpha'' = \nabla_{\alpha'}\alpha'$ of the curve.

GEODESICS

Straight lines in Euclidean space generalize to geodesics in a semi-Riemannian manifold M. A curve γ in M is a *geodesic* provided its acceleration is zero: $\gamma'' = 0$. In coordinate terms the preceding formula gives the *geodesic equations*

$$x^{k\prime\prime} + \sum \Gamma^k{}_{ij} x^{i\prime} x^{j\prime} = 0 \qquad \text{for } k = 1, \ldots, n.$$

Here the prime denotes derivative with respect to the parameter s of γ, and as before, x^i is short for $x^i \circ \gamma$. Differential equations theory guarantees that for any tangent vector v to M there is a unique geodesic $\gamma_v \colon I \to M$ with initial velocity $\gamma_v'(s_0) = v$. Here we can arrange that the interval I is as large as possible. If $I = \mathbf{R}$ for all $v \in TM$, that is, if every geodesic can be extended (as a geodesic) over the entire real line, M is said to be *geodesically complete*.

A reparametrization $\beta(s) = \gamma(h(s))$ of a geodesic is again a geodesic only if h has the form $h(s) = as + b$. Thus geodesic parametrizations have geometric significance. A curve that admits a reparametrization as a geodesic is sometimes called a *pregeodesic*.

Using Lagrangian methods the geodesic equations can be expressed in a different way that is often superior to the formulation above. Let $L \colon TM \to \mathbf{R}$ be a smooth function on the tangent bundle of M. A curve α in M is an *extremal* for L provided that for enough coordinate systems $x^1, \ldots, x^n, \dot{x}^1, \ldots, \dot{x}^n$ to cover TM the following Euler equations hold:

$$\frac{d}{ds}\left[\frac{\partial L}{\partial \dot{x}^j}(\alpha')\right] = \frac{\partial L}{\partial x^i}(\alpha') \quad \text{for } i = 1, \ldots, n.$$

For a semi-Riemannian manifold M, let $L: TM \to \mathbf{R}$ be $L(v) = \frac{1}{2}<v, v>$. In terms of coordinates, $L = \frac{1}{2}\Sigma g_{ij}(x^1, \ldots, x^n)\dot{x}^i\dot{x}^j$. Note that $\dot{x}^i(\alpha') = x^{i\prime}$, where as usual $x^{i\prime}$ means $d(x^i \circ \alpha)/ds$. The resulting Euler equations are

$$\frac{d}{ds}\left[\Sigma g_{k\ell}\, x^{\ell\prime}\right] = \frac{1}{2}\sum \frac{\partial g_{ij}}{\partial x^k}x^{i\prime}x^{j\prime} \quad \text{for } k = 1, \ldots, n.$$

These equations are equivalent to the geodesic differential equations. To see this carry out the differentiation on the left, and use the classical formula (above) for $\Gamma^k_{\ ij}$ in terms of derivatives of g_{ij}.

Lagrangian methods will be used to advantage in Chapter 4.

TYPE-CHANGING

On a semi-Riemannian manifold, the metric tensor can be used to change a tensor of type (r, s) of a tensor to any other type (r', s') such that $r' + s' = r + s$. The new tensor contains the same information as the original and is often considered to be merely a different manifestation of the same entity. (Thus the distinction between covariant and contravariant becomes unimportant.)

We describe type-changing in terms of frame fields. It is sufficient to tell how to change type (r, s) to $(r - 1, s + 1)$ or to $(r + 1, s - 1)$. Let A be, say, a $(2,2)$ tensor field with components $A^{ij}_{\ k\ell}$ relative to a frame field $\{E_i\}$. To change A to type $(1,3)$ there are two choices: lower i or lower j. Take the former. Then the new tensor has components $A_i^{\ j}_{\ k\ell} = \Sigma g_{im}A^{mj}_{\ k\ell}$, where g_{ij}, as always, denotes the components $<E_i, E_j>$ of the metric tensor relative to this frame field.

To change to type $(3,1)$, operate similarly with the inverse matrix (g^{ij}) of (g_{ij}). Hence, raising the index k of $A^{ij}_{\ k\ell}$ gives $A^{ijk}_{\quad \ell} = \Sigma g^{km}A^{ij}_{\ m\ell}$.

Thus we can think of a tensor, classically, as its name $A, B, \ldots$ followed by a sequence of $r + s$ switches that can each be flipped *up* or *down*. (Such a flip is called *raising* or *lowering* an index.)

Note that in the Euler equation above, $\partial L/\partial \dot{x}^i = \Sigma g_{i\ell}\dot{x}^\ell$ is the ith tensor component of the covector field along α that is metrically equivalent to its (contravariant) tangent vector field α'. When α is a relativistic particle, this is the covariant expression of its energy-momentum.

A crucial advantage of the frame fields over coordinates is as follows:

Lemma 1.3.2 *Let* $\{E_i\}$ *be an* orthonormal *frame field, so* $g_{ij} = \delta_{ij}\varepsilon_j$, *with* $\varepsilon_i = \pm 1$. *The effect on components of raising of lowering an index* i *is merely multiplication by* ε_i.

Proof. For instance, suppose we lower the index i of $A^i{}_{jk}$. The new tensor has components:

$$A_{ijk} = \Sigma g_{im} A^m{}_{jk} = \Sigma_m \delta_{im} \varepsilon_i A^m{}_{jk} = \varepsilon_i A^i{}_{jk}.$$

$\square$

This last expression is not summed; by the *weak* Einstein convention in Section 1.2, Σ must appear explicitly in index summations.

Recall from Section 1.2 that contraction of an (r,s) tensor eliminates two component indices, one contravariant and one covariant, to give a tensor of type $(r - 1, s - 1)$. On a semi-Riemannian manifold, this contraction operation can be extended to any two indices of the same variance—one need only raise (or lower) one of them, then use the previous natural contraction. For example, contraction on the first and third covariant indices of a (1,4) tensor A gives the (1,2) tensor $C_{13}A$ whose components are $\Sigma g^{ab} A^i{}_{ajbk}$. Furthermore, if the frame field is *orthonormal*, the double sum in last expression reduces to the single sum $\Sigma \varepsilon_m A^i{}_{mjmk}$.

CURVATURE

The notion of Gaussian curvature of a surface has a far-reaching generalization to semi-Riemannian manifolds.

Definition 1.3.3 *The* Riemannian curvature tensor *of a semi-Riemannian manifold M is the function $R: \mathfrak{X} \times \mathfrak{X} \times \mathfrak{X} \to \mathfrak{X}$ given by*

$$R_{XY}Z = \nabla_X(\nabla_Y Z) - \nabla_Y(\nabla_X Z) - \nabla_{[X,Y]}Z,$$

where, as usual, ∇ is the Levi-Civita connection of M.

It is easily verified that R is, in fact, a tensor, although its ingredients ∇ and [,] are not. If vector fields X and Y have bracket zero (as in the case of coordinate vector fields), then the vector field $R_{XY}Z$ measures the failure of ∇_X and ∇_Y to commute. Thus curvature is likely to appear in any geometric computation involving more than one covariant derivative; it is the dominant invariant of differential geometry.

The definition of curvature is sometimes given with opposite sign; the choice above is the traditional one.

Proposition 1.3.4 (Symmetries of curvature). *For* $x, y, v, w \in T_p(M)$,
1. $R_{xy} = -R_{yx}$
2. $<R_{xy}v, w> = -<R_{xy}w, v>$,
3. $R_{xy}z + R_{yz}x + R_{zx}y = 0$,
4. $<R_{xy}v, w> = <R_{vw}x, y>$.

In terms of a frame field $\{E_i\}$ the components of the curvature tensor are

$$R^i_{\ jk\ell} = \omega^i(R_{E_k E_\ell} E_j),$$

hence the duality formula $X = \Sigma \omega^m(X)E_m$ from Section 1.2 gives

$$R_{\partial_k \partial_\ell}(\partial_j) = \Sigma R^m_{\ jk\ell} \partial_m.$$

Because of the symmetries of curvature, there is—but for sign—only one nonzero contraction of the curvature tensor: the *Ricci curvature tensor* Ric. The sign, by firm convention, is specified as follows:

Lemma 1.3.5 *In terms of an arbitrary frame field the components of* Ric *are* $R_{ij} = Ric(E_i, E_j) = \Sigma R^m_{\ imj}$. *In the orthonormal case,* $Ric(V, W) = \Sigma \varepsilon_m <R_{V E_m} E_m, W>$.

Ricci curvature is of crucial importance in relativity theory, because it is Ric, not the full curvature tensor R, that appears (slightly disguised) in the Einstein equation.

Contraction of Ric yields the *scalar curvature* $S = \Sigma g^{ij} R_{ij}$. Also the *Kretschmann curvature invariant* $k = <R, R> = \Sigma R^{ijk\ell} R_{ijk\ell}$ is sometimes useful.

The information contained in the curvature tensor R of M can also be expressed in terms of *sectional curvature* K, a real-valued function on the set of all nondegenerate 2-planes Π tangent to M. For any basis v, w for Π,

$$K(\Pi) = \frac{-<R_{vw}v, w>}{<v, v><w, w> - <v, w>^2}.$$

This expression is easily shown to be independent of the choice of basis. (That Π is nondegenerate means that the denominator above is nonzero; see Section 1.5).

If S is a surface, that is, a 2-dimensional Riemannian manifold, then for $p \in S$ the sectional curvature K of the 2-plane $T_p(S)$ is the classical *Gaussian curvature* of S at p.

The Riemannian curvature tensor R has a remarkable symmetry property involving its covariant derivatives. For any vector field V the covariant derivative operation ∇_V can be extended to apply to arbitrary tensor fields A, with $\nabla_V A$ having the same tensor type as A. For example, in the case of R, the (1,3) tensor field $\nabla_V R$ is given by

$$(\nabla_V R)_{XY} Z = \nabla_V(R_{XY} Z) - R_{\nabla_V X, Y}(Z) - R_{X, \nabla_V Y}(Z) - R_{XY}(\nabla_V Z).$$

Note that this expression expresses a natural Liebnizian product rule for $\nabla_V(R_{XY}Z)$. The *Bianchi curvature identities* assert that for all X, Y, Z the following cyclic relation holds:

$$(\nabla_X R)_{YZ} + (\nabla_Y R)_{ZX} + (\nabla_Z R)_{XY} = 0.$$

For details, see Section 13 of Misner, Thorne, Wheeler (1973) or Chapter 3 of O'Neill (1983). We use these identities only in Chapter 5, where they are reexpressed in the Newman-Penrose formalism.

The various forms of curvature are, of course, geometric invariants; that is, they are preserved (in an appropriate sense) by isometries. For instance, if $\psi: M \to N$ is an isometry, then $Ric_N(d\psi(v), d\psi(w)) = Ric_M(v, w)$ for all tangent vectors v, w on M.

KILLING VECTOR FIELDS

Let X be a vector field on a semi-Riemannian manifold M. Assume for simplicity that X is complete, so its flow $\{\psi_s\}$ is globally defined (Section 1.1). Then X is a *Killing vector field* provided each stage $\psi_s: M \to M$ is an isometry. Thus X is also called an *infinitesimal isometry*. For example, on the circle S^1 any vector field of constant length is Killing since its flow consists of rotations of S^1.

If X is not complete, the definition above remains valid using flows defined only locally.

Proposition 1.3.6 *The following conditions on a vector field X are equivalent:*
(1) *X is a Killing vector field.*
(2) *$X<V, W> = <[X, V], W> + <V, [X, W]>$ for all V, W.*
(3) *$<\nabla_V X, W> = -<\nabla_W X, V>$ for all V, W.*

Note that if X is a coordinate vector field, say ∂_r, the criterion (b) reduces to merely $\partial g_{ij}/\partial x^r = 0$ for all i, j.

As the definition above shows, Killing vector fields are a help in finding isometries $M \to M$. They also have a vital role in the study of geodesics, because of the following *conservation lemma*.

Lemma 1.3.7 *If X is a Killing vector field on M and γ is a geodesic, then the scalar product $<X, \gamma'>$ is constant along γ.*

Proof. Since $\gamma'' = 0, (d/ds)<X, \gamma'> = <\nabla_{\gamma'}(X), \gamma'>$, which, by criterion (3) in the preceding proposition, is zero. □

If M is connected, then any isometry ψ of M is completely determined by its differential map $d\psi_p$ at a single point $p\varepsilon M$. The corresponding Killing fact is that a Killing vector field X is completely determined by its values on any arbitrarily small neighborhood of a single point.

1.4   Extending Manifolds

Differential geometers usually begin work by assuming they have a complete semi-Riemannian manifold at their disposal; however, in relativity theory one is lucky to have a metric tensor defined on some small coordinate neighborhood $\mathcal{U}$. To study global consequences of that metric, then, it is necessary to extend the initial domain $\mathcal{U}$ as far as possible. We describe two such methods.

COORDINATE EXTENSION

Suppose $\mathcal{U}$ is a semi-Riemannian manifold that we want to extend. The following plan seems almost too simple. Find a diffeomorphism ϕ of $\mathcal{U}$ onto an open set $\mathcal{U}_1$ in a smooth (or analytic) manifold M; assign $\mathcal{U}_1$ the metric g_1 that makes ϕ an isometry. Extend the tensor field g_1 to a larger set $\mathcal{U}_2$, and let $\tilde{\mathcal{U}}$ be the open set on which $\tilde{g} = g_2|\tilde{\mathcal{U}}$ is still a (nondegenerate) metric tensor (so $\mathcal{U}_1 \subset \tilde{\mathcal{U}} \subset \mathcal{U}_2$). In favorable cases $\tilde{\mathcal{U}}$ is larger than $\mathcal{U}_1 \approx \mathcal{U}$; furnished with the metric $\tilde{g}$ it is the desired extension.

But where do we find the diffeomorphism $\phi: \mathcal{U} \to M$ and how do we extend g_1? In the following special case these questions answer themselves, and in later applications this case leads to more general extensions.

Let $\mathcal{U}$ be the domain of a coordinate system $x^1, \ldots, x^n$, and let g_{ij} be the components of a metric tensor g on $\mathcal{U}$. The coordinate system constitutes a diffeomorphism $\xi = (x^1, \ldots, x^n)$ of $\mathcal{U}$ onto its image $\mathcal{U}_1$, an open set in $\mathbf{R}^n$. As above, assign $\mathcal{U}_1$ the metric that makes ξ an isometry. Note that the components g_{ij} of this metric have exactly the same coordinate expressions as before, but with $x^1, \ldots, x^n$ now considered as the natural Cartesian coordinate functions of $\mathbf{R}^n$. It may well be possible to extend these functions g_{ij} over an open set $\mathcal{U}_2$ strictly larger than $\mathcal{U}_1$, and in the analytic case such extensions are unique. (In our applications they are gotten merely by continuing to use the same polynomials or trigonometric functions on the larger domain.)

Now if $\tilde{\mathcal{U}}$ is the region on which the extended components $\tilde{g}_{ij}$ define a (nondegenerate) metric tensor; then $\tilde{\mathcal{U}}$ is the desired extension of $\mathcal{U}$.

GLUING TOPOLOGICAL SPACES

Suppose that semi-Riemannian manifolds M and N are given, together with an isometry μ from an open set $\mathcal{U} \subset M$ onto an open set $V \subset N$. We shall describe a natural way to glue M and N together along $\mathcal{U} \approx V$, producing a new manifold Q. We call $\{M, N, \mu: \mathcal{U} \to V\}$ *gluing data* and μ the *matching map*. Only the topology of Q requires care, so assume at first that M and N are merely topological spaces. Manifold structures and metrics will be added later, almost automatically.

Let $M \cup N$ be the *disjoint union* of M and N; on it, let $\sim$ be the equivalence relation such that $p \sim q$ means

$$p = q \quad \text{or} \quad p = \mu(q) \quad \text{or} \quad q = \mu(p).$$

Then $Q = M \cup_\mu N$ will be the *quotient space* $M \cup N / \sim$, in which points $u \in \mathcal{U}$ and $\mu(u) \in V$ are fused into a single point of Q. The points of Q are of three mutually exclusive types:

$$m \in M - \mathcal{U},$$
$$n \in N - V,$$
$$(u, \mu(u)) = (\mu^{-1}(v), v), \quad \text{where } u \in \mathcal{U}, v \in V.$$

Define $i: M \to Q$ to be the map sending $m \in M - \mathcal{U}$ to $m \in Q$ and sending $u \in \mathcal{U}$ to $(u, \mu(u)) \in Q$. Similarly, $j: N \to Q$ sends $n \in N - V$ to n and $v \in V$ to $(\mu^{-1}(v), v)$. Hence, when restricted to $\mathcal{U}$, $i = j \circ \mu$.

The maps i and j are called *natural injections*; they combine to give a map $i \cup j: M \cup N \to Q$ that is just the usual natural projection for a quotient space (see Figure 1.1.) A subset S of Q is open if and only if $i^{-1}(S)$ is open in M and $j^{-1}(S)$ is open in N. Equivalently, a map $\phi: Q \to X$ is continuous if and only both $\phi \circ i$ and $\phi \circ j$ are continuous.

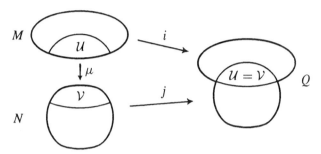

FIGURE 1.1. Gluing M and N along $\mathcal{U} \approx \mathcal{V}$ to form Q.

Lemma 1.4.1 *With notation as above, the natural injection i is a homeomorphism of $\mathcal{U}$ onto an open set $i(\mathcal{U})$ in Q, and j is a homeomorphism of $\mathcal{V}$ onto $j(\mathcal{V})$ in Q.*

Proof. A previous remark implies at once that i and j are continuous. We assert that i carries an open set W of M to an open set of Q. Now $i(W)$ is open if and only if $i^{-1}(i(W)$ and $j^{-1}(i(W))$ are open. Evidently $i^{-1}(i(W) = W$, which is open in M. Next,

$$j^{-1}(i(W)) = j^{-1}(i(W) \cap j(N)) = j^{-1}(i(W \cap \mathcal{U})) = \mu(W \cap \mathcal{U}),$$

which is open in N. In particular, $i(\mathcal{U})$ is open in Q.

Since i is one-one we conclude that it is a homeomorphism onto $i(\mathcal{U})$. The corresponding results hold for $j: N \to j(\mathcal{V})$.   □

We shall often ignore the natural injections, writing simply $i(M) = M$ and $j(N) = N$. Then the space Q is the union of M and N, with $M \cap N = \mathcal{U} \approx \mathcal{V}$.

Lemma 1.4.2 *Let $\{M, N, \mu: \mathcal{U} \to \mathcal{V}\}$ be gluing data. If $\phi_M: M \to P$ and $\phi_N: N \to P$ are continuous maps such that $\phi_M | \mathcal{U} = \phi_N \circ \mu$, then there exists a unique continuous map $\phi: Q \to P$ such that $\phi \circ i = \phi_M | \mathcal{U}$ and $\phi \circ j = \phi_N | \mathcal{V}$.*

Ignoring canonical injections, the assertion is that if ϕ_M and ϕ_N agree on $M \cap N$ they define a map of $Q = M \cup N$.

Proof. We must define ϕ on $i(M)$ by $\phi(i(m)) = \phi_M(m)$ and on $j(N)$ by $\phi(j(n)) = \phi_N(n)$. These definitions are consistent since if $i(m) = j(n)$, then $m \in \mathcal{U}$, $n \in \mathcal{V}$, and $n = \mu(m)$; hence $\phi_N(n) = \phi_N(j(m)) = \phi_M(m)$. By the preceding lemma, ϕ is continuous, since $\phi \circ i$ and $\phi \circ j$ are given as continuous maps. $\qquad\square$

A similar argument gives the following more general result.

Lemma 1.4.3 (Mapping Lemma). *Let M, N, $\mu: \mathcal{U} \to \mathcal{V}$ and M', N', $\mu': \mathcal{U}' \to \mathcal{V}'$ be two sets of gluing data. Let $\phi_M: M \to M'$ and $\phi_N: N \to N'$ be continuous mappings such that $\phi_M(\mathcal{U}) \subset \mathcal{U}'$, $\phi_N(\mathcal{V}) \subset \mathcal{V}'$, and $\mu' \circ \phi_M | \mathcal{U} = \phi_N \circ \mu$. Then there is a unique continuous mapping $\phi: Q \to Q'$ such that $\phi \circ i = i' \circ \phi_M$ and $\phi \circ j = j' \circ \phi_N$. (If natural injections are ignored, the latter conditions are just $\phi | M = \phi_M$ and $\phi | N = \phi_N$.)*

There is just one problem: By definition, a semi-Riemannian manifold must be a Hausdorff space, but even when M and N are Hausdorff, Q need not be. For example, let $M = N = \mathbf{R}^1$ and let $\mathcal{U} = \mathcal{V} = \{t : t < 0\}$, with μ the identity map. Then gluing produces a branched line Q as in Figure 1.2. There are two zeroes 0 and $0'$, and every neighborhood of 0 meets every neighborhood of $0'$, so the Hausdorff condition fails.

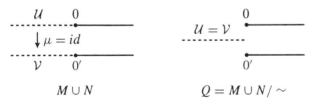

FIGURE 1.2. Non-Hausdorff gluing.

A practical way to prevent this difficulty is described as follows.

Definition 1.4.4 *Gluing data M, N, $\mu: \mathcal{U} \to \mathcal{V}$ satisfy the* Hausdorff condition *if there exists no sequence $\{p_n\}$ in U such that both*

$$\{p_n\} \to p \in M - \mathcal{U} \quad and \quad \{\mu(p_n)\} \to q \in N - \mathcal{V}.$$

Moved to $M \cup_\mu N$, such a sequence would be converging to two different points (as can certainly happen in the example shown in Figure 1.2).

Lemma 1.4.5 *If $M, N, \mu \colon \mathcal{U} \to \mathcal{V}$ satisfy the Hausdorff condition (with M and N Hausdorff), then $Q = M \cup_\mu N$ is a Hausdorff space.*

Proof. Let $x \neq y$ be points of $Q = i(M) \cup j(N)$. If both are in $i(M)$, then x and y have disjoint neighborhoods in M since $i(M)$ is open and M is Hausdorff; similarly for $j(N)$. There remains the case: $x \in i(M) - j(\mathcal{V})$ and $y \in i(N) - i(\mathcal{U})$. Let $\{\mathcal{N}_k \colon k = 1, 2, \ldots\}$ be a decreasing basis for the neighborhoods of x in the open set $i(M)$, with $\{\mathcal{N}'_k\}$ analogous for y. We assert that for some integer k, $\mathcal{N}_k$, and $\mathcal{N}'_k$ are disjoint.

Assume not; then for every $k \geq 1$ there is a point $x_k \in \mathcal{N}_k \cap \mathcal{N}'_k$. Thus, the sequence $\{x_k\}$ converges to both x and y. Applying the homeomorphism $i^{-1} \colon i(M) \to M$ gives a sequence $\{i^{-1}(x_k)\}$ in $\mathcal{U}$ that converges to $p = i^{-1}(x) \in M - \mathcal{U}$. Similarly $\{j^{-1}(x_k)\}$ in $\mathcal{V}$ converges to $q = j^{-1}(y) \in N - \mathcal{V}$. Since $\mu \circ i^{-1} = j^{-1}$, this contradicts the Hausdorff condition.   □

Example 1.4.6 Let $M = N$ be the plane $\mathbf{R}^2$ with rectangular coordinates r, t. Let $\mathcal{U} = \mathcal{V} = \{(r, t) \colon r < 0\}$. If $\mu \colon \mathcal{U} \to \mathcal{V}$ is taken to be the identity map, then, as in the example above, $Q = M \cup_\mu N$ is not Hausdorff. So redefine μ by $\mu(r, t) = (r, t/r)$. Now the Hausdorff condition holds because if a sequence (r_n, t_n) in $\mathcal{U} = \{(r, t) \colon r < 0\}$ converges to a point (r_o, t_o) of $M - \mathcal{U}$, then necessarily $r_o = 0$. But then the sequence $\mu(r_n, t_n) = (r_n, t_n/r_n)$ *diverges* since $t_n/r_n \to \infty$. Thus $Q = M \cup_\mu N$ is a Hausdorff space.

Since this example is a prototype for the constructions in Chapter 3, we consider some of its properties. In Figure 1.3 the planes M and N are drawn as rectangles with horizontal r-axis. Q also is homeomorphic to $\mathbf{R}^2$. However, r and t are no longer coordinate functions; indeed, t is not even well defined on Q, though r is. In Q the level curve $r = r_o$ has one component if $r_o < 0$, but two if $r_o \geq 0$.

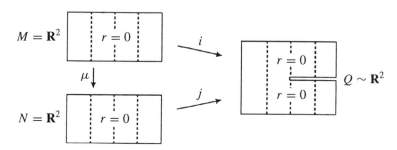

FIGURE 1.3. Q is made of two 2-planes $\mathbf{R}^2(r, t)$ glued along the sets $\{r \colon r < 0\}$.

GLUING SEMI-RIEMANNIAN MANIFOLDS

Given gluing data with M and N manifolds and $\mu\colon \mathcal{U} \to \mathcal{V}$ a smooth map, there is a unique way to make Q a manifold with $\mathcal{U}$ and $\mathcal{V}$ as submanifolds. It suffices to assign Q the atlas consisting of all coordinate systems of $M \approx i(M)$ and all those of $N \approx j(N)$. It is easy to check that those in $i(M)$ overlap smoothly with those in $j(N)$.

The mapping results above remain valid with *continuous* replaced by *smooth*, so homeomorphisms become diffeomorphisms. Furthermore, *smooth* can in turn be everywhere replaced by *analytic*.

Finally, if M and N are semi-Riemannian manifolds and the matching map μ is an isometry, then the metric tensors of M and N agree on $M \cap N = \mathcal{U} \approx \mathcal{V}$, and hence give a metric tensor on Q. Thus Q is a semi-Riemannian manifold that extends both M and N. Such extensions will be used repeatedly in the construction of maximal Kerr spacetimes.

In general, properties of M and N that are preserved by μ will obtain on Q. For example, if M and N are Lorentz 4-manifolds, then Q is a Lorentz 4-manifold (since signature and dimension are preserved by isometries).

Similarly, the mapping results above remain valid with smooth manifolds replaced by semi-Riemannian manifolds and smooth maps replaced by (local) isometries.

1.5 Lorentz Vector Spaces

Relativity theory takes place in Lorentz manifolds, and many of its most striking features derive easily from the linear algebra of the tangent spaces to such manifolds. In this section we examine these tangent spaces abstractly and derive some of their basic properties.

Let V be a real vector space of dimension n. As mentioned in Section 1.2, a *scalar product g* on V is a symmetric, nondegenerate bilinear form on V. Vectors are *orthogonal*, written $v \perp w$ if $g(v, w) = 0$. Consequently, the nondegeneracy of g asserts that only the zero vector is orthogonal to every vector; that is, $v \perp V \Rightarrow v = 0$. A standard criterion for nondegeneracy is that for one (hence every) basis $v_1, \ldots, v_n$ for V the matrix $g_{ij} = g(v_i, v_j)$ has nonzero determinant.

As mentioned previously, $g(v, w)$ is often written as $<v, w>$. The vector space V, furnished with a scalar product g, is called a *scalar product space*. If g is positive definite, it is an *inner product*, and V is then an *inner product space*.

Since $<v, v>$ may well be negative for vectors in V, the *norm* $|v|$ of v is defined to be $|<v, v>|^{1/2}$. If $|u| = 1$, that is, if $<u, u> = \pm 1$, then u is a *unit vector*. A set of mutually orthogonal unit vectors is said to be *orthonormal*.

Proposition 1.5.1   *Every scalar product space $V \neq 0$ has an orthonormal basis. In fact, any orthonormal set $e_1, \ldots, e_k$ in V can be enlarged to an orthonormal basis.*

Proof.   The Gram–Schmidt formulas fail here, but there is an easy inductive proof.

1. There exists a vector $v \in V$ such that $<v, v> \neq 0$, for otherwise, $0 = <v + w, v + w> = 2<v, w>$ for all v, w. Let $e_1 = v/|v|$.

2. If $e_1, \ldots, e_k$ is orthonormal set with $k < n = \dim V$, there is a vector e_{k+1} such that $e_1, \ldots, e_k, e_{k+1}$ is orthonormal.

Proof. The scalar product is nondegenerate on the subspace W consisting of all vectors orthogonal to span $\{e_1, \ldots, e_k\}$, for otherwise some nonzero vector in W would be orthogonal to all of V. Now apply (1) to W. $\square$

Associated with an orthonormal basis $e_1, \ldots, e_n$ are n numbers $\varepsilon_i = \pm 1$ such that $<e_i, e_j> = \delta_{ij}\varepsilon_i$.

Proposition 1.5.2 (Sylvester).   *Every orthonormal basis for V has the same signs $\varepsilon_1, \ldots, \varepsilon_n$ (but for order).*

We call $(\varepsilon_1, \ldots, \varepsilon_n)$ the signature of V—usually listing the negative signs first. The number ν of these negative signs is the *index* of V.

Using the signs ε_i, *orthonormal expansion* is expressed as $v = \Sigma \varepsilon_i <v, e_i> e_i$. To see this, note that v minus the Σ is orthogonal to every e_i, hence to V. Then by nondegeneracy, $v - \Sigma = 0$.

Definition 1.5.3   *A Lorentz vector space V is a scalar product space of dimension $n \geq 2$ and index $\nu = 1$.*

The Lorentz signature is thus $(-1, +1, \ldots, +1)$. The alternative $(+1, -1, \ldots, -1)$ is an equivalent possibility. From now on, V denotes a Lorentz vector space of dimension n.

We now consider some characteristic properties of Lorentz vector spaces not present in inner product spaces. These can be readily tested in the simplest Lorentz vector space: the *Minkowski plane* $\mathbf{R}_1^2$, which is $\mathbf{R}^2$ with scalar product $<(x, t), (x', t')> = xx' - tt'$.

Definition 1.5.4 *A vector v in a Lorentz vector space V is*

- spacelike *if* $<v, v> > 0$ *or* $v = 0$,
- null *(or* lightlike*) if* $<v, v> = 0$ *and* $v \neq 0$,
- timelike *if* $<v, v> < 0$.

The type into which v falls is called its causal character.

Perhaps the most useful single fact about a Lorentz vector space is this:

Lemma 1.5.5 *A vector orthogonal to a timelike vector z in V is spacelike. Thus $z^{\perp} = \{v \in V : v \perp z\}$ is an inner product space, and V is the direct sum $\mathbf{R}z \oplus z^{\perp}$.*

Proof. Since $u = z/|z|$ is a unit vector, Lemma 1.5.5 shows that there is an orthonormal basis $u, e_1, \ldots, e_{n-1}$. If $v \perp z$, so $v \perp u$, then orthonormal expansion gives $v = b_1 e_1 + \cdots + b_{n-1} e_{n-1}$. Hence $<v, v> = \Sigma b_i^2 \geq 0$, and equality here implies $v = 0$, so v is spacelike.

Evidently $u^{\perp}$ is the subspace spanned by $e_1, \ldots, e_{n-1}$, so the preceding argument shows that $u^{\perp}$ is an inner product space (under the scalar product of V). Orthonormal expansion gives $V = \mathbf{R}u \oplus u^{\perp}$, and $u^{\perp} = z^{\perp}$. □

In many proofs it is helpful to express vectors v in terms of this direct sum as $v = au + x$, where u is a timelike unit vector and $x \in u^{\perp}$. Then we have the valuable identity, $<v, v> = -a^2 + |x|^2$.

The set Λ of all null vectors in V is called its *nullcone*. (In general terms, a *cone* in V is a subset closed under multiplication by positive scalars.) Like any n-dimensional real vector space, V has a natural manifold structure whose coordinate systems are the linear isomorphisms $V \to \mathbf{R}^n$.

Lemma 1.5.6 *The nullcone Λ of V has two components Λ^+ and Λ^-, with $\Lambda^- = -\Lambda^+$. Each component is a cone diffeomorphic to $\mathbf{R}^+ \times S^{n-2}$ (see Fig. 1.4).*

Proof. Pick a unit timelike vector u in V, and, as suggested above, write each null vector v as $au + x$, with $x \perp u$. Let Λ^+ consist of those with $a > 0$ and Λ^- those with $a < 0$ (none have $a = 0$). Evidently Λ, Λ^+, and Λ^- are cones, and $\Lambda^- = -\Lambda^+$. Since v is null, $|x|^2 = a^2$.

We now show that Λ^+ is diffeomorphic to $\mathbf{R}^+ \times S^{n-2}$, where S^{n-2} is the unit sphere in the inner product space $u^{\perp} \approx \mathbf{R}^{n-1}$. If $v = au + x$ is in Λ^+, then $|x| = a$, so we define $v \to (a, x/a) \in \mathbf{R}^+ \times S^{n-2}$. A map in the reverse direction is gotten

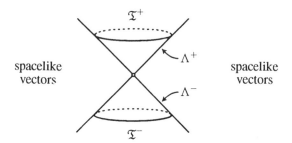

FIGURE 1.4. Nullcones $\Lambda^{\pm}$ and timecones $\mathfrak{T}^{\pm}$ in a Lorentz vector space.

by sending $(r, y) \in \mathbf{R}^+ \times S^{n-2}$ to $r(u + y)$, which is indeed a null vector. It is readily checked that the two maps are inverses.

For $n \geq 3$, the sphere S^{n-2} is connected, hence the disjoint sets Λ^+ and Λ^- are actually the connected components of the full nullcone.

The case $n = 2$ is exceptional since $S^0 = \{\pm 1\}$, so Λ^+ and Λ^- are not connected; each consists of two rays. (Take $\mathbf{R}_1^2$ for concreteness; there Λ is the union of the four $\pm 45°$ rays from the origin.) $\square$

Λ^+ and Λ^- are also referred to as *nullcones*. Despite the notation, there is no invariant way to distinguish one from the other.

The set $\mathfrak{T}$ of all timelike vectors in V is called its *timecone*.

Lemma 1.5.7 *The timecone $\mathfrak{T}$ in V has two components $\mathfrak{T}^+$ and $\mathfrak{T}^-$. Each is an open convex cone, $\mathfrak{T}^- = -\mathfrak{T}^+$, and (choosing $\pm$ consistently)*

$$bdry\ \mathfrak{T}^+ = \Lambda^+ \cup \{0\},$$
$$bdry\ \mathfrak{T}^- = \Lambda^- \cup \{0\}.$$

Proof. Pick a unit timelike vector u and write every vector v as $au + x$, with $x \perp u$. Since $<v, v> = -a^2 + |x|^2$, v is timelike if and only if $a^2 > |x|^2$.

Let $\mathfrak{T}^+$ consist of those $v \in \mathfrak{T}$ for which $a > 0$, and similarly for $\mathfrak{T}^-$ with $a < 0$. Evidently $\mathfrak{T}, \mathfrak{T}^+$, and $\mathfrak{T}^-$ are cones, with $\mathfrak{T}^- = -\mathfrak{T}^+$. Since they are defined by inequalities, all three cones are open sets of V.

To prove that, say, $\mathfrak{T}^+$ is convex, note first that for any vector v in V the set $H_v = \{w \in V: <w, v> < 0\}$ is convex, that is, if $w_1, w_2 \in \mathfrak{T}^+$ then the vector $w_t = tw_1 + (1 - t)w_2$ is also in $\mathfrak{T}^+$ for all $0 < t < 1$. Since an intersection of convex sets is convex, it suffices to show that $\mathfrak{T}^+$ is the intersection of the sets H_{u+e} for the unit vector u and all (spacelike) unit vectors $e \perp u$.

Suppose first that $v \in \mathfrak{T}^+$. Write $v = au + x$ as before, so $a > |x|$. Then for any $e \perp u$, by using the Schwarz inequality we find

$$<v, u + e> = <au + x, u + e> = -a + <x, e> \leq -a + |x| < 0.$$

Hence $v \in \cap H_{u+e}$.

In showing that $v \in \cap H_{u+e}$ implies $v \in \mathfrak{T}^+$ we can suppose $v \neq \lambda u$, so $v = au + x$, with $0 \neq x \perp u$. In particular, $v \in H_{u+x/|x|}$, hence

$$0 > <au + x, u + x/|x|> = -a + |x| < 0. \quad \text{Thus } v \in \mathfrak{T}^+.$$

$\square$

Corollary 1.5.8 *Two timelike vectors z, z' are in the same timecone if and only if $<z, z'> \, < 0$.*

Proof. Consider the function $f: \mathfrak{T} \to R$ given by $f(v) = <v, z>$. Now $f(z) < 0$, and timecones $\mathfrak{T}^{\pm}$ are connected, hence $f < 0$ on the timecone, say $\mathfrak{T}^+$, that contains z. Thus if z' is also in $\mathfrak{T}^+$, $<z, z'> \, = f(z') < 0$. But if $z' \in \mathfrak{T}^-$, then $-z' \in \mathfrak{T}^+$; so $<z, z'> \, = -<z, -z'> \, > 0$. $\square$

A vector v that is not spacelike is said to be *nonspacelike* (or *causal*). The *causal cones* in V are $\mathfrak{T}^+ \cup \Lambda^+$ and $\mathfrak{T}^- \cup \Lambda^-$ (with consistent $\pm$). The following variant of the result above is sometimes needed: A timelike vector z and a causal vector w are in the same causal cone if and only if $<z, w> \, < 0$.

The notion of causal character can be generalized to subspaces W by considering the restriction to W of the scalar product of V.

Definition 1.5.9 *Let V be a Lorentz vector space with scalar product g. A subspace W of V is*

- spacelike *if $g|W$ is positive definite,*
- null *if $g|W$ is degenerate,*
- timelike *if $g|W$ is nondegenerate and has index 1.*

It follows easily from Propositions 1.5.1 and 1.5.2 that each subspace W of V falls into exactly one of these three types—this type called the *causal character* of W. The null case, which at first may seem pathological, is a crucial to Lorentz geometry and hence to general relativity.

If dim $W = 1$, this definition is consistent with Definition 1.5.4 in the sense that an individual vector v has the same causal character as the subspace $\mathbf{R}v$ it spans. In particular, the zero subspace, like the zero vector, is spacelike.

Evidently, W is spacelike if and only if it is an inner product space (the Schwarz inequality is then available in W). Excluding the one-dimensional case, W is timelike if and if only if W is a Lorentz vector space.

As Figure 1.5 illustrates, the causal character of a subspace is determined geometrically by its relation to the nullcone. This is clear enough in the timelike and spacelike cases, so we examine the more interesting null case.

Lemma 1.5.10 *In a Lorentz vector space, orthogonal null vectors are collinear.*

Proof. Let v and w be orthogonal null vectors. If they are independent, then the two-dimensional space N they span consists entirely of null vectors (save for 0). But by Lemma 1.5.5, V always contains an $(n-1)$-dimensional spacelike subspace S. A dimension count shows $N \cap S \neq 0$: contradiction. □

Lemma 1.5.11 *If W is a null subspace of V, then*
(1) *There is a nonzero vector $d \in W$, unique up to nonzero scalar multiplication, such that $d \perp W$. (In particular, d is null).*
(2) *Every vector in $W - \mathbf{R}d$ is spacelike (see Figure 1.5).*

Proof. (1) Since $g|_W$ is, by definition, degenerate, such a vector $d \neq 0$ exists. By Lemma 1.5.10 two such vectors cannot be independent, and nonzero scalar multiplication does not affect the property $d \perp W$.

(2) Let $x \in W - \mathbf{R}d$. In particular, $d \perp x$. If x were timelike, Lemma 1.5.5 would make d spacelike. If x were null, Lemma 1.5.10 would give $x \in \mathbf{R}d$. Since neither is possible, x can only be spacelike. □

For a subspace $W \subset V$, let $W^{\perp} = \{v : v \perp W\}$, as usual. If W is spacelike [timelike], then extending an orthonormal basis for W to an orthonormal basis

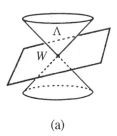

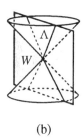

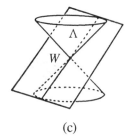

(a)   (b)   (c)

FIGURE 1.5.   Subspaces W that are (a) spacelike, (b) timelike, and (c) null. In the null case, W is tangent to the nullcone Λ.

for V shows that $W^\perp$ is timelike [spacelike] and $V = W \oplus W^\perp$. Since $(W^\perp)^\perp = W$ it follows that W null implies $W^\perp$ null.

1.6 Introduction to General Relativity

This section reviews some of the fundamentals of relativity theory.

As a basis for comparison, let S be a 3-dimensional Riemannian manifold called *space*. A *particle* in space is a curve $\alpha: I \to S$, where naturally we think of the parameter as *time*, so $\alpha(t) \in S$ is the location of the particle at time t. In Newtonian physics, S is usually 3-dimensional Euclidean space $\mathbf{R}^3$, and the motion of the particle is governed by physical laws. The case of importance for us is the classic Newtonian one: a mass M is located at the origin of $S = \mathbf{R}^3$, and particles of various masses $m \ll M$ move in S in obedience to Newton's gravitational law and laws of motion. Let us translate this situation into relativity.

As a first step, form the Riemannian product manifold $S \times \mathbf{R}^1$, considering $\mathbf{R}^1$ (with its usual metric) as a time axis. Call this manifold *space-time*. Each of its points (x, t) is an *event*: an instantaneous happening at position x and time t. A particle α is now viewed as a curve $t \to (\alpha(t), t)$ in space-time (see Figure 1.6).

A second step toward general relativity is to replace the line-element $d\sigma^2 + dt^2$ of the Riemannian product by the Lorentz line-element $d\sigma^2 - dt^2$. The timecones of this *spacetime*, as we shall see, furnish a mathematical way to express the physical observation that, in a vacuum, no material particle travels as fast as light. (Taking the space S to be $\mathbf{R}^3$ and setting mass M $= 0$ gives *Minkowski spacetime*, the arena of special relativity.)

The third step is the substantial one: Replace the preceding spacetime (four-dimensional Lorentz manifold) by one whose geometry has absorbed both Newtonian laws in this sense: *Particles that move solely under the gravitational influence of the central mass* (physical concept) *are geodesics* (geometric concept).

The first such spacetime of this kind was found by Karl Schwarzschild only weeks after Einstein's 1915 formulation of general relativity. The star that produces Schwarzschild gravitation does not rotate, though most astronomical bodies do. (Rotation is gravitationally irrelevant in Newtonian theory but not so in relativity where, in addition to mass, energy of the rotation is a source of gravity.) Almost a half century passed before Roy Kerr discovered the rotating model that now bears his name (see the Introduction).

Discard of the product structure (space) $\times$ (time) forces the fusion of these two concepts in relativity theory. Minkowski said it best: "Henceforth space by itself, and time by itself are doomed to fade away into mere shadows, and only a kind

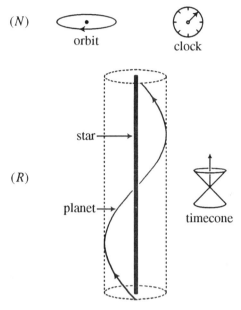

FIGURE 1.6.   A planet in circular orbit around a star. In the Newtonian model (N) the planet travels repeatedly around a fixed circle in space, parametrized by universal Newtonian time. In the relativistic model (R) the planet spirals through spacetime, parametrized by its own proper time.

of union of the two will preserve an independent reality." The main psychological problem in learning relativity is to abandon the entrenched idea that time and space are separate concepts. The difficulty is compounded by that fact that in relativity the two are repeatedly—though not invariantly—being separated.

We have considered only relativistic analogues of a central force field. In general, of course, relativity aims at giving spacetime models for gravitational fields of all kinds; however, few are known that have physical significance.

Now we define more carefully some of the concepts mentioned above.

SPACETIMES

A spacetime begins with a four-dimensional Lorentz manifold M, whose points are called *events*. M is required to be connected, since no interaction would be possible between different components. Additional structure is required to express the natural idea that we are inexorably proceeding into the future. Since each tangent space to M is a Lorentz vector space, for each $p \in M$ the two components of the timecone $\mathfrak{T}$ in $T_p(M)$ are available to distinguish future from past. To select, in a continuous fashion, one of the two at each point of M and is to *time-orient M*. The selected cones are called *future timecones* (or simply *futurecones*), the others *past timecones*. Using partitions of unity it can easily be shown a Lorentz manifold is time-orientable if and only if it admits a nonspacelike vector field (globally defined, continuous). Recall that a curve or vector field is timelike, spacelike, or null, respectively, if each of its tangent vectors has that causal character; similarly for nonspacelike and nontimelike.

Definition 1.6.1 *A* spacetime *is a connected time-oriented four-dimensional Lorentz manifold.*

A timelike tangent vector z is *future-pointing* if it is in a future timecone, otherwise it is past-pointing. The nullcone in the boundary of each future timecone is a *future nullcone*, and its vectors are *future-pointing* (see Figure 1.7). A nonspacelike curve or vector field is *future-pointing* if all its tangent vectors vectors are future-pointing.

The term *spacetime* is also applied when dimension 4 is replaced by 3 or 2. These allow relativistic descriptions of the analogues of Newtonian motion in a surface or on a line, respectively.

Implicit in the definition of spacetime is Einstein's insight that the physical invariants of a gravitational fields are exactly the geometric invariants of spacetimes.

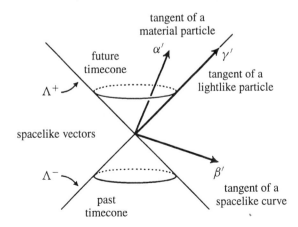

FIGURE 1.7. Tangents and timecones.

This may seem almost to be a truism, but its effect is revolutionary: No concept expressed in terms of coordinates is physically significant unless it can be shown to be independent of the choice of (spacetime) coordinates.

PARTICLES

Informally, a particle is an object that is small compared to the typical distances in its environment: a planet orbiting a star, an astronaut landing on a planet.

Definition 1.6.2 *A material particle* α *in a spacetime* M *is a future-pointing timelike curve* $\alpha: I \to M$. *The proper time* τ *of* α *is its arc length function, and its mass is* $m = |\alpha'| > 0$. *(So both* τ *and* m *are functions on the domain* I *of* α).

Thus, although universal time has been banished from relativity, every material particle α has its own time. Proper time τ is determined only up to an additive "clock-setting" constant since arc length can be measured from any initial point, but time intervals $\Delta\tau$ are well defined: As α proceeds from event $p = \alpha(\tau_0)$ to event $q = \alpha(\tau_1)$ its elapsed time is $\Delta\tau = \tau_1 - \tau_0$.

Many of the novelties of relativity are illustrated in the so-called "Twin Paradox." A pair of twins part at event p. When they meet later on, at event m, they will generally not be the same age—without universal time, no reason presents itself as to why they should be. The phenomenon is a standard feature of particle physics and has even been checked as simply as by carrying one of a pair of atomic clocks around the earth on commercial airlines (see Sachs, Wu 1977).

Each twin's proper time proceeds at its usual pace, no matter how he travels. The time differences at the meeting m cannot be attributed to the twins travelling at different speeds. Speed is undefined in general relativity; indeed, there is no absolute way to decide whether a particle is moving or not, since there is nothing absolute to measure against. The basic dichotomy in relativity is not between motion and rest but between spacetime acceleration and nonacceleration. At their meeting a nonaccelerating twin will be older than his accelerating twin (see examples in Misner, Thorne, Wheeler 1973).

Fundamental to relativity is the exceptional character of light (i.e., electromagnetic radiation), which is carried by photons.

Definition 1.6.3 *A lightlike particle* γ *in a spacetime* M *is a future-pointing null geodesic.*

In this sense it is still geometrically true that light "travels in straight lines." A lightlike particle γ has mass zero, since $|\gamma'| = 0$, so (lacking a clock) it has no proper time.

By contrast, a material particle α may or may not be a geodesic. In general, *freely falling* means moving solely under the influence of gravity; thus, in relativity, a particle is freely falling if and only if it is a geodesic. This in sharp contrast with Newtonian terminology. For example, the (freely falling) orbiting planet mentioned above has nonzero spatial Newtonian acceleration, but no spacetime acceleration. Thus a spacetime realizing Figure 1.6 must have a metric tensor that makes the spiralling orbit a geodesic.

A freely falling particle has constant mass m since, as for any geodesic, $<\alpha', \alpha'> = -m^2$ is constant.

Remark 1.6.4 In physical terminology, the tangent vector $\alpha' = d\alpha/ds$ of a material particle is called its *energy-momentum 4-vector* and denoted by p. As we soon show, it is the relativistic fusion of Newtonian energy and momentum. Often α is reparametrized by its proper time τ (i.e., by arc length.) For such a parametrization $\tilde{\alpha}$ we have $p = d\alpha/ds = m\,d\tilde{\alpha}/d\tau$ since by definition $m = |d\alpha/ds| = d\tau/ds$. The unit vector $d\tilde{\alpha}/d\tau$, called the 4-*velocity* of the particle, is the fusion of Newtonian energy per unit mass and Newtonian velocity (or "3-*velocity*"). In practice a separate notation $\tilde{\alpha}$ is seldom used, the parametrization of α being inferred from context.

For a lightlike particle γ, the *energy-momentum* 4-*vector* is just $p = \gamma'$.

CURVATURE

The central idea of general relativity is that a gravitational field should be expressed by a curved spacetime in which freely falling particles are geodesics. The physical effect of gravity on the particles—expressed directly in Newton's physics by his laws—is now indirectly expressed as the geometric effect of Riemannian curvature on geodesics. In situations that are only mildly relativistic (i.e., slow speeds and weak gravity), it has been shown that, in a properly chosen spacetime, the gravitational effects predicted by relativity are well approximated by those of Newtonian mechanics—at least in the case of material particles. The effect of gravity on light (called "bending" in Newtonian language) is purely relativistic.

Einstein himself used an approximate relativistic model of the solar system to put the finishing touches on general relativity, adjusting it to eliminate the famous shortfall in the Newtonian prediction for the advance of the perihelion of Mercury. The final touch was the precise form of the *Einstein equation*, which links the curvature of the spacetime model to the physical properties of the source of the gravitational field. Three natural geometric levels of curvature are physically significant.

1. In special relativity, the curvature vanishes; that is, the spacetime is *flat*. Thus, the theory applies only in situations where gravity is negligible (e.g., particle physics).

2. A *vacuum* or *empty* spacetime contains no sources of gravity, and this, by the Einstein equation, is equivalent to the vanishing of Ricci curvature: such spacetime is *Ricci flat*. Thus, a model of the gravitational field of a single isolated star—omitting the star itself—must be Ricci flat. The *Riemannian* curvature of such a model need not vanish; indeed, its geometric effect on freely falling particles shows intuititively how gravity can be described by curvature.

3. When the Ricci curvature does not vanish, it describes the motion of the source of the gravity. For example, in cosmology the spacetime models are filled with the flow of the galaxies whose gravitational interaction—expressed as Ricci curvature—accounts for their relative motion.

STATIONARY OBSERVERS

An *observer* in a spacetime is simply a material particle parametrized by proper time. Observers enjoy sending and receiving messages, and they keep close track of their proper time. In special relativity a single freely falling observer can impose his proper time (and space) on the entire Minkowski spacetime, but in general relativity a whole family of observers is needed for analogous results.

An *observer field* U on a spacetime M is a future-pointing, timelike unit vector field. Thus each integral curve of U is an observer, called a U-*observer*.

Definition 1.6.5 *An observer field U on M is* stationary *provided there exists a smooth function $f > 0$ on M such that fU is a Killing vector field. (Then M is* stationary *relative to U.) If U is also hypersurface-orthogonal (i.e., if $U^{\perp}$ is integrable), then U is* static.

The physical rationale is this: If U is stationary, the local flows $\{\psi_s\}$ of fU are isometries that carry each U-observer α to itself (though usually distorting proper time). Thus, for U-observers the local universe is not changing. A spacetime is *absolutely stationary* if it has a unique stationary observer field, and in this case the U-observers are said to be *at rest*.

If U is static, the integral manifolds of $U^\perp$ are three-dimensional, spacelike submanifolds that are isometric under the flow, and hence constitute a common "space" for the U-observers. In this sense the spacetime has been separated into space and time in a significant way.

It is rare for an entire spacetime M to be stationary, but M may well have connected open regions $\mathcal{U}$—spacetimes in their own right—that are stationary. Practically speaking, a spacetime modeling the gravitational field outside a single star is stationary if the physical properties of the star are not changing (e.g., the star is not collapsing or exploding). If the star is also not rotating, the spacetime is static.

In a spacetime suppose $\xi = (x^0, x^1, x^2, x^3)$ is a coordinate system for which the coordinate vector field ∂_0 is timelike future-pointing and span $\{\partial_1, \partial_2, \partial_3\}$ is space-like. Then the (suitably reparametrized) x^0 coordinate curves are observers, and the coordinate slices $x^0 = $ constant are three-dimensional Riemannian manifolds. Such coordinates provide a framework vaguely like Newton's in which one might hope to see relativistic analogues of Newtonian notions. But such coordinates exist plentifully at every point of every Lorentz manifold; invariant results can be deduced only if the coordinate system is significantly related to the geometry of the spacetime. One example is as follows:

Lemma 1.6.6 *If x^0, x^1, x^2, x^3 is a coordinate system such that ∂_0 is timelike future-pointing and $\partial g_{ij}/\partial x^0 = 0$ for all i, j, then the observer field $U = \partial_0/|\partial_0|$ is stationary. Conversely, for a stationary U, such a coordinate system can be found at every point.*

Proof. For such a coordinate system, $\partial g_{ij}/\partial x^0 = 0$ for all i, j implies that ∂_0 is a Killing vector field (Section 1.3). Then $fU = \partial_0$, where $f = |\partial_0| > 0$.

Conversely, let $f > 0$ be a function such that fU is Killing. Since fU is nonvanishing, at each point there is a coordinate system $x^0, \ldots, x^3$ such that $\partial_0 = fU$. But ∂_0 Killing implies $\partial g_{ij}/\partial x^0 = 0$ for all i, j. $\square$

To adjust this lemma to the static case, add the coordinate condition $g_{0j} = 0$ for all $j > 0$.

INSTANTANEOUS OBSERVERS

Separation of space and time is always possible infinitesimally. Define an *instantaneous observer* at an event $p \in M$ to be a future-pointing timelike unit vector u in $T_p(M)$. We imagine that u has the instantaneous information common to every ordinary observer α whose 4-velocity is u as it passes through p. Specifically, u knows the tangent space $T_p(M)$ and breaks it up into time and space by the time-axis $\mathbf{R}u$ in $T_p(M)$ and the space $u^\perp \approx \mathbf{R}^3$. There is a natural way for u to measure the speed of any particle α as it passes through p (say, at proper time τ_0). If α is a material particle, then $\alpha'(\tau_0) = au + x$, where $x \in u^\perp$. As Figure 1.8 suggests, the number a represents the instantaneous rate at which u's time, t, is increasing relative to α's time τ; so we write $a = dt/d\tau$. Similarly, $|x|$ is the rate at which arc length σ in $u^\perp$ is increasing relative to τ, so $|x| = d\sigma/d\tau$. Thus u measures the speed of α at event p as

$$\frac{d\sigma}{d\tau} = \frac{d\sigma/d\tau}{d\tau/dt} = \frac{|x|}{a} < 1$$

The final inequality arises as follows: Since α is timelike, $0 > <\alpha', \alpha'>|_p = -a^2 + |x|^2$, and since both u and α' are future-pointing, $a > 0$. Hence $|x| < a$.

When α is replaced by a lightlike particle γ, then since γ' is null, $0 = <\gamma', \gamma'> = -a^2 + |x|^2$, and thus $d\sigma/ds = 1$.

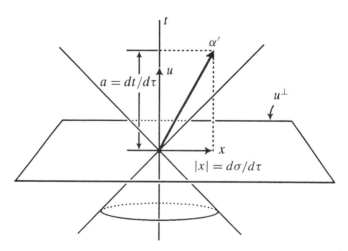

FIGURE 1.8.   Measurements of a particle α by an instantaneous observer u in terms of u's time t and space $u^\perp$.

In conclusion, instantaneous observers measure the speed of light to be $c = 1$ (in geometric units; see Appendix A), and they measure the speed of any material particle to be strictly less than 1.

Though all such observers agree on the speed of light, they seldom agree on anything else. For example, let $u_1 \neq u_2$ be instantaneous observers at $p \in M$. Using the measurement procedure above, u_1 writes $u_1 = 1 \cdot u_1 + 0$, and concludes that, for himself, $d\sigma/dt = |x| = 0$, so he is *at rest*. Then, writing $u_2 = au_1 + x$, he concludes that u_2 is moving since $u_1 \neq u_2$ implies $x \neq 0$.

But u_2 writes $u_2 = 1 \cdot u_2 + 0$ and $u_1 = bu_2 + y$, and hence concludes that she is at rest and that it is u_1 who is moving.

Returning to u_1's view of u_2: With the notations established earlier, the number a must be denoted dt_1/dt_2. But

$$-1 = <u_2, u_2> = -a^2 + |x|^2,$$

hence

$$a = (1 + |x|^2)^{1/2} > 1.$$

Consequently, $dt_1/dt_2 > 1$. Thus u_1 asserts that his time runs faster than u_2's. Since u_1 considers that he is at rest and u_2 is moving, he endorses the slogan *Moving clocks run slow.*

It is probably superfluous to say that u_2—who regards herself as at rest and u_1 as moving—finds the different measurement $dt_2/dt_1 > 1$, but supports the same slogan.

Evidently these disagreements between u_1 and u_2 derive naturally from their different decompositions of $T_p(M)$ into time and space.

SPECIAL RELATIVITY

Special relativity is the general relativity of a spacetime isometric to Minkowski spacetime $\mathbf{R}_1^4$. Thus for an arbitrary spacetime M, special relativity obtains in each tangent space $T_p(M) \approx \mathbf{R}_1^4$, and this provides a link to general relativity in M.

For a material particle α let us consider the physical significance of $p = d\alpha/ds = md\alpha/d\tau$ as viewed by some infinitesimal observer u at some event $p = \alpha(\tau_0)$. (See Remark 1.6.4.) Let $\bar{\alpha}$ be the projection of α into the space $u^\perp$ of the observer. Now identify M near p with its linear approximation the tangent space $T_p(M) = \mathbf{R}u + u^\perp$ (using, for rigor, the exponential map at p). Thus we can write $\alpha(\tau) = t(\tau)u + \bar{\alpha}(\tau)$, where t is the observer's time, τ the proper time of the particle. Hence $d\alpha/d\tau = (dt/d\tau)u + d\bar{\alpha}/d\tau$, and since this is a timelike unit vector we get $-1 = -(dt/d\tau)^2 + |d\bar{\alpha}/d\tau|^2$. Now $v = |d\bar{\alpha}/dt| = d\sigma/dt$ is the Newtonian speed of α as measured by u (so $0 \le v < 1$), and by the chain rule $|d\bar{\alpha}/d\tau| = v \, dt/d\tau$. Thus, algebraic manipulation gives $dt/d\tau = (1 - v^2)^{-1/2}$. Consequently,

$$p = m\frac{d\alpha}{d\tau} = \frac{m}{\sqrt{1 - v^2}}\, u + \frac{m}{\sqrt{1 - v^2}}\frac{d\bar{\alpha}}{dt}.$$

Taking v small, hence $(1 - v^2)^{-1/2}$ near 1, gives a Newtonian comparison. Evidently the second summand corresponds to Newtonian 3-momentum. In the binomial expansion

$$E = m(1 - v^2)^{-1/2} = m + \tfrac{1}{2}mv^2 + O(v^4)$$

kinetic energy $\tfrac{1}{2}mv^2$ appears, and Einstein declared this scalar E to be the *energy* of the particle as measured by u, with the rest-mass $m = E|_{v=0}$ as merely one form of energy. (In arbitary units, with c the speed of light, this is the famous formula $E_0 = mc^2$.)

To gain some intuition about relativistic motion, let us consider the simplest case: motion on a line (as above) with zero gravity. This is special relativity in two dimensions; its spacetime is $M \approx \mathbf{R}_1^2$ (with only coordinate-free results valid). $\mathbf{R}_1^2$ has the same geodesics as the Euclidean plane $\mathbf{R}^2$, namely, straight lines. Light rays have $dx/dt = \pm 1$, that is, "travel at 45°," and material particles are slower: $|dx/dt|$. As in Figure 1.6, the future is upward.

Let α and β be observers moving in their spaceships along a line out in space, far from the nearest galaxy. We draw their *worldlines*, that is, the spacetime routes they follow.

Case 1. Both α and β are freely falling along the same line, and are initially drawing closer to each other. In Figure 1.9 the dashed lines represent messages they exchange. Since they are in free fall, each observer tends to consider that he is at rest with the other observer moving toward him.

Case 2. α is freely falling, but β, initially drawing closer to α, accelerates by firing her rocket engines toward α. After coming close to α, β moves away again, and her speed approaches that of light. (As Figure 1.10 suggests, this statement is meaningful, even though "speed" itself not invariantly defined.)

The dashed lines in Figure 1.10 again represent messages—the odd numbers sent by α, even by β.

- Messages such as (1) sent by α at an early proper time reach β.
- α always receives messages such as (2), (4), (6) sent by β.

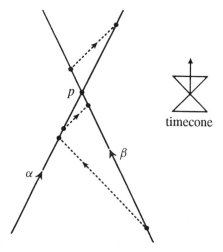

FIGURE 1.9.   Freely falling observers α and β exchange messages. Moving on the same line, they pass each other at event p.

- α replies to β's message (2) by (3), and later receives an answer (4).
- α receives message (4) at $\alpha(a)$, but even if he replies instantly, his return message (5) never reaches β—nor does any later message such as (7).

Radar can be regarded as a message sent and (if received) instantly returned. Thus α eventually gets a complete description of the location of β, even though he can cease transmitting at $\alpha(a)$. By contrast, β locates only a finite open interval of α's worldline. This interval ends at $\alpha(a)$ since none of β's transmissions past $\beta(b)$ are returned; it begins when transmissions from her early past finally begin reaching α.

Note that α, who is freely falling, does not know if he is moving or not. But β is well aware that she is accelerating because if not strapped down she will be pressed up against the wall of her ship.

For detailed treatments of the foundations of general relativity, see for example, Harpaz (1992); Misner, Thorne, Wheeler (1965); Sachs, Wu (1977).

1.7   Submanifolds

We consider submanifolds first of smooth manifolds and then of semi-Riemannian manifolds, especially Lorentz manifolds.

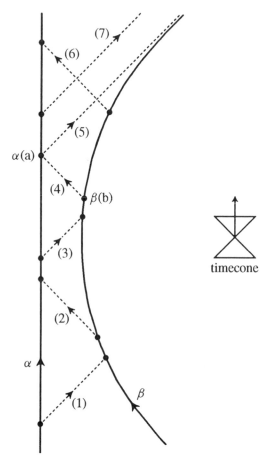

FIGURE 1.10. Freely falling observer α and accelerating observer β exchange messages.

SMOOTH SUBMANIFOLDS

By definition, a manifold P is a submanifold of a manifold M provided

1. P is a subset of M, and the inclusion map $j: P \to M$ is smooth;
2. The differential maps of j are injective (that is, $(dj)(v) = 0 \Rightarrow v = 0$); and
3. P is a topological subspace of M (that is, has the relative topology).

For a submanifold P, it is customary to ignore dj and consider $T_p(P)$ to be a vector subspace of $T_p(M)$. If P has dimension one less than $n = \dim M$ it is called a *hypersurface* of M.

Given a subset of M it can be shown that there is at most one way (topology and atlas) to make it a submanifold. Accordingly, it makes sense to say of a subset of M that it *is* or *is not* a submanifold.

A direct way to get submanifolds is this: Let P be a subset of a smooth n-manifold M. Suppose that for each point $p \in P$ there is a coordinate system $\xi = (x^1, \ldots, x^n)$ of M on a neighborhood $\mathcal{U}$ of p such that the *k-slice* of $\mathcal{U}$ given by

$$x^{k+1} = x^{k+1}(p), \ldots, x^n = x^n(p)$$

is the intersection $\mathcal{U} \cap P$. Then P can be made a k-dimensional submanifold of M as follows. First, assign P the induced topology; then for coordinate systems as above, the functions $x^1, \ldots, x^k$ on $\mathcal{U} \cap P$ will provide coordinate systems making P a manifold—and a submanifold of M.

If P is given as a submanifold of M, then an *adapted coordinate system* (as described above) exists at every $p \in P \subset M$.

Some simple examples of submanifolds of $\mathbf{R}^n$ are the various coordinate axes and planes, and the $(n - 1)$-spheres $r = $ const. It is known that every smooth manifold is diffeomorphic to a submanifold of some Euclidean space.

We say that submanifold of M is *closed* if it is a closed set of M. (This terminology though common is not universal.)

The following result provides an easy way to get closed hypersurfaces.

Lemma 1.7.1 *Let f be a smooth real-valued function on M, and let c be a value of f. If the differential df is nonzero at every point of the set $S = \{p \in M \mid f(p) = c\}$, then S is a closed hypersurface of M.*

For example, on $\mathbf{R}^n$ let $f = \Sigma (u^i)^2$, where $u^1, \ldots, u^n$ are the usual Cartesian coordinates. Then for any $r > 0$, this lemma shows that $\{f = r^2\}$ is a hypersurface, the $(n - 1)$-sphere of radius r.

Since 2-spheres are vital to the construction of Kerr spacetime we now record some coordinate conventions.

Remark 1.7.2 Let S^2 be the sphere of unit radius in $\mathbf{R}^3$. Then spherical coordinates of $\mathbf{R}^3$ restricted to S^2 give coordinates ϑ, φ on S^2. By firm relativistic convention (contrary perhaps to general usage), ϑ denotes *colatitude* and φ *longitude*, as indicated in Figure 1.11. We treat the coordinate φ as circular (see Section 1.1), so the coordinate system ϑ, φ covers all of the sphere except its north and south poles $(0, 0, \pm 1)$. The coordinate function φ is undefined at the poles, but its coordinate vector field ∂_φ is well-defined and smooth on the entire sphere and is zero at the

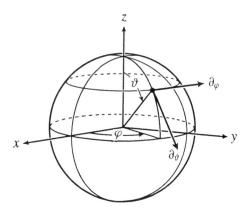

FIGURE 1.11. Spherical coordinates ϑ, φ on the unit 2-sphere.

poles. Defining $\vartheta(0, 0, +1) = 0$ and $\vartheta(0, 0, -1) = \pi$ extends the function ϑ to the entire sphere, with $0 \leq \vartheta \leq \pi$. At the poles, ϑ is only continuous, but $\cos \vartheta$ and $\sin \vartheta$ are smooth (indeed, analytic) everywhere.

FOLIATIONS

Definition 1.7.3 *A* distribution Π *of dimension k on a smooth manifold M is a smooth field of tangent k-planes on M. Explicitly, Π assigns to each $p \in M$ a k-dimensional subspace Π_p of the tangent space $T_p(M)$, and locally Π has a basis of k smooth vector fields.*

In vector bundle terminology, a distribution on M is a subbundle of the tangent bundle TM of M. A submanifold P of M such that $T_p(P) = \Pi_p$ for all $p \in P$ is called an *integral manifold* of Π. If through every point of M there is an integral manifold, then Π is *integrable*.

We say that a vector field V is *in* a distribution Π if $V_p \in \Pi_p$ for all $p \in M$.

Theorem 1.7.4 (Frobenius' Theorem). *A distribution Π on M is integrable if and only if whenever vector fields X, Y are in Π, their bracket $[X, Y]$ is also in Π.*

The necessity of the bracket condition follows from the fact that if two vector fields are tangent to a submanifold so is their bracket. The sufficiency proof shows, in fact, that at each point $p \in M$ there is a coordinate system all of whose k-slices

(k = dim Π) are integral manifolds of Π. Then it follows easily that through each point p there is a unique largest connected integral manifold P. (Note that P may only be an *immersed* submanifold of M; that is, it need not have the induced topology.) The collection of all such maximal integral manifolds of Π is called a *foliation* of dimension k = dim Π. For example, on $\mathbf{R}^n - 0$ the spheres $\{f = r^2\}$ constitute a $(n - 1)$-dimensional foliation, with Π consisting of their tangent spaces.

Every one-dimensional distribution is integrable since the bracket condition in Frobenius' theorem is trivial; such foliations are also called *congruences*. A congruence on M fills it with one-dimensional integral submanifolds; in general, there is no canonical way to parametrize these curves. Evidently any nonvanishing vector field X on M determines a unique congruence on M. For a given one-dimensional distribution Π, such a vector field need not exist; if it does, Π is said to be *orientable*. Then Π supplies a *direction* at each point p of M, that is, an oriented tangent line Π_p in $T_p(M)$.

HYPERSURFACES

If a curve α meets a hypersurface S at a point $\alpha(s_0) \in S$ the meeting is *transverse* if $\alpha'(s_0)$ is not tangent to S. This guarantees that α will cut sharply across S at the meeting point.

Every closed submanifold has a particularly simple kind of neighborhood, said to be tubular. In the case of a closed hypersurface S in a manifold M, a neighborhood $\mathcal{N}$ of S is *tubular* if there is a line bundle $\pi: E \to S$ over S and a diffeomorphism $\phi: E \to \mathcal{N}$ such that $\phi(0_p) = p$ for all $p \in S$. (A *line bundle* is a vector bundle whose fibers are one-dimensional.) Then, $\mathcal{N}$ is foliated by curves that cut transversally across S, as suggested in Figure 1.12.

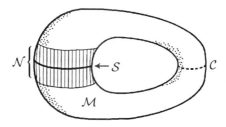

FIGURE 1.12. A normal neighborhood $\mathcal{N}$ of a hypersurface S in a manifold M.

A hypersurface S in a manifold M is said to be *two-sided* in M provided there exists a continuous vector field Z of M on S such that Z is never tangent to S (hence, is never 0). For example, the central circle in an ordinary band $S^1 \times \mathbf{R}^1$ is two-sided but that in a Möbius band is not.

Let $\mathcal{N}$ be a tubular neighborhood of a closed connected hypersurface S in M. It is not hard to show that if S is two-sided, then $\mathcal{N} - S$ has two connected components; otherwise, S is called *one-sided* and $\mathcal{N} - S$ is connected.

A closed hypersurface S in M separates M provided $M - S$ is not connected.

Lemma 1.7.5 *If S is a closed connected two-sided hypersurface of a connected manifold M, then $M - S$ has at most two components.*

Proof. Let $\mathcal{N}$ be a tubular neighborhood of S, so $\mathcal{N} - S$ has two components. Since M is connected, for each $m \in M$ there is a curve from m to S. Because S is closed, this curve meets $\mathcal{N} - S$ before it meets S. This means that every component of $M - S$ contains a component of $\mathcal{N} - S$.   □

The proof also shows that if S is one-sided, then $M - S$ is connected: S does not separate. In the two-sided case, although S separates any tubular neighborhood into two components, Figure 1.11 shows that it need not separate the entire manifold M. This is a global issue, for in the figure if the circle C (far from S) is excised from M, then S does separate the reduced manifold $M' = M - C$.

SUBMANIFOLDS OF LORENTZ MANIFOLDS

Let M be a Lorentz manifold. As with curves in M, submanifolds also may have *causal character.*

Definition 1.7.6 *A submanifold P of M is spacelike, null, or timelike provided that every tangent space $T_p(P)$ of P is spacelike, null, or timelike, respectively, as a vector subspace of $T_p(M)$.*

Of course, most submanifolds do not possess causal character. For example, a circle in $\mathbf{R}^2_1$ has tangent lines of all three causal characters. The restriction of the metric tensor $g = \langle, \rangle$ of M to a submanifold P merely applies g to tangent vectors $v, w \in T_p(P) \subset T_p(M)$. Consequently, a spacelike submanifold is, intrinsically, a Riemannian manifold, and a timelike submanifold (of dim ≥ 2) is a Lorentz manifold. In these two cases the geometries of P and M are related essentially as in the classical Riemannian case, by second fundamental forms (shape tensors), and the Gauss and Codazzi equations (Eisenhart 1926, O'Neill 1983).

On the other hand, if P is a null submanifold, the restriction of g to P is degenerate, so the intrinsic geometry of P is rather meagre. Nevertheless, null submanifolds occur naturally—and are crucial in relativity theory.

A submanifold of a spacetime is *nontimelike* provided it has no timelike tangent vectors. Spacelike and null submanifolds are special cases. A nontimelike sub-manifold N has the key property that if a timelike curve α meets N at $\alpha(s_0)$, then it cuts transversally through N at that point since $\alpha'(s_0)$, being timelike, cannot be tangent to N.

In a spacetime M, *every nontimelike hypersurface S is two-sided* for, being time-orientable, M admits a globally defined timelike vector field Z, and Z cannot be tangent to S.

Proposition 1.7.7 *Let S be a closed connected nontimelike hypersurface in a spacetime M. If $\mathcal{N}$ is a tubular neighborhood of S, then the two components of $\mathcal{N} - S$ can be denoted $\mathcal{N}^+$ and $\mathcal{N}^-$ in such a way that any future-pointing timelike curve α with $\alpha(s_0) \in S$ crosses S transversally from $\mathcal{N}^-$ to $\mathcal{N}^+$.*

Proof. As above let Z be a future-pointing timelike vector field on M, so integral curves of Z starting in S cross S transversally. If β_p is the integral curve of Z starting at $p \in S$, then $\beta_p(s) \in \mathcal{N} - S$ for a largest interval $(0, b_p)$. The set of all such points $\beta_p(s)$ for all p is the continuous image of a connected subspace of $S \times (0,\infty)$, and hence is contained in a single component of $\mathcal{N} - S$. Denote this component by $\mathcal{N}^+$, the other by $\mathcal{N}^-$.

Now let α be a curve that meets S transversally at $p = \alpha(s_0)$. Since $\alpha'(s_0)$ and Z_p are both future-pointing timelike they lie on the same side of the hyperplane $T_p(S)$ in the tangent space $T_p(M)$. Use of an adapted coordinate system shows that, for $s > 0$ sufficiently small, the points $\beta_p(s)$ and $\alpha(s)$ are on the same side of S, that is, $\alpha(s) \in \mathcal{N}^+$. Since the crossing is tranversal, $\alpha(s) \in \mathcal{N}^-$ for $s < 0$ small. $\qquad\square$

The preceding results have the following immediate consequence.

Corollary 1.7.8 *Let S be a closed connected nontimelike hypersurface in a space-time M. If S separates M, then $M - S$ has exactly two components and these can be denoted by C^+ and C^- so that any future-pointing timelike curve α that meets S transversally crosses S from C^- to C^+. (Here C^+ is called the future side of S, and C^- the past side.)*

Here is another source of hypersurfaces. Recall that a nonvanishing vector field X is *hypersurface-orthogonal* if $X^\perp$ is integrable. Then its integral manifolds are hypersurfaces normal to X. For example, the outward radial vector field $X = \Sigma u^i \partial_i$ in $\mathbf{R}^n - 0$ has this property since X is everywhere orthogonal to the spheres $\{p: |p| = \text{const}\}$.

TOTALLY GEODESIC SUBMANIFOLDS

Definition 1.7.9 *A submanifold P of a semi-Riemannian manifold M is* totally geodesic *provided that if vector fields X, Y of M are tangent to P, then the covariant derivative $\nabla_X Y$ is also tangent to P.*

Simple examples are hyperplanes in $\mathbf{R}^n$ and great spheres in S^n. For a proof of the following proposition and related results, see Chapter 4 of O'Neill (1983).

Proposition 1.7.10 *Let P be a totally geodesic submanifold of M. If a geodesic γ of M meets P at a point $\gamma(s_0)$ where $\gamma'(s_0)$ is tangent to P, then $\gamma(s)$ remains in P for all s near s_0.*

However, γ need not stay in P forever. For example, if P is an open disk in a plane of $\mathbf{R}^3$, then a straight line (geodesic) initially tangent to the disk will obviously soon leave it. But when the submanifold is also a closed set, there is no escape.

Corollary 1.7.11 *If a geodesic $\gamma: I \to M$ is tangent to a closed, totally geodesic submanifold P at a single point of I, then γ lies entirely in P.*

Proof. If $\gamma'(b)$ is tangent to P, then the proposition above implies that the nonempty set $I_P = \{s \in I : \gamma(s) \in P\}$ is open. Since P is a closed set, I_P is also closed. But I is connected, so $I_P = I$.   □

The following result gives two prime sources of closed, totally geodesic submanifolds.

Theorem 1.7.12 *If a subset P of a semi-Riemannian manifold M is either*
(1) *the set of fixed points of an isometry $\phi: M \to M$, or*
(2) *the set of zeros of a Killing vector field,*
then P is a closed, totally geodesic submanifold of M.

We merely give the scheme of the proof. (1) Evidently P is a closed set. If $p \in P$, let F_p be the set of vectors v in $T_p(M)$ such that $d\phi(v) = v$. Then F_p is a subspace of $T_p(M)$, and for each $v \in F_p$, the geodesic γ_v lies entirely in P, since

$$\phi(\gamma_v(s)) = \gamma_{d\phi(v)}(s) = \gamma_v(s) \quad \text{for all } s.$$

In fact, P near p is filled with such geodesics. Since this is true for all $p \in P$ it follows (by the converse mentioned above) that P is totally geodesic.

(2) This case follows by applying this argument to the flow of X.   □

A variant of Proposition 1.7.10 shows this: Let α be a curve in a totally geodesic submanifold P of M, and let V be a parallel M vector field along α (that is, $V' = \nabla_{\gamma'} V = 0$). If V is initially tangent to P, then V is always tangent to P.

Consequently, *a connected, totally geodesic submanifold of a Lorentz manifold has causal character*, for if its tangent space is, say, spacelike at one point, then it is spacelike at every point since parallel translation preserves causal character.

Remark 1.7.13 The failure of a submanifold P of M to be totally geodesic can be measured by its *shape tensor* (or *second fundamental form tensor*) S. For vector fields X, Y tangent to P, we define $S(X, Y)$ to be the component of the M covariant derivative $\nabla_X Y$ that is normal to P; thus $S(X, Y) = \text{nor}\nabla_X Y$. Consequently, P is totally geodesic if and only if $S = 0$. Using the properties of ∇ in Theorem 1.3.1 it is easy to check that S is a tensor; that is, it is $\mathfrak{F}(P)$-bilinear.

1.8   Cartan Computations

The Cartan approach to geometry is expressed in terms of the calculus of differential forms (see Appendix B). The introduction given here, though brief, nevertheless provides a powerful method for the computation of covariant derivatives and curvature.

Let $E_1, \ldots, E_n$ be an orthonormal frame field on a semi-Riemannian manifold M. Frame fields need not exist globally, so M will usually be an open set in some larger manifold. Taking the dual one-forms of the frame field gives an *orthonormal coframe field* $\omega^1, \ldots, \omega^n$ (Recall that ω^i is characterized by $\omega^i(E_j) = \delta^i{}_j$). Such coframe fields can be recognized directly from the line-element ds^2 of M as follows.

Lemma 1.8.1   *One-forms $\omega^1, \ldots, \omega^n$ constitute an orthonormal coframe field on M and $(\varepsilon_1, \ldots, \varepsilon_n)$ is the signature of M if and only if*

$$ds^2 = \Sigma \varepsilon_i (\omega^i)^2.$$

Proof.   Suppose $\{\omega^i\}$ and $\{\varepsilon_i\}$ have the stated properties. Let $\{E_j\}$ be the dual frame field. For any tangent vector, $v = \Sigma \omega^m(v) E_m$, hence

$$ds^2(v) = \,<v, v> \,= \sum_{m,k} \omega^m(v) \omega^k(v) <E_m, E_k>.$$

Since $<E_m, E_k> \,= \delta_{mk}\varepsilon_m$, this becomes $\Sigma \varepsilon_m (\omega^m(v))^2 = [\Sigma \varepsilon_m (\omega^m)^2](v)$.

Conversely, suppose $\{\omega^i\}$ are one-forms and $\varepsilon_1, \ldots, \varepsilon_n$ numbers ± 1 such that $ds^2 = \Sigma \varepsilon_m (\omega^m)^2$. Then the metric tensor is $g = \Sigma \varepsilon_m \omega^m \otimes \omega^m$. Let $\{E_i\}$ be the vector fields dual to $\{\omega^i\}$. These form an orthonormal frame field, since

$$<E_i, E_j> \,= g(E_i, E_j) = \Sigma \varepsilon_m \omega^m(E_i)\omega^m(E_j) = \Sigma \varepsilon_m \delta_{mi} \delta_{mj} = \varepsilon_i \delta_{ij}.$$

Thus $(\varepsilon_1, \ldots, \varepsilon_n)$ is the signature and the one-forms ω^i are orthonormal.   □

Definition 1.8.2   *The connection forms of a frame field $E_1, \ldots, E_n$ are the one-forms $\omega^i_j (1 \leq i, j \leq n)$ such that for all tangent vectors, v,*

$$\omega^i_j(v) = \omega^i(\nabla_v E_j) \quad \text{for all } i, j$$

where ∇, as usual, denotes the Levi–Civita covariant derivative.

Hence by the duality formula, $\nabla_v E_j = \Sigma \omega^m_j(v) E_m$. Thus the matrix $(\omega^i_j(v))$ tells how the frame is changing as it moves in the v direction—showing that in the Cartan approach, connection forms replace Christoffel symbols.

Lemma 1.8.3 (The First Structural Equation). $d\omega^i = -\Sigma \omega^i_m \wedge \omega^m$ *for all i.*

Proof.   For $1 \leq a, b \leq n$, the exterior derivative formula (Appendix B) gives

$$(d\omega^i)(E_a, E_b) = E_a \omega^i(E_b) - E_b \omega^i(E_a) - \omega^i[E_a, E_b] = -\omega^i[E_a, E_b],$$

the last equality since $\omega^i(E_j) = \varepsilon_i \delta_{ij}$, constant.

The right-hand side of the asserted equation yields the same result since

$$[-\Sigma \omega^i_m \wedge \omega^m](E_a, E_b) = -\Sigma \omega^i_m(E_a)\omega^m(E_b) + \Sigma \omega^i_m(E_b)\omega^m(E_a)$$

$$= -\omega^i_b(E_a) + \omega^i_a(E_b) = -\omega^i(\nabla_{E_a} E_b) + \omega^i(\nabla_{E_b} E_a)$$

$$= -\omega^i(\nabla_{E_a} E_b - \nabla_{E_b} E_a) = -\omega^j[E_a, E_b].   □$$

Like the Christoffel symbols, the connection forms do not constitute a tensor on M, but their indices can still be raised and lowered in the simple way described in Lemma 1.3.2: for the index i, multiply by $\varepsilon_i = <E_i, E_i>$. For example, $\omega_{ij} = \varepsilon_i \omega^i{}_j$, since

$$\omega_{ij}(v) = <E_i, \nabla_v(E_j)> = <E_i, \Sigma\omega^m{}_j(v)E_m>$$
$$= \Sigma\omega^m{}_j(v)<E_i, E_m> = \varepsilon_i\omega^i{}_j(v).$$

(Recall that by the weak convention in Section 1.2 this last expression is not summed. Indeed, it is the frequent appearance such index patterns that argues against the strong summation convention.) Multiplying the previous formula by ε_i then gives $\omega^i{}_j = \varepsilon_i \omega_{ij}$.

Lemma 1.8.4 $\omega^i{}_j = -\varepsilon_i\varepsilon_j\omega^j{}_i$, *hence, in particular,* $\omega^i{}_i = 0$.

Proof. Since $<E_i, E_j>$ is constant, $<\nabla_v E_i, E_j> + <\nabla_v E_j, E_i> = 0$ for every tangent vector v. By the formula above, this means $\omega_{ij}(v) + \omega_{ji}(v) = 0$ for all v, so ω_{ij} is skew-symmetric in i and j. Raising the first indices of each gives $\varepsilon_j\omega^j{}_i(v) + \varepsilon_i\omega^i{}_j(v) = 0$, and the result follows. □

Consequently, all the connection forms are determined by $\omega^i{}_j$ for $i < j$.

Corollary 1.8.5 *For a coframe field* $\omega^1, \ldots, \omega^n$, *the connection forms* $\omega^i{}_j$ *are the unique one-forms that satisfy the equations in Lemmas 1.8.3 and 1.8.4.*

In view of this result, the connection forms can sometimes be found by cut-and-try methods. However, we will derive an explicit formula for them in terms of the coframe fields and signature. This analogue of the classical formula for Christoffel symbols will produce connection forms when guessing fails.

Proof. Apply the first structural equation (Lemma 1.8.3) to E_j, E_k and then multiply by ε_i to get

$$\varepsilon_i d\omega^i(E_j, E_k) = -\varepsilon_i\omega^i{}_k(E_j) + \varepsilon_i\omega^i{}_j(E_k).$$

Next, cyclically permute the indices i, j, k and use Lemma 1.8.4 to get

$$\varepsilon_j d\omega^j(E_k, E_i) = \varepsilon_i\omega^i{}_j(E_k) + \varepsilon_j\omega^j{}_k(E_i).$$

Analogously, another permutation yields

$$\varepsilon_k d\omega^k(E_i, E_j) = \varepsilon_j\omega^j{}_k(E_i) - \varepsilon_i\omega^i{}_k(E_j).$$

Now subtract this last equation from the sum of the first two to get

$$\varepsilon_i d\omega^i(E_j, E_k) + \varepsilon_j d\omega^j(E_k, E_i) - \varepsilon_k d\omega^k(E_i, E_j) = 2\varepsilon_i \omega^i{}_j(E_k).$$

This gives the required formula

$$\omega^i{}_j(E_k) = \tfrac{1}{2}\varepsilon_i[\varepsilon_i d\omega^i(E_j, E_k) + \varepsilon_j d\omega^j(E_k, E_i) - \varepsilon_k d\omega^k(E_i, E_j)].$$

$\square$

In the Cartan approach, differential forms also describe curvature.

Definition 1.8.6 *For tangent vectors v, w let $\Omega^i{}_j(v, w)$ be the matrix of the curvature operator R_{vw} relative to the frame field $\{E_i\}$. Explicitly,*

$$R_{vw}(E_j) = \sum_i \Omega^i{}_j(v, w) E_i$$

for all j.

Here $\Omega^i{}_j$ is a two-form, since R is a tensor and $R_{vw} = -R_{wv}$. The forms $\Omega^i{}_j$ are called the curvature forms *of the frame field.*

Proposition 1.8.7 (The Second Structural Equation).

$$\Omega^i{}_j = d\omega^i{}_j + \Sigma \omega^i{}_m \wedge \omega^m{}_j \quad \text{for all } i, j.$$

Proof. For $1 \le a, b \le n$, the definition of Ω gives

$$R_{E_a E_b}(E_j) = \Sigma R^i{}_{jab}(E_i) = \Sigma \Omega^i{}_j(E_a, E_b) E_i.$$

We recompute $R_{E_a E_b}(E_j)$ using the definition of R; then comparison of the two results will give the required formula. Now

$$R_{E_a E_b}(E_j) = \nabla_{E_a}(\nabla_{E_b} E_j) - \nabla_{E_b}(\nabla_{E_a} E_j) - \nabla_{[E_a, E_b]} E_j.$$

Since $\nabla_{E_a}(E_j) = \Sigma \omega^i{}_j(E_a) E_i$, we find for the three terms on the right

$$\nabla_{E_a}(\nabla_{E_b} E_j) = \Sigma E_a[\omega^i{}_j(E_b)] E_i + \Sigma \omega^m{}_j(E_b) \nabla_{E_a} E_m$$
$$= \Sigma E_a[\omega^i{}_j(E_b)] E_i + \Sigma \omega^m{}_j(E_b) \cdot \omega^i{}_m(E_b) E_i,$$
$$\nabla_{E_b}(\nabla_{E_a} E_j) = \text{ the same with } a \text{ and } b \text{ reversed, and}$$
$$\nabla_{[E_a, E_b]} E_j = \Sigma \omega^i{}_j([E_a, E_b]) E_i.$$

Substituting these expressions into the formula for $R_{E_a E_b}(E_j)$ yields

$$R_{E_a E_b}(E_j) = \sum_i \left[E_a \omega^i{}_j(E_b) - E_b \omega^i{}_j(E_a) - \omega^i{}_j[E_a, E_b] \right] E_i$$
$$+ \sum_{m,i} \left[\omega^i{}_m(E_a)\omega^m{}_j(E_b) - \omega^i{}_m(E_b)\omega^m{}_j(E_a) \right] E_i.$$

But the terms within square brackets are the values on E_a, E_b of the differential forms $d\omega^i{}_j$ and $\Sigma \omega^i{}_m \wedge \omega^m{}_j$, respectively. Hence,

$$R_{ab}(E_j) = \sum_i [d\omega^i{}_j + \sum \omega^i{}_m \wedge \omega^m{}_j](E_a, E_b) E_i.$$

Comparison with the first equation of the proof shows that $\Omega^i{}_j$ and $d\omega^i{}_j + \Sigma \omega^i{}_m \wedge \omega^m{}_j$ and agree on E_a, E_b. $\qquad\square$

As with the connection forms, lowering an index i of Ω, gives $\Omega_{ij} = \varepsilon_i \Omega^i{}_j$, which, by a symmetry of curvature, is skew-symmetric in i and j. Hence

Corollary 1.8.8 $\Omega^i{}_j = -\varepsilon_i \varepsilon_j \Omega^j{}_i$. *In particular,* $\Omega^i{}_i = 0$.

This corollary and Lemma 1.8.4 say that $(\Omega^i{}_j)$ and $(\omega^i{}_j)$, considered as matrix-valued forms, take their values in the appropriate Lie algebra (see Lemma 9.3 in O'Neill, 1983).

Comparison of definitions shows that the components of the curvature tensor R with respect to the frame field $\{E_i\}$ are

$$R^i{}_{jk\ell} = \Omega^i{}_j(E_k, E_\ell).$$

Hence $R_{ijk\ell} = \varepsilon_i R^i{}_{jk\ell}$. The sectional curvatures of the frame field 2-planes are

$$K(\Pi_{ij}) = K(E_i, E_j) = \varepsilon_i \varepsilon_j R_{ijij} = \varepsilon_j R^i{}_{jij},$$

and the curvature invariant k can be written as $k = \Sigma \varepsilon_i \varepsilon_j \varepsilon_k \varepsilon_\ell (R_{ijk\ell})^2$.

Remarks 1.8.9 Let $\{E_i\}$ be an orthonormal frame field on M.

1. If M is a Riemannian manifold, the signs ε_i are all $+1$. Thus, raising and lowering of indices does not even require a sign change. Hence the matrices $(\omega^i{}_j)$ and $(\Omega^i{}_j)$ are skew-symmetric.

2. If M is a Lorentz manifold, then a sign change is involved only for the "timelike" index $0 \approx 4$. The skew-symmetry of $\omega^i{}_j$ leads to

$$\omega^i{}_0 = \omega^0{}_i,$$

but

$$\omega^i{}_j = -\omega^j{}_i \quad \text{for } i, j > 0,$$

and similarly for $\Omega^i{}_j$.

Our basic application of Cartan methods is in Sections 2.6 and 2.7, but here is a useful example.

Corollary 1.8.10 *Let u, v be orthogonal coordinates in a semi-Riemannian surface S. Write $<\partial_u, \partial_u> = E = \varepsilon_1 e^2$, $<\partial_v, \partial_v> = G = \varepsilon_2 g^2$, where ε_1 and ε_2 are ± 1, and $e > 0$, $g > 0$. The Gaussian curvature of S is given in these terms as*

$$K_S = \frac{-1}{eg} \left[\varepsilon_1 \left(\frac{e_v}{g} \right)_v + \varepsilon_2 \left(\frac{g_u}{e} \right)_u \right].$$

Proof. Define vector fields $E_1 = \partial_u/e$, $E_2 = \partial_v/g$. Clearly this is an orthonormal frame field. By definition in Section 1.2 the Gaussian curvature K_S is $<R_{E_1 E_2} E_2, E_1> = <\Omega^1{}_2(E_1, E_2)E_1, E_1> = \varepsilon_1 \Omega^1{}_2(E_1, E_2)$.

The dual coframe is evidently $\omega^1 = e\,du$, $\omega^2 = g\,dv$. We find the connection form $\omega^1{}_2 = -\varepsilon_1 \varepsilon_2 \omega^2{}_1$ by exploiting its uniqueness in the first structural equation. Thus we need

$$d\omega^1 = e_v\,dv \wedge du = -(e_v/g)\,du \wedge \omega^2$$

and

$$d\omega^2 = g_u\,du \wedge dv = -(g_u/e)\,dv \wedge \omega^1.$$

The first of these equations shows that $\omega^1{}_2 = (e_v/g)\,du + (?)\,dv$ and the second that $\omega^2{}_1 = (?)du + (g_u/e)\,dv$. Hence $\omega^1{}_2 = (?)\,du - \varepsilon_1 \varepsilon_2(g_u/e)\,dv$. These two formulas for $\omega^1{}_2$ determine it uniquely as

$$\omega^1{}_2 = (e_v/g)\,du - \varepsilon_1 \varepsilon_2(g_u/e)\,dv.$$

In dimension 2 the second structural equation reduces to just $\Omega^1{}_2 = d\omega^1{}_2$. Here

$$d\omega^1{}_2 = (e_v/g)_v\,dv \wedge du - \varepsilon_1 \varepsilon_2(g_u/e)_u\,du \wedge dv$$
$$= -[(e_v/g)_v + \varepsilon_1 \varepsilon_2(g_u/e)_u]\,du \wedge dv.$$

Hence

$$K_S = \varepsilon_1 \Omega^1{}_2(E_1, E_2) = \varepsilon_1 d\omega^1{}_2(E_1, E_2) = \varepsilon_1 d\omega^1{}_2(\partial_u/e, \partial_v/g)$$
$$= -\varepsilon_1[(e_v/g)_v + \varepsilon_1\varepsilon_2(g_u/e)_u]/(eg),$$

which gives the required formula for K_S.   □

1.9   Overview of a Kerr Black Hole

Kerr spacetime is the standard relativistic model of the gravitational field of a rotating star. The spacetime is fully revealed only when the star collapses, leaving a black hole—otherwise the bulk of the star blocks exploration. The qualitative character of Kerr spacetime depends on its mass and its rate of rotation, the most interesting case being when the rotation is fairly slow. (If the rotation stops entirely, Kerr spacetime reduces to Schwarzschild spacetime.)

The existence of black holes in our universe is generally accepted—by now it would be hard for astronomers to run the universe without them. Everyone knows that no light can escape from a black hole, but convincing evidence for their existence is provided by observable effects around binary stars whose companions are invisible.

Suppose that, traveling in our spacecraft, we approach an isolated, slowly rotating black hole. It can then be observed as a black disk against the stars of the background sky. Explorers familiar with the Schwarzschild black holes will refuse to cross its boundary horizon. First of all, return trips through a horizon are never possible, and in the Schwarzschild case, there is a more immediate objection— after the passage, any material object will, in a fraction of a second, be devoured by a singularity in spacetime.

If we dare to penetrate the horizon of this Kerr black hole we will find ... another horizon. Behind this one, the singularity in spacetime finally appears, not as a central focus but as a *ring*—a circle of infinite gravitational forces. Fortunately, this singularity is not as dangerous as the Schwarzschild one; it is possible to avoid it and enter a new region of spacetime, passing through either of two "throats" bounded by the ring. In the new region, escape from the ring singularity is easy because the gravitational effect of the black hole is reversed—it now repels rather than attracts. As distance increases, this negative gravity weakens, as on the positive side, until its effect becomes negligible.

A quick departure may be prudent but will prevent discovery of something strange: The ring singularity is the outer equator of a spatial solid torus that is, precisely, a time machine. Traveling within it, one can reach arbitrarily far back into the past of any entity inside the double horizons. In principle you can arrange a bridge game, with all four players being yourself at different ages. But there is no way to meet Julius Caesar or your (predeparture) childhood self since these lie on the other side of two impassable horizons.

This rough description is reasonably accurate within its limits, but its apparent completeness is deceptive. The Kerr black hole is vaster and more symmetrical. Outside the horizons, it turns out that the model described above (and pictured in Figure 2.2) lacks a distant past and, on the negative-gravity side, a distant future. Harder to imagine are the deficiencies of the spacetime region between the two horizons. This region definitely does not resemble the Newtonian three-space between two concentric spheres, plus a clock to tell time. In it, space and time are turbulently mixed. Pebbles dropped experimentally can simply vanish in finite time—and new objects magically appear. The complete model of Kerr spacetime built in Chapter 3 adds two more horizons to each such interhorizon region (there will be many regions), and shows that Kerr spacetime is organized symmetrically around the spatial 2-spheres at which these horizons intersect.

BEGINNING KERR SPACETIME

To present the Kerr metric in intuitive terms, let us picture a distant, spherically symmetric star rotating in space about a vertical axis through its center. To describe this situation it is natural to use spherical coordinates r, ϑ, φ on $\mathbf{R}^3$, and a time coordinate t on $\mathbf{R}^1$, interpreting r as distance to the center of the star, ϑ as colatitude, and φ as longitude—the star rotating in the positive φ direction. Only the gravitational field of the star is to be modeled, not the star itself. (The word star should be interpreted liberally, to also include the *black hole* possibility.)

These coordinates could be used for a Newtonian description, where it is irrelevant whether the star is rotating or not. They serve also (initially, at least) for the two relativistic cases: Schwarzschild spacetime, where the star does not rotate, and Kerr spacetime, where it does. When the Kerr metric is presented in terms of these familiar coordinates, they are called *Boyer–Lindquist coordinates*.

A major purpose of this chapter is to extend the domain on which the Kerr metric is defined. This is readily done in the case of an ordinary star since only its exterior is modeled. Black holes are another story. With the star's bulk gone, the spacetime must extend from the distant vantage point at which our description begins, on into the physically unobservable regions at its core and beyond. If the black hole rotates rapidly enough, this spacetime is readily constructed. However, in the principal case, where the rotation is not too rapid, the extension comes in stages. To know how to construct these we need to understand some of the geometry of the initial

Boyer–Lindquist domain—notably, its two remarkable families of *principal null geodesics*. The basic geometric invariant, Riemannian curvature, can already be computed at this level. Also, we notice some special submanifolds, notably the *horizons*, that will grow as the initial domain is extended. The next chapter will complete the construction of the (slowly rotating) maximal Kerr black hole.

2.1   The Kerr Metric

The simplest description of the Kerr metric tensor is in terms of the coordinates t, r, ϑ, φ on $\mathbf{R}^4 = \mathbf{R}^3 \times \mathbf{R}^1$ mentioned above: a time coordinate t on $\mathbf{R}^1$ and spherical coordinates r, ϑ, φ on $\mathbf{R}^3$ (as in Chapter 1). To see how rotation changes the Schwarzschild metric to the Kerr metric we shall give the components g_{ij} of both—and of Minkowski spacetime as well—in terms of these coordinates.

Kerr spacetime depends on two parameters: $\mathrm{M} > 0$, its *mass*, and $a \neq 0$, its *angular momentum per unit mass*. Halt the rotation by setting $a = 0$ and Kerr spacetime reverts to Schwarzschild spacetime; then remove the mass by setting $\mathrm{M} = 0$ and only (empty) Minkowski spacetime remains.

The following two functions pervade the study of Kerr spacetime:

$$\rho^2 = r^2 + a^2 \cos^2 \vartheta$$
$$\Delta = r^2 - 2\mathrm{M}r + a^2.$$

Their Schwarzschild meaning can be seen by setting $a = 0$. Thus, ρ^2 obviously generalizes r^2, while Δ is, in effect, a generalization of the Schwarzschild horizon function $h(r) = 1 - 2\mathrm{M}/r$, since

$$\frac{\Delta}{r^2} = 1 - \frac{2\mathrm{M}}{r} + \frac{a^2}{r^2}.$$

Table 2.1 gives the tensor components of the three metrics in terms of the coordinates t, r, ϑ, φ. From this table we observe the following:
1. When $a = 0$ the Kerr metric does reduce to the Schwarzschild metric.
2. Boyer–Lindquist coordinates fail on the z-axis (where $\sin \vartheta = 0$) as spherical coordinates always do. More seriously, the formulas above show that the Kerr metric as here presented fails when either $\rho^2 = 0$ or $\Delta = 0$. Points where $\Delta = 0$ will turn out, as in the Schwarzschild case, to give the *horizons* of Kerr spacetime.

TABLE 2.1.

Minkowski (1908)		Schwarzschild (1916)	Boyer–Lindquist (1966)
g_{tt}	-1	$-1 + 2M/r$	$-1 + 2Mr/\rho^2$
g_{rr}	$+1$	$[1 - (2M/r)]^{-1} =$ $r/(r - 2M)$	ρ^2/Δ
$g_{\vartheta\vartheta}$	r^2	r^2	$\rho^2 = r^2 + a^2 \cos^2 \vartheta$
$g_{\varphi\varphi}$	$r^2 \sin^2 \vartheta$	$r^2 \sin^2 \vartheta$	$\left[r^2 + a^2 + \dfrac{2Mra^2 \sin^2 \vartheta}{\rho^2} \right] \sin^2 \vartheta$
$g_{ij} (i \neq j)$	all zero	all zero	all zero except $g_{\varphi t} =$ $g_{t\varphi} = -2Mra \sin^2 \vartheta / \rho^2.$

3. Since the coordinates t and φ do not appear in the preceding metric formulas, the coordinate vector fields ∂_t and ∂_φ are Killing vector fields (Section 1.3). The flow ∂_t consists of the coordinate translations that send t to $t + \Delta t$, leaving the other coordinates fixed. Roughly speaking, these isometries express the time-invariance of the model. For ∂_φ the flow consists of coordinate rotations $\varphi \to \varphi + \Delta\varphi$; these isometries express its axial symmetry.

4. The form of the metric shows that the double sign change $t \to -t$, $\varphi \to -\varphi$ gives an isometry: Running time backward reverses the rotation. (Since $g_{t\varphi} \neq 0$, reversing t or φ separately is not an isometry.)

5. Kerr spacetime is *asymptotically flat*. When the coordinate r is large we can think of it as "distance to the star." The formulas above show that as $r \to \infty$ the Kerr metric becomes nearly Minkowskian. This expresses the natural idea that far from the star its gravitational field is weak. In general the notion of asymptotic flatness is nontrivial (Wald 1984), but the special properties of r, developed in Chapter 3, make Kerr spacetime a model for the concept (Ashtekar, Hansen 1978).

The metric components above are complicated enough to account for the difficulty in discovering them, but inspection reveals some useful relations among them. In expressing these we use abbreviations that will simplify many subsequent calculations:

$$S = \sin \vartheta, \qquad C = \cos \vartheta.$$

Lemma 2.1.1 *The Boyer–Lindquist metric components satisfy the following* metric identities:

(m1) $g_{\varphi\varphi} + aS^2 g_{\varphi t} = (r^2 + a^2)S^2,$

(m2) $g_{t\varphi} + aS^2 g_{tt} = -aS^2,$

(m3) $ag_{\varphi\varphi} + (r^2 + a^2)g_{t\varphi} = \Delta aS^2,$

(m4) $ag_{t\varphi} + (r^2 + a^2)g_{tt} = -\Delta.$

Since the formulas for $g_{\varphi\varphi}$ and $g_{\varphi t}$ are rather cumbersome, it is good tactics to use the metric identities as early and as often as possible. Because $g_{ij} = \;<\partial_i, \partial_j>$ they can be interpreted as scalar products of vector fields as follows.

Definition 2.1.2 *The* canonical Kerr vector fields *are*

$$V = (r^2 + a^2)\partial_t + a\partial_\varphi, \quad and$$
$$W = \partial_\varphi + aS^2\partial_t.$$

In terms of V and W the metric identities can be written more simply as

$$<V, \partial_\varphi> = \Delta aS^2 \qquad <W, \partial_\varphi> = (r^2 + a^2)S^2$$
$$<V, \partial_t> = -\Delta \qquad\quad <W, \partial_t> = -aS^2.$$

Lemma 2.1.3 $<V, V> = -\Delta\rho^2, \quad <W, W> = \rho^2 S^2, \quad <V, W> = 0.$

Proof. This is an easy computation; for example,

$$<V, V> = \;<V, (r^2 + a^2)\partial_t + a\partial_\varphi> = (r^2 + a^2)<V, \partial_t> + a<V, \partial_\varphi>$$
$$= (r^2 + a^2)(-\Delta) + a(\Delta aS^2) = -\Delta(r^2 + a^2 - a^2S^2) = -\Delta\rho^2.$$

$\square$

Now we want to show that components above actually constitute a metric tensor.

Lemma 2.1.4 *For Boyer–Lindquist coordinates,*

1. $g_{tt}g_{\varphi\varphi} - g_{t\varphi}{}^2 = -\Delta \sin^2 \vartheta,$ *and*
2. $\det(g_{ij}) = -\rho^4 \sin^2 \vartheta.$

Proof. (1) The expressions for g_{ij} show at once that $g_{t\varphi}{}^2$ cancels a term in $g_{tt}g_{\varphi\varphi}$ leaving

$$-g_{\varphi\varphi} + (2Mr\rho^{-2})(r^2 + a^2)S^2 = -g_{\varphi\varphi} - a^{-1}(r^2 + a^2)g_{t\varphi}.$$

But by metric identity (m3) this equals $-a^{-1}(\Delta a S^2) = -\Delta S^2$. Then assertion (2) follows.   □

Equation (2) in the preceding lemma shows that the Boyer–Lindquist formulas—at least when ρ^2, Δ, and $\sin \vartheta$ are all nonzero—give a nondegenerate metric tensor having Lorentz signature (up to a sign reversal). The causal character of the coordinate vector fields is by no means regular and is examined in the next section.

We do not attempt to impose a fixed order on the Boyer–Lindquist coordinates. For comparison with the usual Schwarzschild formula, the ordering t, r, ϑ, φ is appropriate. But geometrically the pairings $(r, \vartheta)(\varphi, t)$ are superior since only r and ϑ are involved in the metric components. On the other hand, from a topological viewpoint the pairings $(r, t)(\vartheta, \varphi)$ turn out to be more natural (Section 3.6).

There is a modification of the Kerr metric for a source that carries an electric charge e. Replacing the preceding definition of Δ by

$$\Delta = r^2 - 2Mr + a^2 + e^2$$

leads to the *Kerr–Newman metric* (Newman, et al, 1965). See also Section 12.3 of Wald (1984).

2.2   Boyer–Lindquist Blocks

The Kerr metric tensor contains parameters M and a, so strictly speaking it defines the two-parameter *Kerr family of spacetimes*. If we think of the mass M > 0 as constant and vary the parameter a (angular momentum per unit mass), a profound change occurs at the *extreme value* $a^2 = $ M^2, where the family splits into two subfamilies.

To avoid countless exceptions we shall not consider Schwarzschild spacetime to be in the Kerr family even though we frequently invoke it by setting $a = 0$. If a sign choice is needed, $a \neq 0$ is assumed to be positive. Fix the following terminology:

$0 = a^2$ gives Schwarzschild spacetime,

$0 < a^2 < $ M^2 gives *slowly rotating Kerr spacetime* ("slow Kerr"),

$a^2 = $ M^2 gives *extreme Kerr spacetime,* and

M$^2 < a^2$ gives *rapidly rotating Kerr spacetime* ("fast Kerr").

A first indication of the differences between these *rotation types* is given by the horizon function $\Delta = r^2 - 2Mr + a^2$.

For Schwarzschild spacetime, Δ has roots 0, 2M (see Figure 2.1).
For slow Kerr, Δ has two roots: $0 < r_{\pm} = M \pm (M^2 - a^2)^{1/2} < 2M$.
For extreme Kerr, $\Delta = (r - M)^2$, so $r = M$ is a double root.
For fast Kerr, Δ has no real roots.

As mentioned above, zeros of Δ will give horizons, a crucial feature of relativistic gravitation. But since the Boyer–Lindquist form of the metric fails when $\Delta = 0$, horizons can appear only after our initial patch of spacetime has been enlarged. As a first step in that direction, we look for the largest region on which the current formulas for g_{ij} give a valid metric tensor.

In sharp contrast to the Schwarzschild case, *the Kerr metric does not fail when* $r = 0$, so the usual spherical-coordinate condition $r > 0$ can be dropped, letting r run over the whole real line. Thus r and t are cartesian coordinates for an entire plane $\mathbf{R}^2$, and—with spherical coordinates ϑ, φ on the 2-sphere S^2—we consider the product manifold $\mathbf{R}^2 \times S^2$. In it there are just three subsets on which the present coordinates or metric fail. To specify them, we recall that the colatitude function ϑ is defined on the whole sphere, including the poles $(0, 0, \pm 1)$, where φ is undefined (see Remark 1.7.2).

1. On the *horizon* $H \colon \Delta = 0$. The cases above show that for fast Kerr spacetime, there is no horizon; for extreme Kerr spacetime there is a single horizon $H \colon r = M$; and for slow Kerr spacetime there are *two* horizons $H_+ \colon r = r_+$ and $H_- \colon r = r_-$.

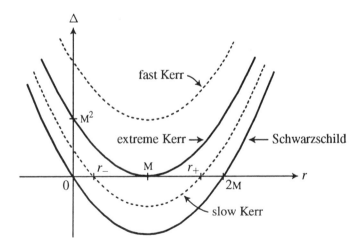

FIGURE 2.1. The horizon function Δ for the various rotation types.

2. On the *ring singularity* Σ: $\rho^2 = 0$. This terminology comes from the fact that $\rho^2 = 0$ if and only if both $r = 0$ and $\cos\vartheta = 0$. Thus, Σ is the cartesian product of a time axis $\mathbf{R}^1(t)$ and a circle S^1, namely, the equatorial circle $\vartheta = \pi/2$ in S^2 at radius $r = 0$. Informally, the circle itself is sometimes called the ring singularity, with $\Sigma = S^1 \times \mathbf{R}^1$ its history through time.
3. On the *axis A*: $\sin\vartheta = 0$. For spherical coordinates on $\mathbf{R}^3$, $\sin\vartheta = 0$ gives the z-axis, but since r is now allowed to take negative values, there are *two* such axes (enduring through time). In the sphere S^2, $\sin\vartheta = 0$ picks out the north pole p: $\vartheta = 0$ and the south pole $-p$: $\vartheta = \pi$, and then in the $\mathbf{R}^2 \times S^2$ each pole gives a complete "z-axis" enduring through time, namely, $\pm p \times R^2(r, t)$.

Figure 2.2 gives a slice of $\mathbf{R}^2 \times S^2$ at $t = $ const showing the relative positions of the three subsets discussed above. The question is this: Can the Kerr metric can be extended over any of the three subsets?

Consider first the axis A. Spherical coordinates always fail on the z-axis, but often the failure is not serious. For example, given spherical symmetry (as in the

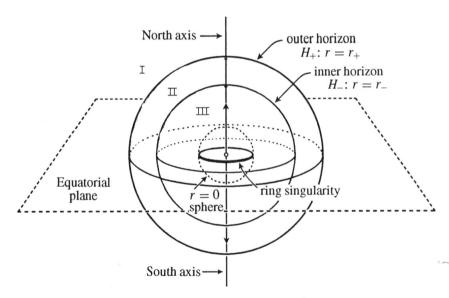

FIGURE 2.2.   An exponential picture of a slice $t = $ const of $\mathbf{R}^2 \times S^2$ (the radius r is drawn as e^r, so $r = -\infty$ is at the center of the figure). The two z-axes each run from $r = -\infty$ to $r = +\infty$. The ring singularity Σ (that is, its t-slice) lies on the central 2-sphere $r = 0$, but does not obstruct passage between the $r > 0$ and $r < 0$. Though it appears smaller, the $r < 0$ side is just as extensive as the $r > 0$ side.
This is the slow case where two horizons separate $\mathbf{R}^2 \times S^2$ into three regions I, II, and III. Both horizons lie on the $r > 0$ side, where they conceal the ring singularity from distant observers. Warning: block II in this figure is not "space"; on it r measures *time* not distance.

Schwarzschild case), it suffices to rotate the old coordinates. But in the Kerr case only axial symmetry prevails, hence the axis A is physically and geometrically unique. Nevertheless, we will now see that the Kerr metric *can* be analytically extended over A.

The Boyer–Linquist form of the line-element is

$$ds^2 = g_{tt}dt^2 + g_{rr}dr^2 + g_{\vartheta\vartheta}d\vartheta^2 + g_{\varphi\varphi}d\varphi^2 + 2g_{\varphi t}d\varphi\, dt,$$

and the formulas in Section 2.1 for the metric components show that they are nonsingular on the axis, since $\rho^2 = r^2 + a^2 > 0$ there. The differentials dt and dr are well defined on $\mathbf{R}^2$, hence over all $\mathbf{R}^2 \times S^2$. By contrast, both $d\varphi$ and $d\vartheta$ are singular at the poles of the 2-sphere and hence on the axis in $\mathbf{R}^2 \times S^2$. However, the combination $S^2d\varphi = \sin^2\vartheta d\varphi$ is extendible since it is expressed in Cartesian coordinates as $x\,dy - y\,dx$. And $d\vartheta$ can be dealt with by using the line-element $d\sigma^2$ of the 2-sphere.

Lemma 2.2.1 *On the unit 2-sphere with poles $(0, 0, \pm 1)$ deleted,*

$$\rho^2d\vartheta^2 + g_{\varphi\varphi}d\varphi^2 = \rho^2d\sigma^2 + (1 + 2Mr\rho^{-2})a^2(S^2d\varphi)^2.$$

Proof. Substituting the formula for $g_{\varphi\varphi}$ into the left side above gives

$$\rho^2d\vartheta^2 + (r^2 + a^2)S^2d\varphi^2 + 2Mr\rho^{-2}a^2 S^4d\varphi^2.$$

The useful Kerr identity, $r^2 + a^2 = \rho^2 + a^2S^2$, turns this into

$$\rho^2(d\vartheta^2 + S^2d\varphi^2) + a^2S^4(d\varphi^2 + 2Mr\rho^{-2}d\varphi^2),$$

and $d\vartheta^2 + S^2d\varphi^2$ is just the line-element $d\sigma^2$ of the (unit) 2-sphere. □

On the right side of this equation, all the terms are extendible over the poles, hence over the axis in $\mathbf{R}^2 \times S^2$.

It follows that although Boyer–Lindquist coordinates do not cover the axis, the Kerr metric is well defined there. This is shown by the following explicit formula, which, by the remarks above, is valid on the axis (horizon points excepted).

Lemma 2.2.2 *On $\mathbf{R}^2 \times S^2 - (H \cup \Sigma)$ the Kerr line-element can be written as*

$$ds^2 = (\rho^2/\Delta)dr^2 + \rho^2d\sigma^2 + a^2S^4d\varphi^2 - dt^2 + 2Mr\rho^{-2}(dt - aS^2d\varphi)^2.$$

Proof. On $\mathbf{R}^2 \times S^2 - (H \cup \Sigma \cup A)$, we express the Boyer–Lindquist formula for ds^2 as a sum

$$[g_{\vartheta\vartheta}d\vartheta^2 + g_{\varphi\varphi}d\varphi^2] + [g_{tt}dt^2 + g_{rr}dr^2 + 2g_{\varphi t}d\varphi \, dt].$$

Lemma 2.2.1 shows that the first summand equals

$$\rho^2 d\sigma^2 + (1 + 2\mathrm{M}r\rho^{-2})a^2 S^4 d\varphi^2.$$

The second summand is

$$(-1 + 2\mathrm{M}r\rho^{-2}) \, dt^2 + (\rho^2/\Delta) \, dr^2 - 4\mathrm{M}r\rho^{-2}aS^2 d\varphi \, dt$$
$$= -dt^2 + (\rho^2/\Delta) \, dr^2 + 2\mathrm{M}r\rho^{-2}[dt - 4\mathrm{M}r\rho^{-2}aS^2 d\varphi] \, dt.$$

Combining these, we recognize the square $(dt - aS^2 d\varphi)^2$ and hence obtain the required formula. We must check that it still gives a (nondegenerate) metric at points of $A - H$. On the axis, $S = 0$ and $C = 1$; hence

$$ds^2|_A = (-1 + 2\mathrm{M}r\rho^{-2})dt^2 + \rho^2\Delta^{-1}dr^2 + \rho^2 d\sigma^2$$
$$= -(r^2 + a^2)^{-1}\Delta dt^2 + (r^2 + a^2)\Delta^{-1}dr^2 + (r^2 + a^2) \, d\sigma^2.$$

Evidently this is nondegenerate when $\Delta \neq 0$. $\square$

The Boyer–Lindquist form of the Kerr metric can be pushed no farther, so to express current results in terms of *spacetimes* (which, by definition, are connected) we must look at the connected components of $\mathbf{R}^2 \times S^2 - (H \cup \Sigma)$. This domain is cut into components by *horizons*.

Definition 2.2.3 *The* Boyer–Lindquist blocks I, II, III *are the following open subsets of* $\mathbf{R}^2 \times S^2 - \Sigma$:

1. *For slow Kerr, where there are two horizons* $r = r_\pm$,

$$\mathrm{I}: r > r_+,$$
$$\mathrm{II}: r_- < r < r_+,$$
$$\mathrm{III}: r < r_-.$$

2. *For extreme Kerr, where there is a single horizon* $r = \mathrm{M}$.

$$\mathrm{I}: r > \mathrm{M}$$
$$\mathrm{III}: r < \mathrm{M}.$$

3. *For fast Kerr, since there is no horizon,* $\mathbf{R}^2 \times S^2 - \Sigma$ *itself can be regarded as a single block* I = III.

These blocks are connected Lorentz 4-manifolds, and they are indeed Kerr building blocks because, as the next chapter will show, *maximally extended Kerr spacetimes are constructed by gluing Boyer–Lindquist blocks together along horizons.*

In the fast Kerr case, it turns out that $\mathbf{R}^2 \times S^2 - \Sigma$ is already a maximal spacetime, hence it can often be neglected. Typically, we discuss slow Kerr spacetime in detail; then eliminating block II gives the extreme case.

To develop some feeling for the geometry of the blocks, it helps to consider the causal characters of the Boyer–Lindquist coordinate vector fields and the canonical vector fields V and W (see Figure 2.3).

As Figure 2.1 illustrates, we have $\Delta > 0$ on I∪III and $\Delta < 0$ on II; however, ρ^2 is always strictly positive. Then Lemma 2.1.3 and the formulas for $g_{ii} = <\partial_i, \partial_i>$ show that

1. ∂_ϑ and W are always spacelike;
2. ∂_r is spacelike on I ∪ III and timelike on II;
3. V is timelike on I ∪ III and spacelike on II;
4. ∂_φ is spacelike when $r > 0$, hence on I ∪ II. On block III its causal character varies, though inspection of $g_{\varphi\varphi}$ shows that it is spacelike for $r \ll -1$; and
5. ∂_t is spacelike on block II since it is orthogonal to ∂_r (timelike on II), but its causal character varies on both I and III. However, it is timelike if $r > 2$M or $r < 0$, as is evident when g_{tt} is written as $\rho^{-2}[-r(r - 2\text{M}) - a^2C^2]$. In fact, ∂_t is still timelike for $r = 0$ since $C = 0$ then gives the ring singularity.

These variable cases will be examined in Section 2.4.

The causal characters of the coordinate vector fields support the following interpretations of the Boyer–Lindquist blocks.

Block I, called the *Kerr exterior*, is the "astronomical" block, the most reasonable of the three. Provided r is not too small, its gravitational field bears comparison with a Newtonian central force field, and the Boyer–Lindquist coordinates t, r, ϑ, φ can be vizualized in a more or less Newtonian way: t as time, r as distance from the center of the black hole, ϑ as colatitude, φ as longitude. Since the t-coordinate vector field ∂_t is timelike for large r, it is natural to *time-orient* block I by declaring these vectors ∂_t to be future-pointing.

Block II, the spacetime region between the two horizons is utterly relativistic. The intuitive meaning of the coordinates is lost in causal turbulence; it is now the coordinate r that is timelike, while t, like the other two, is spacelike! Strictly speaking, II is not yet a spacetime, for although it is time-orientable, there is no

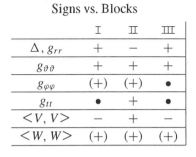

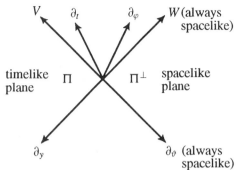

FIGURE 2.3.   The vector fields V, W, ∂_r, ∂_ϑ are mutually orthogonal. The plane $\Pi = \mathrm{span}\{V, \partial_r\}$ is always timelike, since $\Pi^\perp = \mathrm{span}\{W, \partial_\vartheta\}$ is always spacelike. In the table of signs, (+) means +, except 0 on the axis, where ∂_φ and W vanish (but remain spacelike). The vector fields V and W span the same plane as ∂_t and ∂_φ.

natural way to time-orient it. (Once II is attached to I, the time-orientation of I can be extended over II.)

For $r \ll -1$, block III may seem to be about the same as block I, with similar natural interpretations of Boyer–Lindquist coordinates, except that $-r$ is distance to the center. However, as we shall see, its gravity repels rather than attracts, and the region near block II is stranger than block II itself because it contains not only the ring singularity, but also a time machine.

2.3   Special Submanifolds

Our goal is to construct maximal Kerr spacetimes as promptly as is reasonably possible, but the construction will involve the geometry of simpler Kerr spacetimes. Here is a working definition:

Definition 2.3.1   *A Kerr spacetime is an analytic spacetime K such that*

1. *A disjoint union of (canonically isometric copies of) Boyer–Lindquist blocks is dense in K.*

2. *There are (unique) analytic functions r and C on K that restrict, on each Boyer–Lindquist block of condition (1), to the Boyer–Lindquist functions r and $C = \cos\vartheta$.*

3. *There is a (unique) isometry $\epsilon\colon K \to K$ called the* equatorial isometry *whose restriction to each Boyer–Lindquist block sends ϑ to $\pi - \vartheta$, leaving the other coordinates unchanged.*

4. *There are (unique)* Killing vector fields $\tilde{\partial}_t$ *and* $\tilde{\partial}_\varphi$ *on K that restrict, on each block, to the Boyer–Lindquist coordinate vector fields ∂_t and ∂_φ.*

In particular, the Boyer–Lindquist blocks themselves (when time-oriented) are the simplest Kerr spacetimes. Note that on any K the functions ρ^2 and Δ, and the canonical vector fields V and W are well defined.

This definition allows us to distinguish certain submanifolds that are important throughout Kerr geometry and physics.

Definition 2.3.2   *In any Kerr spacetime K,*

1. *The* Axis $A = Ax(K)$ *is the set of zeroes of the Killing vector field $\tilde{\partial}_\varphi$ as in condition 4 in Definition 2.3.1.*

2. *The* Equatorial plane $Eq = Eq(K)$ *is the set of fixed point of the* equatorial isometry ϵ. *(We often call Eq merely the* equator.*)*

By Theorem 1.7.12, the axis is a closed, totally geodesic submanifold; it is two-dimensional with Lorentz signature. For $|r|$ large, A is a relativistic version of the axis of rotation of the star—a spatial axis enduring through time. As Kerr spacetime is enlarged, its axis grows correspondingly. A always has two connected components: north axis, $\vartheta = 0$, and south axis, $\vartheta = \pi$ (see Figure 2.2), and these are isometric under the equatorial isometry ϵ. Further geometric properties of A are considered in Section 2.5 where its relation to horizons can be shown.

In any Kerr spacetime K, the above-mentioned result, Theorem 1.7.12 shows that the equatorial plane Eq is also a closed, totally geodesic submanifold. Intrinsically, it is a three-dimensional Lorentz manifold that represents—at least for $|r|$ large—the spatial equatorial plane $\vartheta = \pi/2$ enduring through time (see Figure 2.2). Because of its symmetric location, Eq is less influenced by rotational effects than are regions to the north or south; hence it is a convenient three-dimensional testing ground for various features of four-dimensional Kerr geometry.

The Boyer–Lindquist coordinates provide a whole family of totally geodesic surfaces as follows:

Definition 2.3.3   *The* polar plane $P = P(t_0, \varphi_0)$ *is the surface $t = t_0$, $\varphi = \varphi_0$ in* $\mathbf{R}^2 \times S^2 - \Sigma$.

On P we remove the restriction $0 < \vartheta < \pi/2$ and let ϑ be a circular coordinate (as in Section 1.1). Then P has single coordinate system (r, ϑ), though it is disconnected by horizons when these exist. Polar planes have (nondegenerate) line-element $ds^2 = \rho^2(\Delta^{-1}dr^2 + d\vartheta^2)$, and the sign of Δ shows that they are Riemannian in $\mathrm{I} \cup \mathrm{III}$ but Lorentz in II. Evidently, φ-rotations and t-translations give isometries of polar planes (see Section 3.7 and 3.8).

For $|r|$ large we can picture P as a plane in space passing through the axis of rotation. However, Figure 2.2 is deceptive in this case, and the remarkable global configuration of P becomes evident only in the extended Kerr spacetimes in Chapter 3.

$P(t_0, \varphi_0)$ is the set of fixed points of the mapping

$$r \to r, \quad \vartheta \to \vartheta, \quad \varphi - \varphi_0 \to -\varphi + \varphi_0, \quad t - t_0 \to -t + t_0,$$

and this mapping is clearly an isometry because the Boyer–Lindquist metric components g_{ij} do not involve φ or t. Thus polar planes are closed and totally geodesic.

We now consider another natural family of surfaces present in any Kerr spacetime. Definition 2.3.1 guarantees that extensions $X = \tilde{\partial}_t$ and $Y = \tilde{\partial}_\varphi$ of the Killing vector fields ∂_t and ∂_φ exist on K. On the axis, ∂_φ, and hence Y, vanish, but elsewhere span $\{X, Y\}$ is two-dimensional. On Boyer–Lindquist blocks the Lie bracket $[\partial_t, \partial_\varphi]$ is zero because t and φ are coordinate functions. Hence, by continuity, $[X, Y] = 0$ on K. Thus, the Frobenius theorem (Section 1.7) implies that $K - A$ is foliated by surfaces—called Killing orbits—to which X and Y are everywhere tangent. On Boyer–Lindquist blocks such an orbit F is just a coordinate slice $r = r_0, \vartheta = \vartheta_0$. Thus, F is a cylinder $\mathbf{R}^1 \times S^1$ and has global coordinates t, φ (with φ circular).

Since the Boyer–Lindquist metric depends only on r and ϑ, holding these coordinates constant to get the Killing orbits means that each F has line-element of the form $ds^2 = A\, dt^2 + B\, d\varphi^2$, with A and B constant. Thus, *Killing orbits are flat.* However, it will soon appear that they are not totally geodesic.

In short, Boyer–Lindquist coordinates express the off-axis part of the each block as a coordinate product of polar planes times Killing orbits.

Another distinctive Kerr submanifold shown in Figure 2.2 is the *throat T* that separates the regions $r < 0$ and $r > 0$. We take T to be the coordinate hypersurface $r = 0$ in Boyer–Lindquist block III; thus T is the *central sphere* $S^2\langle 0 \rangle = \{r = 0, t = 0\}$ enduring through time: $T \approx S^2\langle 0 \rangle \times \mathbf{R}$. If a black hole can be said to have a center, this is it. Note that $S^2\langle 0 \rangle$ is not a topological 2-sphere since, as Figure 2.2 shows, it is cut by the (missing) ring singularity $r = 0, \vartheta = \pi/2$ into two disks, north and south.

To find the line-element of T we set $r = 0$ in general Boyer–Lindquist expression, getting $d\sigma^2 = a^2 C^2 d\vartheta^2 + a^2 S^2 d\varphi^2 - dt^2$. The first two terms give the line-element of the sphere $S^2 \langle 0 \rangle$, and classical formulas show that its Gaussian curvature is zero (O'Neill 1966). In fact, each of the two components of T is isometric to the interior $D \times \mathbf{R}$ of a cylinder $S^1 \times \mathbf{R}^1$ of radius $r = a$ in Minkowski three-space $\mathbf{R}_1^3 \approx \mathbf{R}^2 \times \mathbf{R}_1^1$. Such an isometry ψ is given by

$$x = aS \cos \varphi, \qquad y = aS \sin \varphi, \qquad t = t.$$

Though the throat is flat, it is not symmetrically placed in Kerr spacetime (the local map $r \to -r$ is not an isometry), and a computation in Section 4.15 shows that it is not totally geodesic.

2.4   Ergosphere and Time Machine

The Boyer–Lindquist coordinate vector fields maintain constant causal character on each block with only three exceptions: ∂_t on blocks I and III, and ∂_φ on block III. A closer look at these cases produces remarkable results. Both ∂_t and ∂_φ are Killing vector fields, and in Chapter 3 we see that every Killing vector field on any Boyer–Lindquist block B is a constant-coefficient linear combination of ∂_t and ∂_φ.

The causal character of ∂_t is physically important because, as discussed in Section 1.6, if a Killing vector field Z is timelike, the corresponding observer field $U = Z/|Z|$ is stationary. For these observers, the spacetime is not changing with time. Furthermore, if this U is unique, the spacetime is absolutely stationary.

Proposition 2.4.1

(1) *Boyer–Lindquist blocks are not stationary; that is, there exists no timelike Killing vector field defined on an entire block.*

(2) *On the regions $|r| \gg 1$ in blocks I and III the only timelike Killing vector fields are constant multiples of ∂_t. (So these regions are absolutely stationary.)*

Proof. (1) By the result quoted above, if X is a Killing vector field, then the vectors X_p lie in span $\{\partial_t, \partial_\varphi\}$ at each point p. It follows from Lemma 2.1.4 that on block II, where $\Delta < 0$, this plane is spacelike. Thus, there can be no timelike Killing field on II—or even on a small open set in II. Hence, ∂_t is the only candidate for a timelike Killing vector field on either block I or block III. But we have already seen that ∂_t does not remain timelike on either block.

To prove assertion (2), let X be a Killing vector that is timelike for $|r|$ large. We must show that $X = A\partial_t$ for some $A \neq 0$. By the result mentioned previously, we can write $X = A\partial_t + B\partial_\varphi$ for constants A and B. Now A cannot be zero because ∂_φ is zero (spacelike!) on the axis, so we can set $A = 1$. Then

$$<X, X> = g_{tt} + 2Bg_{t\varphi} + B^2 g_{\varphi\varphi}.$$

The metric identities in Lemma 2.1.1 show that as $|r| \to \infty$,

$$g_{tt} \to -1,$$
$$g_{t\varphi} \to 0, \quad \text{and}$$
$$g_{\varphi\varphi} \to +\infty \text{ (except on the axis)}.$$

Thus, $<X, X> \to +\infty$ (off-axis) unless $B = 0$. $\square$

Block I was time-oriented by requiring that where ∂_t is timelike, it is future-pointing. Thus, $U = \partial_t/|\partial_t|$ is a stationary observer field on the open set $\{g_{tt} < 0\}$; its integral curves are called *Kerr stationary observers*. For large $|r|$, these observers are unique in viewing the surrounding spacetime as stationary—with themselves as at rest in it. However, the following result shows that the Kerr observers are not static.

Lemma 2.4.2 *The coordinate vector field ∂_t is not hypersurface-orthogonal on any open set in a Kerr spacetime.*

Proof. We apply Lemma B.1 in Appendix B. Let θ be the one-form metrically dual to ∂_t; explicitly, $\theta(v) = <v, \partial_t>$ for all vectors v. Then $\partial_t^\perp$ is integrable if and only if $\theta \wedge d\theta = 0$. In Boyer–Lindquist coordinates,

$$\theta = \Sigma\, \theta(\partial_i)dx^i = \Sigma\, g_{0i}dx^i = g_{tt}dt + g_{t\varphi}d\varphi.$$

Since $d\theta = dg_{tt} \wedge dt + dg_{t\varphi} \wedge d\varphi$, we find

$$\theta \wedge d\theta = [g_{t\varphi}dg_{tt} - g_{tt}dg_{t\varphi}] \wedge d\varphi \wedge dt.$$

Because the metric components depend only on r and ϑ, the one-form β within the square brackets is a linear combination of dr and $d\vartheta$, and is thus independent of $d\varphi$ and dt. Consequently, $\theta \wedge d\theta = 0$ if and only if $\beta = 0$. But $\beta = (g_{t\varphi})^2 d(g_{tt}/g_{t\varphi})$, and we compute (neglecting nonvanishing factors):

$$(\partial/\partial r)(g_{tt}/g_{t\varphi}) \approx S^2(r^2 - a^2C^2)$$

and

$$(\partial/\partial\vartheta)(g_{tt}/g_{t\varphi}) \approx rSC\Delta.$$

Evidently these functions vanish simultaneously on no open set.   □

Physically, this result is a consequence of the Kerr rotation because in Schwarz-schild spacetime, where $a = 0$, the Boyer–Lindquist coordinates become ortho-gonal, and thus the hypersurfaces $t = $ const are orthogonal to ∂_t.

Boyer–Lindquist coordinates can be regarded as the way the Kerr observers impose their notions of time and space on the spacetime. Far from the star, these coordinates are quite reasonable, since Kerr geometry approaches the Minkowski metric of empty spacetime as $|r| \to \infty$. But they become less reasonable as $|r|$ decreases. In fact, first the "time coordinate" t stops measuring time, then the "radial distance" coordinate r stops measuring distance.

Lemma 2.4.3 *In slow Kerr spacetime the* stationary limit $L = \{g_{tt} = 0\}$ *is a (disconnected) hypersurface in* $\mathrm{I} \cup \mathrm{III}$ *that separates the stationary region* $\{g_{tt} < 0: \partial_t$ *timelike*$\}$ *and the region* $\mathcal{D} = \{g_{tt} > 0: \partial_t$ *spacelike*$\}$. $\mathcal{D}$ *lies between* $r = 0$ *and* $r = 2M$. *Horizons excluded,* L *is timelike.*

Proof. Write $g_{tt} = -1 + 2Mr\rho^{-2}$ as $\ell\rho^{-2}$; thus, $\ell = r(2M - r) - a^2C^2$. Recall that, as mentioned earlier, if either $r < 0$ or $r > 2M$, then ℓ and hence g_{tt} are negative.

To show that L is a hypersurface it suffices to check that $d\ell \neq 0$ at all points of L: $\ell = 0$. Now $d\ell = 2(r - M)\,dr - 2a^2SC\,d\vartheta$, so if $d\ell = 0$, then ℓ becomes $M^2 - a^2C^2$, which is cannot be zero for slow Kerr spacetime. Furthermore, the fact that $d\ell \neq 0$ on L shows that ℓ changes signs across L, proving the common boundary assertion.

It remains to show that L is timelike, that is, has Lorentz signature. On a Boyer–Lindquist block, L cannot meet the axis A because there ℓ becomes $-\Delta$. At any point p of L, the vector $z = a^2SC\partial_r + (r - M)\,\partial_\vartheta$ is tangent to L since $d\ell(z) = 0$. Because $\Delta > 0$ on $\mathrm{I} \cup \mathrm{III}$, it follows that z is spacelike. Evidently z is orthogonal to ∂_t and ∂_φ, which are also tangent to L and (by Lemma 2.1.4) span a timelike 2-plane. Thus, $T_p(L)$ is timelike.   □

Once horizons are incorporated in Kerr spacetime, L will be *null* where it meets the axis, since there it is tangent to a (null) horizon, as shown in Figure 2.4.

The region $\mathcal{E}$ in block I where ∂_t is spacelike is called the *ergosphere*, and we apply the same term to the corresponding region $\mathcal{E}'$ in block III. These form a enveloping zones around block II that becomes thinner at higher latitudes, leaving the poles uncovered (Figure 2.4.).

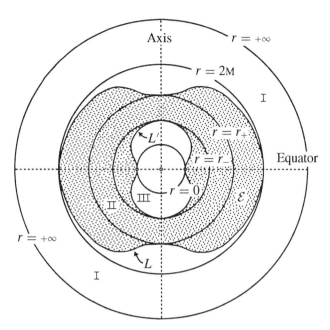

FIGURE 2.4.   The region $\mathcal{D}$ (shaded) on which ∂_t is spacelike in a slow Kerr black hole. (Rotating this polar-plane figure about the axis gives a t-slice as in Figure 2.2, with r exponential.) $\mathcal{D}$ contains all of block II and both ergospheres $\mathcal{E}$ and $\mathcal{E}'$. It can be considered as the region on which Kerr geometry is too relativistic for reasonable Newtonian analogies.

Ergospheres, first noticed by C.V. Vishveshwara (1968), are another consequence of rotation, nonexistent in Schwarzschild spacetime (Remark 2.5.8). They have an essential role in black hole theory and, in principle, enable energy to be extracted from the black hole (Section 4.1).

Lemma 2.4.4

(1) *On any timelike curve α in $\mathcal{E} \cup \mathcal{E}'$, the coordinate φ is strictly monotonic.*

(2) *With the natural time-orientation on $\mathcal{E}$, if α is future-pointing, then φ is increasing. Furthermore,*

(3) *If α meets the horizon, then $t' \to \infty$ and $\varphi' \to \infty$.*

Assertion (2) says that any traveler in the ergosphere $\mathcal{E}$, regardless of how powerful and well directed his rockets are, is forced (according to the Kerr observers) to rotate in the direction of rotation of the black hole.

Proof. Each hypersurface $N: \varphi = $ const is coordinatized by r, ϑ, t. Thus, the mutually orthogonal vector fields $\partial_r, \partial_\vartheta, \partial_t$ span $T_p(N)$ at each point $p \in N$. Both ∂_r and ∂_ϑ are spacelike throughout blocks I and III. But ∂_t is spacelike in $\mathcal{E} \cup \mathcal{E}'$, hence N is spacelike there. Thus the curve α can never be tangent to N, so by the Basis Theorem (Section 1.1), $d(\varphi \circ \alpha)/ds = \alpha'[\varphi]$ is never zero. This proves assertion (1).

The sign of $d(\varphi \circ \alpha)/ds$ depends on the time-orientation of the spacetime (and the choice of sign a). So far, we have only time-oriented the block I, requiring that ∂_t be timelike for $|r| \gg 1$. Let v be a nonspacelike vector at a point $p \in \mathcal{E} \subset \text{I}$, and write $v = \Sigma v^i \partial_i$, where i runs through r, ϑ, φ, t. Then

$$0 > <v, v> = \Sigma g_{ii} (v^i)^2 + g_{\varphi t} v^\varphi v^t.$$

Of the diagonal components, g_{rr} and $g_{\vartheta\vartheta}$ are positive on all Boyer–Lindquist blocks, $g_{\varphi\varphi} \geq 0$ holds on block I, and $g_{tt} > 0$ on $\mathcal{E}$. Thus, $g_{\varphi t} v^\varphi v^t < 0$. But $g_{\varphi t} = -2Mr\rho^{-2}aS^2 < 0$, since $a > 0$ and $\mathcal{E}$ does not meet the axis. Hence $v^\varphi v^t > 0$. The proof following Proposition 2.4.6 shows (independently) that if v is future-pointing, then $v^t > 0$ holds off-axis in block I. We conclude that $v^\varphi > 0$. Then setting $v = \alpha'(s)$ gives $d(\varphi \circ \alpha)/d\tau = v^\varphi > 0$.

For assertion (3), suppose that the r-coordinate of α approaches r_+ as $s \to s_0$, and r' approaches a nonzero limit (the particle passes through the horizon). Then as $s \to s_0$, their formulas show that $g_{rr} \to +\infty$, while $g_{\vartheta\vartheta}$ and (off the axis) $g_{\varphi\varphi}$ and $-g_{\varphi t}$ all approach (finite) positive limits. Hence $\varphi' t' \to +\infty$. It will follow that both φ' and t' must approach $+\infty$ because in the expression above for $<v, v> = <\alpha', \alpha'>$, the only negative term is $g_{\varphi t}\varphi' t'$. Thus, if, say, only $t' \to \infty$, the positive term $g_{tt} t'^2$ alone would suffice to contradict $<\alpha', \alpha'> < 0$. $\square$

If a pebble is dropped from $r > r_+$ we expect (correctly) that it will fall through the horizon $r = r_+$ in finite proper time. But according to the Kerr stationary observers, it never gets there because, as their time t runs on toward $+\infty$, the particle spirals ever faster with $\varphi' \to \infty$ as $r \to r_+$. Thus, they deny that r actually reaches r_+. Furthermore, even if the pebble is replaced by a well-directed spacecraft, their opinion is unchanged. The Kerr observers (who measure in Boyer–Lindquist coordinates) are not "wrong"—they will never see the horizon penetrated—but near the horizon their measurements differ radically from those of local observers.

One of the disturbing possibilities opened up by general relativity is *violation of causality*. If a timelike curve α is closed, then α represents a material particle that, starting from a certain event p, proceeds through an interval of its proper time only to find itself back once more at exactly the same event p. Thus, a person, returning from such a trip to the original departure event, could prevent himself from traveling after all. This certainly spells trouble for causality. No general physical principle is known that will prevent causality violations, but their appearance is unsettling. We shall see now that Kerr spacetime always contains such violations—though not in the intuitive exterior block I or even in the interhorizon block II.

Definition 2.4.5 *A spacetime M is* chronological [causal] *if there exist no closed timelike [nonspacelike] curves in M.*

Evidently, causal spacetimes are chronological, but examples show that the converse is not true. In applications, the spacetime M will often be merely a connected open set in a larger spacetime $\widetilde{M}$—and then M can be causal (or chronological) even though $\widetilde{M}$ is not.

Proposition 2.4.6 *Boyer–Lindquist blocks* I *and* II *are causal.*

Proof. First we show that in block I, *the hypersurfaces N*: $t = t_0$ *are spacelike.* (Note that N includes axis points since holding t constant gives a hypersurface in $\mathbf{R}^2 \times S^2$.) At each off-axis point $p \in N$, the vectors ∂_r, ∂_ϑ, ∂_φ form a basis for $T_p(N)$, and these vectors are spacelike and mutually orthogonal, so $T_p(N)$ is spacelike. If $p = (t_0, r, q) \in N \subset \mathbf{R}^2 \times S^2$ is in the axis, so $q = (0, 0, \pm 1)$, it suffices to replace ∂_ϑ and ∂_φ by any basis for $T_q(S^2)$. Hence N is spacelike.

It follows that on any nonspacelike curve α, the coordinate t is strictly monotonic. For if

$$(d(t \circ \alpha)/ds)(s_0) = \alpha'(s_0)[t] = 0,$$

then $\alpha'(s)$ is tangent to the hypersurface $t = t(\alpha(s_1))$—an impossibility, because $\alpha'(s_1)$ is not spacelike. Thus, α cannot be a closed curve.

On block II, the same argument works, with t and r reversed, since in *block* II, *the hypersurfaces $r = r_0$ are spacelike.* In fact, at off-axis points of N: $r = r_0$, $T_p(N) = \text{span}\{\partial_\vartheta, \partial_\varphi, \partial_t\}$, hence $\partial_r \perp T_p(N)$—and by continuity this still holds at axis points. Since ∂_r is timelike throughout II, $T_p(N)$ is spacelike. $\square$

Although t does not measure time everywhere on block I, nevertheless it is strictly increasing on any future-pointing particle. To verify this sign, note that grad t is timelike on the Kerr exterior since its orthogonal hypersurfaces t const are spacelike. Now $<\text{grad } t, \partial_t> = dt/dt = 1 > 0$, and ∂_t is future-pointing (when timelike), so grad t is past-pointing everywhere on block I. Thus, if nonspacelike vector v is future-pointing, $v[t] = <v, \text{grad } t> > 0$, which for $v = \alpha'$ means $d(t \circ \alpha)/ds > 0$.

On block III, however, the argument for block I and II fails radically. Consider first the Boyer–Lindquist coordinate φ. The integral curves of the vector field ∂_φ are closed curves circling around the axis of rotation (except on the axis, where $\partial_\varphi = 0$). Although ∂_φ is spacelike on blocks I and II, it becomes timelike on a region $\mathfrak{T}$ in III near the ring singularity. $\mathfrak{T} = \{g_{\varphi\varphi} < 0\}$ is called the *Carter time machine*. By its use, causality can be violated in a severe way.

For concreteness, time-orient block III by the canonical vector field V (timelike on III). In the following proofs we often denote a point (alias event) by its Boyer–Lindquist coordinates.

Proposition 2.4.7 *Block III is* vicious, *that is, given any events $p, q \in \text{III}$ there exists a timelike future-pointing curve in III from p to q.*

Hence, reversing p and q, there also exists a past-pointing timelike curve from p to q. In preparation for the proof, note first that *any point with $\vartheta = \pi/2$ and $r < 0$ sufficiently small is in $\mathfrak{T}$.* In fact, $\vartheta = \pi/2$ implies $S = 1$, $C = 0$, and $\rho^2 = r^2$. Thus, $g_{\varphi\varphi}$ reduces to $r^2 + a^2 + 2Ma^2 r^{-1}$, which is negative for small r.

Next we show that inhabitants of block III can freely go to and from the time machine as indicated in Figure 2.5.

Lemma 2.4.8 *Given $p = (r_0, \vartheta_0, \varphi_0, t_0) \in \text{III}$ there is a number $c > 0$ such that for any numbers $\Delta t \geq c$ and φ, there is*

1. *A future-pointing timelike curve α from p to $(r^*, \pi/2, \varphi, t_0 + \Delta t) \in \mathfrak{T}$, and*
2. *A past-pointing timelike curve β from p to $(r^*, \pi/2, \varphi, t_0 - \Delta t) \in \mathfrak{T}$.*

Proof. By the remark above we can choose r^* small enough so that every $(r^*, \pi/2, \varphi, t)$ is in $\mathfrak{T}$. Let $\alpha_1(s) = (r(s), \vartheta(s), \varphi_0, t_0)$ be a curve in block III from p to $(r^*, \pi/2, \varphi_0, t_0)$. For a number $A > 0$ consider the curve

$$\alpha(s) = (r(s), \vartheta(s), \varphi_0 + Aas, t_0 + At(s)),$$

where $t(s)$ satisfies $dt/ds = r^2 + a^2$. Now

$$\alpha' = \alpha_1' + A[a\partial_\varphi + (r^2 + a^2)\partial_t] = \alpha_1' + AV,$$

where the canonical vector field V is restricted to α. Since $V \perp \alpha_1'$,

$$<\alpha', \alpha> = <\alpha_1, \alpha_1'> + A^2 <V, V> = <\alpha_1', \alpha_1'> - A^2 \Delta \rho^2.$$

The curves are defined on a closed interval, so $<\alpha_1', \alpha_1'>$ is bounded above, and on α the positive functions Δ and r^2 are bounded away from 0. Thus, for A large,

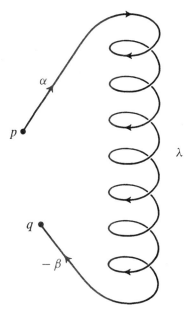

FIGURE 2.5. By using the time machine, inhabitants of block Ⅲ can travel to any event, including those in their own past.

α is timelike. Furthermore, α is future-pointing since V is future-pointing on Ⅲ and $\langle \gamma', V \rangle = A\langle V, V \rangle < 0$.

Extending α by an integral curve of the future-pointing timelike vector fields V or $-\partial_\varphi$ (future-pointing timelike) lets us freely increase the final t coordinate of α and adjust the φ coordinate to any value. Thus, the required number c exists.   □

Proof of Proposition 2.4.7. The required curve will be constructed by joining three segments. The first is a curve α as in the preceding lemma. Then for a constant $B > 0$, consider the curve

$$\lambda(s) = (r^*, \pi/2, \bar{\varphi} - Bs, \bar{t} - s) \quad \text{for } s \geq 0$$

that starts at the final point $(r^*, \pi/2, \bar{\varphi}, \bar{t})$ of α. Then λ is in the time machine and has velocity $\lambda' = -B\partial_\varphi - \partial_t$. Thus,

$$\langle \lambda', \lambda' \rangle = B^2 g_{\varphi\varphi} - 2Bg_{\varphi t} + g_{tt}.$$

Since the metric components depend only on r and ϑ, they are constant along λ. But $g_{\varphi\varphi} < 0$, hence for B large, λ is timelike. Furthermore, it is future-pointing since a short computation using the metric identities shows that $<\lambda', V> = \Delta(1 - aB)$, which is negative for $B > 1/a$ since $\Delta > 0$ on $\amalg\!\!\amalg$.

What makes the time machine work is that, on the curve λ, $dt/ds = \lambda'(s)[t] = -1$. Thus, as λ proceeds into the future, its Boyer–Lindquist coordinate t is steadily *decreasing*. Let λ run until it reaches a point, say, $(r^*, \pi/2, \varphi_1, t_1)$ that can be reached from event q by a past-pointing timelike curve β. Then reversing the parametrization of β gives a curve $\alpha \cup \lambda \cup (-\beta)$, as required (see Figure 2.5). $\qquad\square$

The composite curve takes its time in going from p to $\mathfrak{T}$ and from $\mathfrak{T}$ to q, but, as Brandon Carter has remarked, its segment within $\mathfrak{T}$ literally makes up for lost time.

Lemma 2.4.9 *The time machine $\mathfrak{T}$ lies in the region $-\max\{M, a\} < r < 0$. A t-slice of $\mathfrak{T}$ is a solid torus whose outer equator is the ring singularity.*

Proof. The causal character of ∂_φ is determined by the sign of

$$g_{\varphi\varphi} = [r^2 + a^2 + 2Mra^2\rho^{-2}S^2]\,S^2.$$

But since ρ^2 is always positive, this sign is also determined by the simpler function

$$f(r, \vartheta) = \rho^2(r^2 + a^2) + 2Mra^2S^2 = r^4 + a^2(1 + C^2)r^2 + 2Ma^2S^2r + a^4C^4.$$

Since all coefficients here are positive, ∂_φ is spacelike when $r \geq 0$. It can be checked directly (and we will see below) that $r < -\max\{M, a\}$ also implies $f(r, \vartheta) > 0$.

On the equator $\vartheta = \pi/2$, the function $f(r, \vartheta)$ becomes $r(r^3 + a^2r + 2Ma^2)$, which is negative for $k < r < 0$, where k is the unique negative root of the polynomial $r^3 + a^2r + 2Ma^2$.

Expressed in terms of r and $C = \cos\vartheta$,

$$f(r, C) = r^4 + a^2(1 + C^2)r^2 + 2Ma^2r - 2Ma^2C^2r + a^4C^4.$$

In the (r, C)-plane, $f(r, C) = 0$ is an oval, symmetric about its diameter $[k, 0]$ in the r-axis. (The point $r = 0$, $C = 0$ is missing; it is in the ring singularity.) On the interior of this oval, ∂_φ is timelike, and, of course, it remains so for all values of φ and t. $\qquad\square$

2.5   Kerr-Star Spacetime

The Kerr metric will now be extended over horizons by means of *Kerr coordinates*. This extension welds adjacent Boyer–Lindquist blocks together along horizons, which become hypersurfaces in the resulting spacetime.

Kerr coordinates derive from two remarkable families of null geodesics, said to be *principal*. These represent light rays, one family going in toward the star, the other out from it. We can picture an ingoing principal null geodesic as a beam of light aimed directly at the star by a distant observer in the Kerr exterior (the outgoers in reverse). These curves are given in terms of Boyer–Lindquist coordinates on block I, by

$$r' = \pm 1, \qquad \vartheta' = 0, \qquad \varphi' = a/\Delta, \qquad t' = (r^2 + a^2)/\Delta,$$

where the sign $+1$ gives the *outgoing* principal null geodesics, the sign -1 the ingoing ones. Here we can see directly that as such a curve approaches a horizon, so $\Delta \to 0$, it exhibits the infinite spiraling and slowing that signals the failure of Boyer–Lindquist coordinates.

The principal null geodesics are deeply imbedded in the structure of Kerr spacetime; they not only stand out as a distinctive class of geodesics (Chapter 4) but can actually be derived directly from the curvature of Kerr spacetime (Chapter 5). That these curves γ are in fact geodesics will be checked in the next section. We verify now that γ is null, future-pointing, and—in the limit as $r \to \infty$—aimed directly toward (or away from) the star. The tangent of γ is

$$\gamma' = \pm \partial_r + \Delta^{-1} a \partial_\varphi + \Delta^{-1}(r^2 + a^2)\partial_t = \pm \partial_r + \Delta^{-1} V,$$

where V is the canonical vector field of Definition 2.1.2. Since $\Delta = r^2 - 2Mr + a^2$, the first of these expressions shows at once that, as $r \to \infty$, γ' approaches $\pm \partial_r + \partial_t$. This is direct radial motion in the nearly Minkowskian region $r \gg 1$ of block I. Next, Lemma 2.1.3 and the formula for g_{rr} give

$$<\gamma', \gamma'> = g_{rr} + \Delta^{-2}<V, V> = 0,$$

so γ is a null curve. Finally, it is future-pointing since ∂_t is future-pointing on block I when r is large, and, by the metric identity (m4) in Lemma 2.1.1

$$<\gamma', \partial_t> = \Delta^{-1}[a\, g_{t\varphi} + (r^2 + a^2)\, g_{tt}] = -1.$$

There are two types of Kerr coordinates corresponding to the two families of principal null geodesics. To avoid sign complications we describe now only the type based on *ingoers*, that is, the ingoing principal null geodesics. (Sign changes will suffice to handle the other type.)

The physical motivation is this: Of all particles passing through a horizon, ingoers surely cut through most directly. To build a coordinate system that covers horizons we turn the ingoers into coordinate curves. Thus, we must (infinitely) untwist φ and slow t, building new φ and t coordinates that are constant on ingoers. These goals are easily accomplished by defining, on all three blocks, the *Kerr-star coordinate functions*

$$t^* = t + T(r), \qquad \varphi^* = \varphi + A(r),$$

where the *time function* $T(r)$ and *angle function* $A(r)$ are any functions such that

$$dT/dr = (r^2 + a^2)/\Delta$$

and

$$dA/dr = a/\Delta.$$

Each ingoing principal null geodesic γ now satisfies $r' = -1$, $\vartheta' = 0$ as before, and by the formulas above,

$$(\varphi^*)' = \varphi' + dA/dr\, r' = 0, \qquad (t^*)' = t' + dT/dr\, r' = 0.$$

Thus, assuming the result below, γ is an r-coordinate curve parametrized by *decreasing r*.

Lemma 2.5.1 *For each Boyer–Lindquist block B = I, II, or III, the mapping* $\xi^* = (r, \vartheta, \varphi^*, t^*)$ *is a coordinate system on B − (axis A).* ξ^* *is called a* Kerr-star coordinate system.

Proof. To show that ξ^* is one-one, suppose $\xi^*(p) = \xi^*(q)$. Then $r(p) = r(q)$, $\vartheta(p) = \vartheta(q)$, hence

$$t^*(p) = t^*(q) \;\Rightarrow\; t(p) + T(r(p)) = t(q) + T(r(q)) \;\Rightarrow\; t(p) = t(q),$$

and similarly for φ^* (regarded as a circular coordinate). Since p and q have the same Boyer–Lindquist coordinates, they are equal.

The Jacobian determinant $\partial(r, \vartheta, \varphi^*, t^*)/\partial(r, \vartheta, \varphi, t)$ is immediately found to be nonzero (in fact, $+1$), hence ξ^* is a diffeomorphism and thus a coordinate system. $\square$

As smooth manifolds, the Boyer–Lindquist blocks fit together naturally in $\mathbf{R}^2 \times S^2$, but the Boyer–Lindquist form of the metric fails at their interfaces. Indeed, we saw in Section 2.4 that the simplest fall through a horizon was registered as an infinite spiral taking infinite time. The functions $T(r)$ and $A(r)$—with Δ in denominators—approach infinity as r approaches a horizon radius $r_\pm$, and provide the explosive coordinate change needed to fit the blocks together *geometrically*.

How are the coordinate vector fields of Kerr-star coordinates ξ^* related to those for Boyer–Lindquist coordinates ξ? Because the coordinate functions differ solely by additive functions of r, *the coordinate vector fields* ∂_t, ∂_ϑ, ∂_φ *are the same for both systems*, except, of course, that in K^* they extend over the horizons. Let ∂_r^* denote the coordinate vector field of r in Kerr-star coordinates. Then the Basis Theorem gives

$$\partial_r = \partial_r^* + \Delta^{-1} a\, \partial_\varphi + \Delta^{-1}(r^2 + a^2)\, \partial_t = \partial_r^* + \Delta^{-1} V,$$

where V, as usual, is the canonical vector field. Thus, $\partial_r^* = \partial_r + \Delta^{-1} V$ extends over horizons though neither of its summand does separately.

To find the components g_{ij}^* of the Kerr metric relative to Kerr-star coordinates, the only changes from Boyer–Lindquist coordinates are those resulting from $\partial_r \to \partial_r^*$. By construction, the ingoing principal null geodesics are integral curves of $-\partial_r^*$, and since this vector is null, $g_{rr}^* = 0$. Also, $g_{r\vartheta}^* = <\partial_r^*, \partial_\vartheta> = <\partial_r - \Delta^{-1} V, \partial_\vartheta> = 0$. However,

$$g_{r\varphi}^* = <\partial_r^*, \partial_\varphi> = <\partial_r - \Delta^{-1} V, \partial_\varphi> = -\Delta^{-1}<V, \partial_\varphi> = -aS^2,$$
$$g_{rt}^* = <\partial_r^*, \partial_t> = <\partial_r - \Delta^{-1} V, \partial_t> = -\Delta^{-1}<V, \partial_t> = 1.$$

The following formula for ds^2 results:

Lemma 2.5.2 *The Kerr metric, expressed in terms of Kerr-star coordinates, has line-element*

$$ds^2 = g_{tt}dt^{*2} + 2g_{t\varphi}\,dt^*d\varphi^* + g_{\varphi\varphi}\,d\varphi^{*2} + \rho^2 d\vartheta^2$$
$$+ 2dt^*dr - 2aS^2d\varphi^*dr.$$

The four components g_{tt}, $g_{t\varphi}$, $g_{\varphi\varphi}$, $g_{\vartheta\vartheta}$ above are the same functions of r and ϑ as before. The most striking change is that the Boyer–Lindquist component $g_{rr} = \rho^2/\Delta$ has been replaced by $2(dt^* - aS^2d\varphi^*)\,dr$, where, as always, $S = \sin\vartheta$. But g_{rr} is the *only* part of the Boyer–Lindquist form that fails when $\Delta = 0$, and the two new terms in ds^2 above are well behaved there. Thus, the Kerr metric is extendible over $\Delta = 0$, and it remains nondegenerate there (off-axis) since a short computation gives $\det(g^*_{ij}) = -\rho^4 S^2$.

The axis is dealt with as before: Kerr-star coordinates fail there just as Boyer–Lindquist coordinates do, but the Kerr metric extends over A in the same way as for Boyer–Lindquist coordinates (Lemma 2.2.2).

Definition 2.5.3   Kerr-star spacetime K^* *has*

1. *Analytic manifold* $R^2(r, t^*) \times S^2(\vartheta, \varphi^*) - \Sigma$, *where* Σ *is the ring singularity* $r = 0$, $\vartheta = \pi/2$,

2. *Metric tensor as in Lemma 2.5.2 (extended over the axis), and*

3. *Time-orientation such that the globally defined null vector field* $-\partial_r^*$ *is future-pointing.*

This time-orientation is consistent with our previous intuitive choice of ∂_t as future-pointing when $r \gg 1$, since $<\partial_t, \partial_r^*> = g^*_{tr} < 0$. This sign is the "standard" choice; it will sometimes be reversed in the construction of maximal Kerr spacetimes in Chapter 3. For these constructions it is important to specify just how Kerr-star spacetime contains the Boyer–Lindquist blocks.

Lemma 2.5.4   *For any Boyer–Lindquist block B the coordinate map $(r, \vartheta, \varphi, t) \rightarrow (r, \vartheta, \varphi^*, t^*)$ has a (unique) analytic extension to an isometry j^* of B onto an open set B^* of K^*.*

Proof.   The coordinate map j is by construction an isometry, but it is defined only off-axis in B. Since the Kerr metric extends over the axis, if we can also analytically extend j over the axis, then by continuity the extension will be an isometry.

For any number α, let R_α be rotation of S^2 around the z-axis through angle α. A matrix expression shows that the resulting map $(\alpha, q) \rightarrow R_\alpha(q)$ from $\mathbf{R} \times S^2$ to S^2 is analytic. Now define j^* to be the map sending each point $(r, t, q) \in B \subset \mathbf{R}^2 \times S^2$ to

$$(r, t^*, R_{A(r)}(q)) \in K^* \subset \mathbf{R}^2 \times S^2.$$

Then j^* is analytic, and it is an extension of j since if $q \neq (0, 0, \pm1)$, the rotation preserves colatitude ϑ and increases longitude φ by $A(r)$ to φ^*. $\square$

Having established this *canonical imbedding* of B in K^*, we ordinarily regard B as a subset of K^*. To summarize: in slow Kerr spacetime, the three Boyer–Lindquist blocks have been welded into a single spacetime, with the horizons $H_+: r = r_+$ and $H_-: r = r_-$ as interfaces (see Figure 2.7). In extreme Kerr spacetime, its two blocks I and III are attached along $H: r = M$. And in the fast case there are no horizons, so K^* is merely an isometric copy of the single block I = III.

Unless the contrary is stated, the Boyer–Lindquist blocks I, II, and III are *time-oriented as subsets of K^**. Then ∂_t is future-pointing wherever it is timelike, e.g., if $r > 2M$ or $r \leq 0$ (see Figure 2.4.) On block II, where ∂_r is timelike, it is past-pointing, since $<\partial_r, -\partial_r^*> > 0$.

Now that horizons have been brought into a spacetime we can examine their remarkable geometric properties.

Proposition 2.5.5 *In K^*, each horizon H is a closed, connected, totally geodesic null hypersurface, with futurecones on the $-\partial_r^*$ side. The restriction to H of the canonical vector field V is the unique null vector field on H that is tangent to H, hence also normal to H. The integral curves of V in H are null pregeodesics called* restphotons *that are said to* generate *H.*

Proof. As an analytic manifold, K^* is $\mathbf{R}^2 \times S^2 - \Sigma$, with r, t^* as coordinates on $\mathbf{R}^2$ (lifted to K^*). Thus, setting r equal to any nonzero constant produces a closed hypersurface diffeomorphic to $\mathbf{R}^1 \times S^2$. In particular, for $r = r_\pm$ (or $r = M$ in the extreme case) each horizon H is a sphere (Figure 2.2) enduring through time.

On any slice $r = $ const, the functions ϑ, φ^*, t^* form a coordinate system for all but axis points. In view of Lemma 2.1.4, for these coordinates we have

$$\det \begin{bmatrix} \rho^2 & 0 & 0 \\ 0 & g_{\varphi\varphi} & g_{\varphi t} \\ 0 & g_{t\varphi} & g_{tt} \end{bmatrix} = -\rho^2 \Delta S^2.$$

Thus, the metric is degenerate when $\Delta = 0$, showing that the horizons H are null. By continuity they remain null at axis points.

Each tangent space $T_p(H)$ is a null subspace of $T_p(K^*)$, hence by Lemma 1.5.11, every vector in $T_p(K^*)$ is spacelike except for those on the line at which the nullcone is tangent to $T_p(H)$. (See Figure 1.5(c).) Since the null vector $-\partial_r^*$ is future-pointing and not tangent to H, it indicates the future side of the horizon.

The canonical vector field V is a linear combination of ∂_φ and ∂_t, so it is tangent to H. But $<V, V> = -\Delta\rho^2$, so V is null on horizons. Thus, it lies in the unique null line tangent to H. Then Lorentz linear algebra (Lemma 1.5.11, again) shows that V is also *normal* to the horizon; in fact, $V^\perp = T_p(H)$.

To show that horizons are totally geodesic, note first that the vector fields $V = (r^2 + a^2)\partial_t + a\partial_\varphi$ and $V_\pm = (r_\pm^2 + a^2)\partial_t + a\partial_\varphi$ agree on the horizon $r = r_\pm$.

Since V_+ and V_- are constant linear combinations of the Killing vector fields ∂_φ and ∂_t, they are themselves Killing vector fields.

Given vector fields X and Y tangent to $H = H_\pm$, we must show that $\nabla_X Y$ is also tangent to H (see Section 1.7). On the submanifold H, where $V = V_\pm$ and $X, Y \perp V$,

$$\langle \nabla_X Y, V \rangle = X\langle Y, V \rangle - \langle Y, \nabla_X V \rangle = -\langle Y, \nabla_X(V_\pm) \rangle.$$

Since $V_\pm$ is Killing, reversing X and Y here reverses signs, so $\langle \nabla_X Y, V \rangle = -\langle \nabla_Y X, V \rangle$. The bracket $[X, Y]$ is also tangent to H, hence orthogonal to V. But $[X, Y] = \nabla_X Y - \nabla_Y X$, so $\langle \nabla_X Y, V \rangle = \langle \nabla_Y X, V \rangle$. We conclude that $\langle \nabla_X Y, V \rangle = 0$, showing that H is totally geodesic.

It remains to show that (when suitably parametrized) the integral curves of V are geodesics. If $p \in H$, let γ be the geodesic of K^* with initial velocity V_p. Since H is closed and totally geodesic, γ remains always in H. And since γ is a geodesic, it maintains the same causal character: initially null, it is always null. Hence by the uniqueness of the null line tangent to H, γ' is always collinear with V. But then γ is a reparametrization of the integral curve of V starting at p. Thus, the integral curves of V are pregeodesics. $\qquad\qquad\square$

Because all futurecones lie entirely on the $-\partial_r{}^*$ side of the horizon H, it is a kind of one-way membrane: *Every future-pointing timelike curve that meets H must cut through transversally in the direction of decreasing r*, and the same is true for null curves (restphotons excepted). Thus, in slow Kerr spacetime K^*, material particles and photons can pass freely from block I through H_+ to block II, and on through H_- to block III. Since these horizons separate the blocks, particles can never return from III to II, or from II to I. (As K^* is maximally extended in Chapter 3 these restrictions are maintained.)

It will soon be clear that the restphotons are in fact *outgoing* principal null geodesics. Thus, each horizon $H_\pm$ is the locus of all photons radiating directly outward—at the speed of light—but held firmly at radius $r_\pm$ by the gravitational attraction of the black hole (see Figure 2.6). This relativistic phenomenon, present also in Schwarzschild spacetime, had its classical antecedent in the imagination of Lewis Carroll, whose Red Queen explained, "Now, here, you see, it takes all the running you can do, to keep in the same place."

Nevertheless, the restphotons do at least circulate in the following sense: on $H_\pm$, where $V = V_\pm = (r_\pm^2 + a^2)\partial_t + a\partial_\varphi$, the flow of the (spacelike!) vector field ∂_t carries integral curves of $V_\pm$, that is, restphotons, to restphotons. Since $V_\pm$ and ∂_t are independent except at the poles, the only fixed restphotons are the parametrizations of $H_\pm \cap A$.

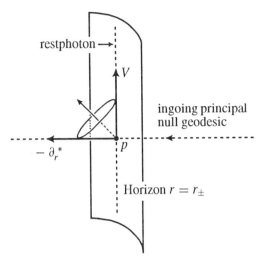

FIGURE 2.6.   Through each point p of a horizon in K^* the ingoing principal null geodesic is the r-parameter curve through p, an integral curve of $-\partial_r{}^*$. The restphoton through p is an *outgoing* principal null geodesic trapped in the horizon. Otherwise, every future-pointing nonspacelike curve through p must cross the horizon in the $-\partial_r{}^*$ direction (given the standard time-orientation of K^*).

We began with ingoing principal null geodesics; now let us see what happens to the outgoing ones in K^*. The outgoers were described initially by the equations

$$r' = +1, \qquad \vartheta' = 0, \qquad \varphi' = a/\Delta, \qquad t' = (r^2 + a^2)/\Delta,$$

which, in Kerr-star coordinates, become

$$r' = +1, \qquad \vartheta' = 0, \qquad \varphi^{*\prime} = 2a/\Delta, \qquad t^{*\prime} = 2(r^2 + a^2)/\Delta.$$

Thus, these curves are undefined on horizons $\Delta = 0$. Because the original definitions of ingoers and outgoers were symmetrical, we can hope that the outgoers too can be extended over all of K^*. This is easy to do—provided their geodesic parametrization is abandoned. The previous formulas present the outgoers as integral curves of the vector field

$$\partial_r{}^* + 2a/\Delta\, \partial_\varphi + 2(r^2 + a^2)/\Delta\, \partial_t = \partial_r{}^* + (2/\Delta)\, V.$$

Scalar multiplication by $\Delta/2$ gives the vector field $N^* = (\Delta/2)\, \partial_r{}^* + V$, which is well defined on all of K^*, including the axis. Furthermore, N^* is always future-pointing because on block I it points in the same direction as the original outgoers. Now we redefine the *outgoing principal null geodesics* on K^* to be geodesic parametrizations of the integral curves of N^*. Then there exists a unique out-goer and unique ingoer (up to parametrization) through each point of Kerr-star spacetime.

This change of parametrization is not a mere technicality; it has two important consequences.

1. There are now outgoing principal null geodesics that meet horizons H. These new outgoers are exactly the restphotons contained in H because the latter are (by definition) geodesic reparametrizations of integral curves of V, and when $\Delta = 0$, N^* is just V.

2. Between the two horizons (i.e., in block II, where $\Delta < 0$) the initial parametrizations of outgoers would have been past-pointing, contrary to the mandate that particles are always future-pointing. In the corrected parametrization the *outgoing* principal null geodesics in block II have *decreasing* r coordinates. This expresses the causal necessity of these photons to proceed into the future, because between the horizons, r measures time not distance, and it is $-\partial_r{}^*$ that is future-pointing.

The ambiguities of parametrization involved in extending the outgoers over horizons make it clear that ingoers and outgoers are not families of parametrized curves, but are rather two *congruences* on K^* (Section 1.7), each of which fills K^* with totally geodesic, one-dimensional submanifolds. There is exactly one such integral curve from each congruence through each point of K^*, and these curves can be geodesically parametrized as needed.

In the constructions of Chapter 3 we need to know how r varies on principal nulls. The ingoers are easy: each runs from limit value $r = +\infty$ to $r = -\infty$. Among the outgoers, the restphotons have r constant at r_+. In block I, $dr/ds = +1$ gives $r(s) = r_0 + s$, so the outgoers run from limit value r_- to $r = +\infty$. In block II the reversal of parametrization discussed previously gives $r(s) = r_0 + s$, so outgoers run from limit value r_+ to r_-. In block III, $r(s) = r_0 - s$ again, so they run from $-\infty$ to r_- (see Figure 2.7). In particular, in III the so-called *out*goers are actually traveling *inward* from $r \ll -1$ toward $r = r_-$. Such misnomers are not uncommon since terminology tends, naturally enough, to derive from the more familiar block I.

We have seen that in K^* that if a particle from block I crosses the outer horizon $H_+: r = r_+$ it can never return . But can an astronaut who comes arbitrarily close to H_+ still escape to $r = \infty$? The *yes* answer proves that H_+ is an *event horizon*: the boundary of the region in K^* for which escape to infinity is possible.

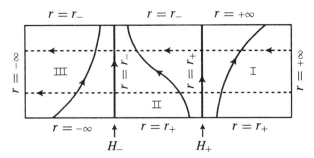

FIGURE 2.7.   Schematic representation of Kerr-star spacetime K^*. Ingoing principal null geodesics, represented by horizontal dashed lines, cross both horizons; outgoers, represented by curved solid lines, cross neither. (The restphotons in the horizons H_+ are also outgoers.)

Lemma 2.5.6   *Let p_0 be a point of slow Kerr-star spacetime K^*. Then p_0 is in block* I *if and only if there exists a material particle (timelike future-pointing curve) α starting at p_0 such that $r(\alpha(\tau)) \to +\infty$.*

Proof.   For points of K^* not in block I, no particle can cross the horizon $r = r_+$. Restphotons in H_+ come close to escaping—at least they do not fall farther into the black hole. So we construct an escaping material particle α by modifying the principal null equations to

$$dr/d\tau = \lambda, \qquad d\vartheta/d\tau = 0, \qquad d\varphi/d\tau = a/\Delta, \qquad dt/d\tau = (r^2 + a^2)/\Delta$$

where λ is a constant, $0 < \lambda < 1$. Such a curve α, starting at $p_0 \in$ I, is defined for all $\tau > 0$ and has $r \to \infty$. Thus, it remains only to show that for $\lambda < 1$ it is timelike. But since $\alpha' = \lambda \partial_r + \Delta^{-1} V$, and $\Delta > 0$ on I,

$$<\alpha', \alpha'> = \lambda^2 g_{rr} + \Delta^{-2} <V, V> = (\lambda^2 - 1)\rho^2/\Delta < 0.$$

$\square$

We now consider the geometry of the (north or south) axis A in Kerr-star spacetime (see Figure 2.2). Coordinates for A can be derived from Kerr-star coordinates by setting ϑ equal to 0 or π and discarding φ^*. Thus, $A \approx \mathbf{R}^2(t^*, r)$, with line-element

$$dS^2 = g_{tt} dt^{*2} + 2dt^* dr = dt^* \theta,$$

where θ is the one-form $-\Delta/(r^2 + a^2)\, dt^* + 2dr$.

A is a closed, totally geodesic Lorentz surface—a spatial line enduring through time. The equations $dt^* = 0$ and $\theta = 0$ describe two families of null curves in A: the lines t^* constant and the solutions of the differential equation $dt^*/dr = 2\Delta/(r^2+a^2)$. *These curves are just principal ingoing and outgoing null geodesics, respectively*, as becomes clear when they are viewed as the integral curves of $-\partial_r{}^*$ and $N^*|_A = (\Delta/2)\partial_r{}^* + (r^2 + a^2)\partial_t$.

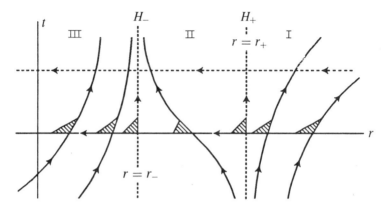

FIGURE 2.8.   Principal null geodesics in the north or south axis A of K^*: horizontal lines represent ingoers, all others (t-axis excepted) are outgoers. Their intersections mark out the cross-hatched future-cones. Each curve $H_\pm \cap A$ is parametrized by a single (outgoing) restphoton. Translation in the t direction is an isometry—preserving future-cones and principal null geodesics.

Thus, in the axis of K^*, as elsewhere, the ingoers move directly inward from $r = \infty$ to $r = -\infty$. In $A \cap \mathrm{I}$ the outgoers also behave "normally," going out toward $r = \infty$. But those in $A \cap \mathrm{III}$ exhibit in sharper form the behavior observed earlier. These geodesics represent photons traveling along the axis *toward* the black hole—an axial beam of light aimed directly at the black hole from block III. Yet, as their differential equation shows, the rate of approach slows and the photons never reach $r = r_-$. In short, they are *repelled* by, not attracted to, the Kerr gravitational field. We conclude that block III is not an analogue of block I but a gravitational reversal of it, a *white hole* in which the mass M > 0 acts as if it were negative.

Vectors tangent to principal null geodesics are called *principal null*; they point in *principal null directions*. At each point p of K^*, the (necessarily timelike) plane Π determined by the two principal directions is called the *principal plane* at p. Thus, $\Pi = \mathrm{span}\{\partial_r{}^*, V\}$ is a two-dimensional distribution on K^*. Its orthogonal distribution $\Pi^\perp = \text{span}\{\partial_\vartheta, W\}$ is spacelike (see Figure 2.3). In Schwarzschild spacetime, Remark 2.5.8 shows that both these distributions are integrable, but here rotation invervenes.

Lemma 2.5.7 *Neither Π nor $\Pi^\perp$ is integrable on any open set of K^*.*

Proof. For Π we have $[\partial_r{}^*, V] = [\partial_r, V] = [\partial_r, (r^2 + a^2)\partial_t + a\partial_\varphi] = 2r\partial_t$. Off-axis in a Boyer–Lindquist blocks, ∂_t is never in Π, since ∂_t, ∂_r, and ∂_φ are independent. Hence Π is not integrable on any open set in K^*.

Similarly, for $\Pi^\perp$, $[\partial_\vartheta, W] = 2aSC\partial_t$ gives the corresponding result.   □

At points of the axis A, both principal directions are tangent to A, so A is an exceptional integral surface of Π. Checking cases in the proof above shows that Π and $\Pi^\perp$ have no others—in K^*. (A highly exceptional integral surface of $\Pi^\perp$ appears in the next chapter.)

Remark 2.5.8 (Schwarzschild comparison) Formal Kerr results usually remain valid for Schwarzschild spacetime when a is set equal to 0; nevertheless the two geometries differ in significant ways. As we saw in Section 2.1, setting $a = 0$ in the Kerr metric produces

$$ds^2 = -h\,dt^2 + h^{-1}dr^2 + r^2(d\vartheta^2 + S^2 d\varphi^2),$$

where

$$h = h(r) = 1 - 2\text{M}/r.$$

This gives a nondegenerate metric on the two *Schwarzschild blocks* I: $r > 2\text{M}$ and II: $0 < r < 2\text{M}$ in the product manifold $\mathbf{R}(t) \times \mathbf{R}^3$. The Kerr-star formulas (with $a = 0$) give a coordinate system covering the unique horizon $H: r = 2\text{M}$. Thus, the two blocks are joined along H, which as in the Kerr case, is a null hypersurface that (with ∂_t future-pointing) allows particles from I to enter II but none to emerge.

There is no ergosphere, since ∂_t is timelike for $h > 0$, that is, for $r > 2\text{M}$. We will see in Section 2.7 that the Schwarzschild origin $r = 0$ is a genuine singularity, so there is no simple extension through it as in the Kerr case. Thus, there is no block III.

With $a = 0$, the Schwarzschild principal null geodesics are given by $r' = \pm 1$, $\vartheta' = 0$, $\varphi' = 0$, $t' = 1/h$. They are simply spacetime parametrizations of light rays running directly out from or into the origin $r = 0$ of $\mathbf{R}^3$. Thus, the distributions Π and $\Pi^\perp$ of the Kerr case become the (trivially integrable) horizontal and vertical tangent planes of a different product structure $\mathbf{R}^2(t, r > 0) \times S^2$. In terms of this structure Schwarzschild spacetime can be expressed geometrically as a *warped product* (O'Neill 1983)—a decisive simplification that fails for Kerr spacetime.

We turn now to the computation of Kerr curvature, which shows that—by contrast with axis and horizons—the ring singularity Σ can never be a subset of a Kerr spacetime.

2.6   Connection Forms

Boyer–Lindquist coordinates suggest a natural frame field on the Boyer–Lindquist blocks. Using Cartan methods as discussed in Chapter 1 we compute the connection forms of this frame field and thereby find the covariant derivatives of its constituent vector fields. A discussion on *curvature* follows in the next section.

We have seen that the vector fields ∂_r, ∂_ϑ, V, and W are mutually orthogonal, where V and W are, as usual, the canonical vector fields (Definition 1.1.2). Taking scalar products gives

$$g_{rr} = \rho^2/\Delta,$$

$$g_{\vartheta\vartheta} = \rho^2,$$

$$<V, V> = -\Delta\rho^2, \text{ and}$$

$$<W, W> = \rho^2 S^2,$$

where, as always, $S = \sin\vartheta$ and $C = \cos\vartheta$. To normalize these vector fields we must use, for example, $|\partial_r| = |g_{rr}|^{1/2} = (\rho^2/|\Delta|)^{1/2}$. Since Δ does not maintain the same sign on all blocks, we write $|\Delta| = \varepsilon\Delta$, where $\varepsilon = \text{sgn } \Delta$ is +1 on I $\cup$ III, -1 on II. Also, let ρ be the positive square root of $\rho^2 = r^2 + a^2 C^2$. Then $|\partial_r| = \rho/\sqrt{\varepsilon\Delta}$.

Definition 2.6.1   *The* Boyer–Lindquist frame field *on* I $\cup$ II $\cup$ III *is*

$$E_1 = \frac{\sqrt{\varepsilon\Delta}}{\rho}\partial_r \qquad E_3 = \frac{W}{|W|} = \frac{1}{\rho S}(\partial_\varphi + aS^2\partial_t)$$

$$E_2 = \frac{1}{\rho}\partial_\vartheta \qquad E_0 = \frac{V}{|V|} = \frac{1}{\rho\sqrt{\varepsilon\Delta}}((r^2 + a^2)\partial_t + a\partial_\varphi)$$

It can be checked directly that this is an orthonormal frame field. Note that the general semi-Riemannian sign $\varepsilon_i = <E_i, E_i>$ here becomes

$$\varepsilon_0 = -\varepsilon, \qquad \varepsilon_1 = \varepsilon, \qquad \varepsilon_2 = \varepsilon_3 = +1.$$

since the timelike vector field in the frame is E_0 on I $\cup$ III, but E_1 on II.

Recall that the corresponding coframe field consists of the one-forms $\{\omega^i\}$ characterized by $\omega^i(E_j) = \delta^i{}_j$. These are readily found.

Lemma 2.6.2 *The* Boyer–Lindquist coframe field *on* $\text{I} \cup \text{II} \cup \text{III}$ *is:*

$$\omega^1 = \frac{\rho}{\sqrt{\varepsilon\Delta}}\,dr \qquad \omega^3 = \frac{S}{\rho}\big((r^2 + a^2)\,d\varphi - a\,dt\big)$$

$$\omega^2 = \rho\,d\vartheta \qquad \omega^0 = \frac{\sqrt{\varepsilon\Delta}}{\rho}(dt - aS^2 d\varphi).$$

Substituting these one-forms into the formula

$$ds^2 = -\varepsilon(\omega^0)^2 + \varepsilon(\omega^1)^2 + (\omega^2)^2 + (\omega^3)^2,$$

gives an new expression for the line-element.

Corollary 2.6.3 *The Kerr line-element can be written in terms of Boyer–Lindquist coordinates as:*

$$ds^2 = -(\Delta/\rho^2)\big[dt - aS^2 d\varphi\big]^2 + (\rho^2/\Delta)\,dr^2$$
$$+\rho^2 d\vartheta^2 + (S^2/\rho^2)\big[(r^2 + a^2)\,d\varphi - a\,dt\big]^2.$$

(Here the mass M *is concealed in* Δ.*)*

To convert Boyer–Lindquist coordinate formulas into coframe formulas, we solve the equations in Lemma 2.6.2 to get

$$dr = \frac{\sqrt{\varepsilon\Delta}}{\rho}\omega^1 \qquad d\varphi = \frac{1}{\rho S}\omega^3 + \frac{a}{\rho\sqrt{\varepsilon\Delta}}\omega^0$$

$$d\vartheta = \frac{1}{\rho}\omega^2 \qquad dt = \frac{aS}{\rho}\omega^3 + \frac{r^2 + a^2}{\rho\sqrt{\varepsilon\Delta}}\omega^0$$

Computation of the connection forms of the Boyer–Lindquist frame field will be easy once we find the exterior derivatives $d\omega^i$ of the dual one-forms ω^i, *expressed in terms of the* ω^i's. For such work a stockpile of partial derivative formulas is a convenience. With notation as above,

$$\frac{\partial\rho}{\partial r} = \frac{r}{\rho}, \quad \frac{\partial}{\partial r}\frac{1}{\rho} = \frac{-r}{\rho^3}, \quad \frac{\partial\rho}{\partial\vartheta} = \frac{-a^2 SC}{\rho}, \quad \frac{\partial}{\partial\vartheta}\frac{1}{\rho} = \frac{a^2 SC}{\rho^3}$$

$$\frac{\partial\Delta}{\partial r} = 2(r - \text{M}) \quad \frac{\partial}{\partial r}\sqrt{\varepsilon\Delta} = \varepsilon\frac{r - \text{M}}{\sqrt{\varepsilon\Delta}}, \quad \frac{\partial}{\partial r}\frac{1}{\sqrt{\varepsilon\Delta}} = -\frac{r - \text{M}}{\Delta\sqrt{\varepsilon\Delta}}.$$

Then the exterior derivatives are

$$d\omega^1 = \frac{a^2 S C}{\rho^3}\omega^1 \wedge \omega^2 \qquad d\omega^2 = \frac{r\sqrt{\varepsilon\Delta}}{\rho^3}\omega^1 \wedge \omega^2$$

$$d\omega^3 = \frac{C}{S}\frac{r^2+a^2}{\rho^3}\omega^2 \wedge \omega^3 - \frac{r\sqrt{\varepsilon\Delta}}{\rho^3}\omega^3 \wedge \omega^1 + \frac{2arS}{\rho^3}\omega^1 \wedge \omega^0$$

$$d\omega^0 = F\omega^1 \wedge \omega^0 - \frac{2aC\sqrt{\varepsilon\Delta}}{\rho^3}\omega^2 \wedge \omega^3 - \frac{a^2 S C}{\rho^3}\omega^2 \wedge \omega^0$$

$$\text{where } F = \frac{\partial}{\partial r}\frac{\sqrt{\varepsilon\Delta}}{\rho} = \varepsilon\frac{(r-\mathrm{M})\rho^2 - r\Delta}{\rho^3\sqrt{\varepsilon\Delta}}.$$

Since the connection forms are semiskew we need only find $\omega^i{}_j$ for $i < j$. Care is required with the "index reversal" sign

$$\omega^i{}_j = \varepsilon_i\omega_{ij} = -\varepsilon_i\omega_{ji} = -\varepsilon_i\varepsilon_j\omega^j{}_i = \pm\omega^j{}_i.$$

As we saw earlier, $\varepsilon_i = <E_i, E_i>$ and $\varepsilon = \mathrm{sgn}\Delta$ are related by

$$\varepsilon_0 = -\varepsilon, \qquad \varepsilon_1 = \varepsilon, \qquad \varepsilon_2 = \varepsilon_3 = +1,$$

hence the reversal signs in the left column below emerge. (The same rules apply later to the curvature forms.)

Proposition 2.6.4 *The connection forms for the Boyer–Lindquist frame field are*

$$-\varepsilon\omega^2{}_1 = \omega^1{}_2 = -a^2 S C\rho^{-3}\omega^1 - \varepsilon r\sqrt{\varepsilon\Delta}\rho^{-3}\omega^2,$$

$$\varepsilon\omega^3{}_0 = \omega^0{}_3 = -\varepsilon arS\rho^{-3}\omega^1 + aC\sqrt{\varepsilon\Delta}\rho^{-3}\omega^2,$$

$$\omega^1{}_0 = \omega^0{}_1 = -\varepsilon arS\rho^{-3}\omega^3 + F\omega^0,$$

$$\varepsilon\omega^2{}_0 = \omega^0{}_2 = -aC\sqrt{\varepsilon\Delta}\rho^{-3}\omega^3 - a^2 S C\rho^{-3}\omega^0,$$

$$-\varepsilon\omega^1{}_3 = \omega^3{}_1 = r\sqrt{\varepsilon\Delta}\rho^{-3}\omega^3 + arS\rho^{-3}\omega^0,$$

$$-\omega^3{}_2 = \omega^2{}_3 = -(C/S)(r^2+a^2)\rho^{-3}\omega^3 - \varepsilon aC\sqrt{\varepsilon\Delta}\rho^{-3}\omega^0.$$

(Here the function F is as defined above.)

Proof. By Corollary 1.8.5, it would suffice merely to verify that these forms satisfy the first structural equations—indeed, this provides a check that they are correct. However, to derive them we use the formula in the proof of that corollary. Take $\omega^3{}_1$, for example.

In the special case $i = k$ the formula becomes $\omega^i{}_j(E_i) = d\omega^i(E_j, E_i)$. This gives two rotation coefficients. First, $\omega^3{}_1(E_3) = d\omega^3(E_1, E_3)$. The formula above for $d\omega^3$ shows that this is $r\sqrt{\varepsilon\Delta}\,\rho^{-3}$. Next, $\omega^3{}_1(E_1) = \pm\omega^1{}_3(E_1) = \pm d\omega^1(E_3, E_1)$, which the formula for $d\omega^1$ shows is zero.

The general formula is needed to find $\omega^3{}_1(E_k)$ for $k = 0, 2$. Since $\varepsilon_3 = +1$ and $\varepsilon_1 = \varepsilon$, it becomes

$$\omega^3{}_1(E_k) = \tfrac{1}{2}\big[d\omega^3(E_1, E_k) + \varepsilon\,d\omega^1(E_k, E_3) - \varepsilon_k d\omega^k(E_k, E_3)\big].$$

For $k = 0$, $d\omega^3(E_1, E_0) = 2arS\rho^{-3}$, but the other two terms are zero. Thus, $\omega^3{}_1(E_k) = arS\rho^{-3}$. For $k = 2$, all three summands vanish, so $\omega^3{}_1(E_k) = 0$. We conclude that

$$\omega^3{}_1 = \Sigma\omega^3{}_1(E_m)\omega^m = r\sqrt{\varepsilon\Delta}\rho^{-3}\omega^3 + arS\rho^{-3}\omega^0.$$

Computation of the other forms is entirely similiar.   □

Remark 2.6.5 (Schwarzschild Comparison.)   The Schwarzschild line-element

$$ds^2 = -h\,dt^2 + (1/h)\,dr^2 + r^2(d\vartheta^2 + \sin^2\vartheta\,d\varphi^2), \quad \text{with } h = 1 - (2\text{M}/r),$$

shows that the coordinate system t, r, ϑ, φ is orthogonal. Thus, from ds^2 we immediately read the *Schwarzschild coframe field*

$$\omega^0 = \sqrt{h}\,dt, \quad \omega^1 = (1/\sqrt{h})\,dr, \quad \omega^2 = r\,d\vartheta, \quad \omega^3 = rS\,d\varphi,$$

where for simplicity we consider only the exterior $r > 2\text{M}$, where $h > 0$. The exterior derivatives are computed in coordinates as

$$d\omega^0 = \text{M}/(r^2\sqrt{h})dr \wedge dt, \quad d\omega^1 = 0, \quad d\omega^2 = dr \wedge d\vartheta,$$
$$d\omega^3 = S\,dr \wedge d\varphi + rC\,d\vartheta \wedge d\varphi$$

In terms of the coframe field these become

$$d\omega^0 = -\text{M}/(r^2\sqrt{h})\omega^0 \wedge \omega^1, \quad d\omega^1 = 0, \quad d\omega^2 = -(\sqrt{h}/r)\omega^2 \wedge \omega^1,$$
$$d\omega^3 = -(\sqrt{h}/(r))\omega^3 \wedge \omega^1 - C/(rS)\omega^3 \wedge \omega^2.$$

The connection forms can now be deduced, as in the preceding proof, by cut-and-try or by formula. We find

$$\omega^0{}_1 = (\text{M}/r^2\sqrt{h})\omega^0, \quad \omega^2{}_1 = (\sqrt{h}/r)\omega^2, \quad \omega^3{}_1 = (\sqrt{h}/r)\omega^3, \quad \omega^3{}_2 = (C/(Sr)\omega^3,$$

with the others ($\pm$'s excepted) equal to zero (see Lemma 1.8.4). To check Proposition 2.6.4 against these, set $a = 0$, $\varepsilon = 1$. Eight of the twelve summands in the proposition vanish immediately. Since $\rho^2 \to r^2$ and $\Delta \to r(r - 2M) = r^2 h$, the complicated function F reduces to $M/(r^2\sqrt{h})$, and resulting forms agree with the previously listed Schwarzschild connection forms.

The connection forms in Proposition 2.6.4 are actually rather simple. With six independent forms, each involving four dimensions, there could have been 24 different rotation coefficients $\omega^i{}_j(E_k)$, but only 12 appear, and—ignoring constant factors—only six of these are different. A partial explanation for this simplicity can be deduced directly from the geometry of Boyer–Lindquist blocks—with connection form information expressed (equivalently) in terms of covariant derivatives.

Lemma 2.6.6 *For the Boyer–Lindquist frame field, the covariant derivative $\nabla_{E_i} E_j$ is in*

1. *Span $\{E_1, E_2\}$ if either $i, j \in \{1, 2\}$ or $i, j \in \{3, 0\}$,*
2. *Span $\{E_3, E_0\}$ if $i \in \{1, 2\}$ and $j \in \{3, 0\}$ or vice versa.*

Proof. Recall that the Boyer–Lindquist blocks are coordinate products of polar planes P and Killing fibers F. Since polar planes are totally geodesic, if vector fields X and Y are tangent to every P, then so is $\nabla_X Y$. And if Z is normal to each plane P, then so is $\nabla_X Z$. But P and F are orthogonal, so normal to P means tangent to F.

Because E_1, E_2 are tangent to polar planes and E_3, E_0 are tangent to fibers, these remarks imply the first alternative in both assertions (1) and (2).

Since the (φ, t) coordinate system on Killing fibers F has constant metric coefficients, the vector fields E_3 and E_0—which are tangent to F—are parallel in the intrinsic geometry of F. Thus, if $i, j \in \{3, 0\}$, the covariant derivative $\nabla_{E_i} E_j$ has no F component, hence is in span $\{E_1, E_2\}$. $\square$

The Cartan approach exhibits this qualitative information as a byproduct of an explicit computation: Substitute the rotation coefficients, read from Proposition 2.6.4, into the covariant derivative formula $\nabla_{E_i} E_j = \Sigma \omega^m{}_j(E_i) E_m$.

Corollary 2.6.7 *The covariant derivatives of the Boyer–Lindquist vector fields* E_1, E_2, E_3, E_0 *are*

$$\nabla_{E_1} E_1 = \varepsilon a^2 SC\rho^{-3} E_2, \qquad \nabla_{E_1} E_2 = -a^2 SC\rho^{-3} E_1,$$

$$\nabla_{E_2} E_1 = r\sqrt{\varepsilon\Delta}\,\rho^{-3} E_2, \qquad \nabla_{E_2} E_2 = -\varepsilon r\sqrt{\varepsilon\Delta}\,\rho^{-3} E_1.$$

$$\nabla_{E_0} E_0 = FE_1 - \varepsilon a^2 SC\rho^{-3} E_2,$$

$$\nabla_{E_0} E_3 = \nabla_{E_3} E_0 = -\varepsilon a r S\rho^{-3} E_1 - \varepsilon a C\sqrt{\varepsilon\Delta}\,\rho^{-3} E_2,$$

$$\nabla_{E_3} E_3 = -\varepsilon r\sqrt{\varepsilon\Delta}\,\rho^{-3} E_1 - (C/S)(r^2 + a^2)\rho^{-3} E_2$$

$$\nabla_{E_0} E_1 = FE_0 + arS\,\rho^{-3} E_3, \qquad \nabla_{E_0} E_2 = -a^2 SC\,\rho^{-3} E_0 + \varepsilon a C\sqrt{\varepsilon\Delta}\,\rho^{-3} E_3,$$

$$\nabla_{E_3} E_1 = -\varepsilon a r S\,\rho^{-3} E_0 + r\sqrt{\varepsilon\Delta}\,\rho^{-3} E_3,$$

$$\nabla_{E_3} E_2 = -a C\sqrt{\varepsilon\Delta}\,\rho^{-3} E_0 + (C/S)(r^2 + a^2)\rho^{-3} E_3.$$

$$\nabla_{E_1} E_0 = -arS\,\rho^{-3} E_3, \qquad \nabla_{E_1} E_3 = -\varepsilon a r S\,\rho^{-3} E_0,$$

$$\nabla_{E_2} E_0 = \varepsilon a C\sqrt{\varepsilon\Delta}\,\rho^{-3} E_3, \qquad \nabla_{E_2} E_3 = a C\sqrt{\varepsilon\Delta}\,\rho^{-3} E_0.$$

With these formulas can verify directly that the principal null curves are in fact geodesics (though Chapter 4 gives simpler methods). They are the integral curves of the vector fields $N_\pm = \pm\partial_r + \Delta^{-1} V$, which, in terms of the Boyer–Lindquist frame field, become $(r/\sqrt{\varepsilon\Delta})[E_0 \pm E_1]$. Thus, reading from above, we find

$$\nabla_{E_0 \pm E_1}(E_0 \pm E_1) = \nabla_{E_0} E_0 + \nabla_{E_1} E_1 \pm [\nabla_{E_0} E_1 + \nabla_{E_1} E_0]$$
$$= FE_1 \pm FE_0 = \pm F(E_0 \pm E_1),$$

where, as before, $F = \partial_r(\sqrt{\varepsilon\Delta}/r)$. Then

$$\nabla_{N_\pm}(N_\pm) = (\rho^2/\Delta)\nabla_{E_0 \pm E_1}(E_0 \pm E_1) + (\rho/\sqrt{\varepsilon\Delta})(E_0 \pm E_1)\big[r/\sqrt{\varepsilon\Delta}\big](E_0 \pm E_1).$$

The first summand here is $\pm(\rho^2/\varepsilon\Delta)F(E_0 \pm E_1)$, and in the second, $E_0\big[\rho/\sqrt{\varepsilon\Delta}\big]$ $= 0$ and

$$E_1\big[\rho/\sqrt{\varepsilon\Delta}\big] = (\sqrt{\varepsilon\Delta}/\rho)\partial_r(\rho/\sqrt{\varepsilon\Delta})$$
$$= -(\rho/\sqrt{\varepsilon\Delta})\partial_r(\sqrt{\varepsilon\Delta}/\rho) = -(\rho/\sqrt{\varepsilon\Delta})F.$$

Thus, the second summand becomes $\mp(\rho^2/\varepsilon\Delta)F(E_0 \pm E_1)$, cancelling the first.

Brackets follow easily from covariant derivatives, by using the formula $[X, Y] = \nabla_X Y - \nabla_Y X$.

Corollary 2.6.8   *The brackets of the Boyer–Lindquist vector fields are*

$$[E_0, E_1] = FE_0 + 2arS\rho^{-3}E_3, \quad [E_1, E_2] = -a^2SC\rho^{-3}E_1 - r\sqrt{\varepsilon\Delta}\,\rho^{-3}E_2,$$

$$[E_0, E_2] = -a^2SC\rho^{-3}E_0, \quad\quad\quad [E_1, E_3] = -r\sqrt{\varepsilon\Delta}\,\rho^3 E_3.$$

$$[E_0, E_3] = 0, \quad\quad\quad\quad\quad\quad\quad [E_2, E_3] = 2aC\sqrt{\varepsilon\Delta}\,\rho^{-3}E_0$$

$$- (C/S)(r^2 + a^2)\rho^{-3}E_3.$$

2.7   Kerr Curvature à la Cartan

Using the results of the preceding section we proceed to find the Riemannian curvature of Kerr spacetime in terms of the Boyer–Lindquist frame field. In the Cartan approach this is accomplished by computing the curvature forms

$$\Omega^i{}_j = d\omega^i{}_j + \Sigma\omega^i{}_m \wedge \omega^m{}_j.$$

For Schwarzschild spacetime (as Remark 2.7.3 shows) this is an easy computation, but here it is more serious. However, it is decisively simpler than a classical coordinate computation via Christoffel symbols. It would be tedious to record full details of the calculations, but we now outline their structure and give details for the worst case: $\Omega^0{}_1$.

The sum $\Sigma\omega^i{}_m \wedge \omega^m{}_j$ always has just two summands, and in Proposition 2.6.4 each connection form has two summands, $\omega^i{}_j = A\omega^r + B\omega^s$, for appropriate rotation coefficients A and B. Thus, there will be three *types* of two-forms to be found; these are all expressed in terms of wedge products $\omega^i \wedge \omega^j$.

$$\Omega^0{}_1 = \begin{cases} d\omega^0{}_1 \\[4pt] + \\[4pt] \sum\omega^0{}_m \wedge \omega^m{}_1 \end{cases} = \begin{cases} dA \wedge \omega^0 + dB \wedge \omega^3 & \text{(type 1)} \\[4pt] +A\,d\omega^0 + B\,d\omega^3 & \text{(type 2)} \\[4pt] +\omega^0{}_2 \wedge \omega^2{}_1 + \omega^0{}_3 \wedge \omega^3{}_1 & \text{(type 3)} \end{cases}$$

For type 3 terms, simply wedge the formulas in Proposition 2.6.4. For type 2 terms, use the formulas for $d\omega^i$ preceding that proposition. For type 1 terms, we build a small machine that differentiates rotation coefficients—these all being of the form $N\rho^{-3}$.

Lemma 2.7.1 $dA = C_1(A)\omega^1 + C_2(A)\omega^2$, *where, if* $A = N\rho^{-3}$,

$$C_1(A) = \frac{\sqrt{\varepsilon\Delta}}{\rho^6}\left[\rho^2\frac{\partial N}{\partial r} - 3rN\right], \quad C_2(A) = \frac{1}{\rho^6}\left[\rho^2\frac{\partial N}{\partial \vartheta} + 3a^2SC\,N\right].$$

There are only six essentially different rotation coefficients; to find $\Omega^0{}_1$ we must apply the lemma to the two appearing in $\omega^0{}_1 = F\omega^0 - \varepsilon\, arS\, \rho^3\omega^3$. Using the partial derivative formulas in Section 2.6 gives

$$C_1(F) = \varepsilon\rho^{-6}[\rho^2(\rho^2 - \Delta - 2r^2 + 2Mr) - \Delta^{-1}\rho^4(r - M)^2 + 3r^2\Delta],$$

$$C_2(F) = \varepsilon\frac{a^2 SC}{\rho^6\sqrt{\varepsilon\Delta}}[\rho^2(r - M) - 3r\Delta].$$

It is these formulas from the most complicated rotation coefficient F that make $\Omega^0{}_1$ a worst case. For $B = -\varepsilon\, arS\, \rho^{-3}$, things are simpler:

$$C_1(B) = -\varepsilon\, aS\sqrt{\varepsilon\Delta}\,\rho^{-6}(-2r^2 + a^2C^2), \quad C_2(B) = -\varepsilon\, arC\,\rho^{-6}(\rho^2 + 3a^2S^2).$$

Now we are ready to assemble $\Omega^0{}_1$. The type 1 terms come from

$$C_1(F)\omega^1 \wedge \omega^0 + C_2(F)\omega^2 \wedge \omega^0 + C_1(B)\omega^1 \wedge \omega^3 + C_2(B)\omega^2 \wedge \omega^3.$$

Rather than write out all the terms of $\Omega^0{}_1$ simultaneously, we look first at the coefficient of $\omega^0 \wedge \omega^1$, which again is the most complicated one. The contributions of the three types are as follows:

type 1 $C_1(F)\omega^1 \wedge \omega^0$

type 2 $+ F(F\omega^1 \wedge \omega^0) - \varepsilon\, arS\, \rho^{-3}(2arS\, \rho^{-3})\omega^1 \wedge \omega^0$

type 3 $+ (-a^2 SC\rho^{-3}\omega^0) \wedge (\varepsilon\, a^2 SC\, \rho^{-3}\omega^1)$

            $+ (-\varepsilon\, arS\, \rho^{-3}\omega^1) \wedge (arS\, \rho^{-3}\omega^0).$

Since $F = \varepsilon\rho^{-6}\Delta^{-1/2}[\rho^2(r - M) - r\Delta]^2$, factoring out $\varepsilon\rho^{-6}\omega^0 \wedge \omega^1$ leaves

$$-\rho^2[\rho^2 - \Delta - 2r^2 + 2Mr] + \underline{\Delta^{-1}\rho^4(r - M)^2} - 3r^2\Delta$$
$$-\Delta^{-1}[\underline{\rho^4(r - M)^2} - 2r\Delta(r - M)\rho^2 + r^2\Delta^2] + 3a^2r^2S^2 - a^4S^2C^2.$$

The welcome cancellation (underlined) reduces this expression to

$$-\rho^2(\rho^2 - \Delta - 4r^2 + 4Mr) - 4r^2\Delta + 3a^2r^2S^2 - a^4S^2C^2.$$

Replacing the functions ρ^2 and Δ by their definitions, $r^2 + a^2C^2$ and $r^2 - 2Mr + a^2$, respectively, yields

$$-(r^2 + a^2C^2)(6Mr - 4r^2 - a^2S^2) - 4r^2(r^2 + a^2 - 2Mr) + 3a^2r^2S^2 - a^4S^2C^2,$$

which collapses trigonometrically to $2Mr(r^2 - 3a^2C^2)$. Thus, the $\omega^0 \wedge \omega^1$ term in $\Omega^0{}_1$ is $\varepsilon\, 2Mr\rho^{-6}(r^2 - 3a^2C^2)$.

All other $\omega^i \wedge \omega^j$ terms are zero, except for $\omega^2 \wedge \omega^3$. For it we find, by a simpler version of the above computation, the following coefficients:

type 1 $\quad - \varepsilon\, 2aC\rho^{-6}[\rho^2(r - M) - r\Delta] - \varepsilon\, arC\,\rho^{-6}(r^2 + a^2)$

type 2 $\quad - \varepsilon\, arC\rho^{-6}(\rho^2 + 3a^2S^2)$

type 3 $\quad + \varepsilon\, 2arC\Delta\rho^{-6}$

Extracting the common factor $\varepsilon aC\rho^{-6}$ leaves

$$-2\rho^2(r - M) + 2r\Delta - r(r^2 + a^2) - r(\rho^2 + 3a^2S^2) + 2r\Delta,$$

Again, replacing ρ^2 and Δ by their definitions produces extensive cancellation, with outcome $-2M(3r^2 - a^2C^2)$. We conclude that

$$\Omega^0{}_1 = \varepsilon\frac{2Mr}{\rho^6}(r^2 - 3a^2C^2)\,\omega^0 \wedge \omega^1 - \varepsilon\frac{2MaC}{\rho^6}(3r^2 - a^2C^2)\,\omega^2 \wedge \omega^3.$$

This nicely balanced formula proves to be a model for *all* the curvature forms, as resolute computation of the remaining (simpler) cases shows. The organization used here is well adapted to a labor-saving computation by computer.

Theorem 2.7.2 *The curvature forms of the Boyer–Lindquist frame field on Kerr spacetime, expressed in terms of $\varepsilon = \text{sgn}\Delta$ and the functions*

$$\mathsf{I} = \frac{Mr}{\rho^6}(r^2 - 3a^2C^2), \qquad \mathsf{J} = \frac{MaC}{\rho^6}(3r^2 - a^2C^2),$$

are as follows:

$\Omega^0{}_1 = \varepsilon\, 2\mathsf{I}\,\omega^0 \wedge \omega^1 - \varepsilon\, 2\mathsf{J}\,\omega^2 \wedge \omega^3 \qquad \Omega^2{}_3 = 2\mathsf{J}\,\omega^0 \wedge \omega^1 + 2\mathsf{I}\,\omega^2 \wedge \omega^3$

$\Omega^0{}_2 = -\mathsf{I}\,\omega^0 \wedge \omega^2 + \varepsilon\,\mathsf{J}\,\omega^3 \wedge \omega^1 \qquad \Omega^3{}_1 = -\mathsf{J}\,\omega^0 \wedge \omega^2 - \varepsilon\,\mathsf{I}\,\omega^3 \wedge \omega^1$

$\Omega^0{}_3 = -\mathsf{I}\,\omega^0 \wedge \omega^3 - \varepsilon\,\mathsf{J}\,\omega^1 \wedge \omega^2 \qquad \Omega^1{}_2 = -\varepsilon\mathsf{J}\,\omega^0 \wedge \omega^3 - \mathsf{I}\,\omega^1 \wedge \omega^2$

(The effect of index reversal is the same as in Proposition 2.6.4; the index arrangements here are convenient for Chapter 5.)

Remark 2.7.3 (Schwarzschild comparison.) The connection forms in Remark 2.6.5 are simple enough so that the second structural equation easily yields the Schwarzschild curvature forms.

$$\Omega^0{}_1 = 2Mr^{-3}\omega^0 \wedge \omega^1, \qquad \Omega^2{}_3 = 2Mr^{-3}\omega^2 \wedge \omega^3,$$

$$\Omega^0{}_2 = -Mr^{-3}\omega^0 \wedge \omega^2, \qquad \Omega^3{}_1 = -Mr^{-3}\omega^3 \wedge \omega^1,$$

$$\Omega^0{}_3 = -Mr^{-3}\omega^0 \wedge \omega^3, \qquad \Omega^1{}_2 = -Mr^{-3}\omega^1 \wedge \omega^2,$$

Setting $a = 0$ and $\varepsilon = +1$ in these Kerr formulas gives $\mathbf{I} = Mr^{-3}$ and $\mathbf{J} = 0$, so they correctly reduce to the Schwarzschild ones. Then easy calculations show that the functions $\pm Mr^{-3}$ and $\pm 2Mr^{-3}$ provide the sectional curvatures of all Schwarzschild coordinate planes, that is all tangent planes spanned by any two of $\partial_t, \partial_r, \partial_\vartheta, \partial_\varphi$.

This comparison shows explicitly the effect of rotation on curvature, notably in producing the off-diagonal terms. In Chapter 5 we find that the description of Kerr curvature in the preceding theorem is as simple as possible. Thus, the advantage of Cartan methods over coordinates is not just ease of computation: In the rotating case, the orthogonal vector fields $\partial_r, \partial_\vartheta, V, W$ that lead to the Boyer–Lindquist coframe field fit the geometry well; the coordinate vector fields $\partial_r, \partial_\vartheta, \partial_\varphi, \partial_t$ do not, as a machine computation of $R_{ijk\ell}$ will show. (The frame vector fields, or scalar multiples of them, can never be coordinate vector fields since, as we saw in Section 2.5, the distribution $\Pi^\perp = \mathrm{span}\{\partial_\vartheta, W\}$, for instance, is not integrable.)

The coefficients in the formulas of Theorem 2.7.2 are just the curvature components $R^i{}_{jk\ell} = \Omega^i{}_j(E_k, E_\ell)$. Thus, for example, $R^0{}_{123} = -\varepsilon 2\mathbf{J}$. Note that if there are exactly three distinct integers in $\{i, j, k, \ell\}$, then $R^i{}_{jk\ell} = 0$. For frame fields, we have seen that raising or lowering an index i involves merely multiplying by $\varepsilon_i = <E_i, E_i>$. Because E_0 is timelike for $\varepsilon = +1$ and E_1 is timelike for $\varepsilon = -1$, this means that

$$\varepsilon_0 = -\varepsilon, \quad \varepsilon_1 = \varepsilon, \quad \text{and} \quad \varepsilon_2 = \varepsilon_3 = +1.$$

Since curvature is the dominant geometric invariant, we can be sure that the curvature of Kerr spacetime has profound geometric-physical consequences.

Corollary 2.7.4 *Kerr spacetime is Ricci flat.*

Proof. We use the formula $\mathrm{Ric}(E_i, E_j) = \Sigma R^m{}_{imj}$. If $i \neq j$, then $\mathrm{Ric}(E_i, E_j) = 0$, since all three-index curvature components are zero.

When $i = j$, we first rearrange the indices, using the values of ε_i so that Theorem 2.7.2 may conveniently be applied. If $i = j = 1$, for example,

$$\mathrm{Ric}(E_1, E_1) = R^0{}_{101} + R^2{}_{121} + R^3{}_{131}.$$

But $R^2{}_{121} = R_{2121} = R_{1212} = \varepsilon R^1{}_{212}$, and similarly for $R^3{}_{131}$. Hence,

$$\mathrm{Ric}(E_1, E_1) = R^0{}_{101} + \varepsilon R^1{}_{212} + \varepsilon R^1{}_{313} = \varepsilon 2\mathbf{l} + \varepsilon(-\mathbf{l}) + \varepsilon(-\mathbf{l}) = 0.$$

$\square$

As noted in Chapter 1 this means that any Kerr spacetime is a *vacuum*; that is, any matter in K is negligible as a source of gravitation.

Corollary 2.7.5 *For Kerr spacetime the Kretschmann curvature invariant $k = <R, R>$ is* $48(\mathbf{l}^2 - \mathbf{J}^2)$.

Proof. We saw in Section 1.8 that in terms of an orthonormal frame field, $k = \Sigma \varepsilon_i \varepsilon_j \varepsilon_k \varepsilon_\ell (R_{ijk\ell})^2$. Since these curvature terms are squared, $R_{ijk\ell}$ can be replaced by $\pm R^i{}_{jk\ell}$, read directly from Theorem 2.7.2. By the symmetries of curvature, each special term with $i < j$ and $k < \ell$ represents four equal terms in the sum. The special terms consist of

- Six main diagonal terms: 0101, 0202, 0303, 1212, 1313, 2323, with $\varepsilon_i \varepsilon_j \varepsilon_k \varepsilon_\ell$ $= +1$;
- Six off-diagonal terms: 0123, 0213, 0312, 1203, 1302, 2301, with $\varepsilon_i \varepsilon_j \varepsilon_k \varepsilon_\ell$ $= -1$,

Thus, working from the theorem, we find

$$k = 4\big[\{4\mathbf{l}^2 + 4(\mathbf{l}^2) + 4\mathbf{l}^2\} - \{4\mathbf{J}^2 + 4(\mathbf{J}^2) + 4\mathbf{J}^2\}\big] = 48(\mathbf{l}^2 - \mathbf{J}^2).$$

$\square$

Our first look at the Kerr metric revealed three problem subsets: axis A, horizons H, and ring singularity Σ. The Kerr metric has been extended over A and H; can it be extended over Σ ? The factor ρ^{-6} in the curvature formulas (and the term *singularity*) certainly suggest that the answer is *no*. However, this is not a proof, since the ρ^{-6} derives from just one frame field, in terms of another frame field, the curvature form might be better behaved. A valid proof must use geometric invariants, independent of a particular choice of frame field or coordinates.

A priori, there may be no need to extend over Σ if the ring singularity is actually located "at infinity," that is, if (like $r = \infty$ in the Euclidean plane) geodesics require an infinite parameter interval to approach Σ . Such considerations suggest:

Definition 2.7.6 *A geodesic γ in a Kerr spacetime* hits *(or* meets*) the ring singularity Σ if (with affine parametrization) there is a finite parameter value b such that $\rho^2(\gamma(s)) \to 0$ as $s \to b$. Equivalently, both r and* $\cos\vartheta$ *approach 0 as $s \to b$.*

For example, as noted earlier, ingoing principal null geodesics have r coordinate $r(s) = r_0 - s$, hence those in the equator $\vartheta = \pi/2$ all hit the ring singularity. Thus, the ring singularity is not "at infinity" in block III.

Furthermore, the Kerr metric cannot be extended—in any very satisfactory way—over Σ. To say that the Kerr metric *can* be extended over Σ should mean at least this: There exists a spacetime $\widetilde{M}$ that contains an isometric copy of block III and such that every geodesic $\gamma\colon [s_0, s_1) \to \mathrm{III} \subset \widetilde{M}$ that hits Σ can be extended, as a geodesic, over a larger interval $[s_0, s_1 + \delta)$. (We could even hope that the points $\tilde{\gamma}(s_1)$ fill a surface in $\widetilde{M}$ diffeomorphic to $\Sigma \approx S^1 \times \mathbf{R}^1$.) However, these goals cannot be achieved.

Corollary 2.7.7 *No equatorial geodesic $\gamma\colon [s_0, s_1) \to \mathrm{III}$ that hits Σ can be geodesically extended past s_1 in any extension $\widetilde{M}$ of III.*

Proof. Let $\widetilde{M}$ be an extension of III and let $\gamma\colon [s_0, s_1) \to \mathrm{III}$ be an equatorial geodesic that hits Σ. In particular, $r \to 0$ on γ as $s \to s_1$. Since γ is equatorial, $C = 0$ and $\rho^2 = r^2$, hence the formula in Corollary 2.7.5 reduces to $k = 48\mathrm{M}^2/r^6$. Thus, $k(\gamma(s)) \to \infty$ as $s \to s_1$.

Assume that γ has an extension $\gamma_{\widetilde{M}}$ in $\widetilde{M}$ past s_1. Since the invariant k is preserved by isometries, $k_{\widetilde{M}}$ restricted to the isometric copy III is just k as above. Thus, $k_{\widetilde{M}}(\gamma(s)) \to \infty$ as $s \to s_1$. Since $\tilde{\gamma}$ is a geodesic of $\widetilde{M}$ defined on a larger interval, $\tilde{\gamma}(s_1)$ is an ordinary point of $\widetilde{M}$, hence $k_{\widetilde{M}}(\gamma(s)) = k_{\widetilde{M}}(\tilde{\gamma}(s))$ approaches the finite limit $k_{\widetilde{M}}(\tilde{\gamma}(s_1)) < \infty$, a contradiction. □

Thus the ring singularity Σ deserves to be called a singularity in spacetime—in view of the argument above, a *curvature singularity* (Ellis, Schmidt 1977, Tipler, Clark, Ellis 1980). In Chapter 4 further knowledge of Kerr geodesics allows us to replace the word *equatorial* in the preceding result by *nonspacelike.* (Note that in Schwarzschild spacetime, where $\mathbf{I} = \mathrm{M}/r^3$ and $\mathbf{J} = 0$, the curvature invariant k is $48\mathrm{M}^2/r^6$, showing that $r = 0$ is a curvature singularity.)

Since there are Kerr geodesics that hit Σ, Corollary 2.7.7 has the following consequence:

Corollary 2.7.8 *No Kerr spacetime containing block III (as all maximal models do) is geodesically complete.*

In the maximal Kerr spacetimes constructed in the next chapter, it turns out that the only incomplete geodesics are those that hit Σ.

In physical terms the fact that spacetime curvatures become arbitrarily large near the ring singularity means that a material object approaching it will be destroyed. But in comparison with Schwarzschild spacetime, Kerr spacetime seems benign. In the Schwarzschild case, *every* particle that crosses from the exterior through the horizon must continue on to destruction at the central singularity $r = 0$ (if it does not perish earlier). By contrast, in Kerr spacetime we picture the ring singularity Σ spatially as just an equatorial circle on the 2-sphere at $r = 0$. In fact, in the only case we have considered so far, almost all of the ingoing principal null geodesics (photons aimed directly inward) miss Σ, passing through $r = 0$ out through the milder regions of block III to $r = -\infty$.

Remark 2.7.9

1. For the Boyer–Lindquist frame field, let K_{ij} be the sectional curvature of the plane $P_{ij} = \text{span}\{E_i, E_j\}$. Using results from Theorem 2.7.2 in the formula $K_{ij} = \varepsilon_i \varepsilon_j R_{ijij} = \varepsilon_j R^i{}_{jij}$ (Section 1.8) gives

$$K_{01} = K_{23} = 2\mathsf{l}, \quad \text{and} \quad K_{02} = K_{03} = K_{12} = K_{13} = -\mathsf{l}.$$

2. For any totally geodesic surface S in M, the (intrinsic) Gaussian curvature K_S of S agrees with the M sectional curvature on the tangent planes of S (see O'Neill 1983). This fact applies to polar planes P and the axis A in any Kerr spacetime.

   a. The tangent planes to P are $\text{span}\{\partial_r, \partial_\vartheta\} = \text{span}\{E_1, E_2\}$; hence, by the preceding remark, $K_P = K_{12} = -\mathsf{l} = -\mathsf{M}r\rho^{-6}(r^2 - 3a^2C^2)$.

   b. The tangent planes to A are limits of the planes $\text{span}\{V, \partial_r\} = \text{span}\{E_0, E_1\}$. Hence

$$K_A = \lim_{C^2 \to 1} K_{01} = K_{01}|_{C^2=1} = 2\mathsf{l}|_{C^2=1} = 2\mathsf{M}r(r^2 - 3a^2)/(r^2 + a^2)^3.$$

   Note that the curvatures of both P and A approach 0 at infinity.

We have now seen that—at least in block I—Kerr spacetime is

(1) Axially symmetric,

(2) Stationary (on $r \geq 2\mathsf{M}$),

(3) Ricci-flat,

(4) Asymptotically flat,

(5) Furnished (except in the fast case) with an event horizon enclosing its ring singularity, and

(6) Determined entirely by parameters M and a (reducing to Schwarzschild space-time when $a = 0$).

These properties constitute impressive evidence that the Kerr metric provides a spacetime model for the region outside a rotating star of radius $r > 2$M. For instance, property (1) is the result of axial rotation; (2) means that the star is not expanding or collapsing; (3) asserts that the star is isolated, the only source of gravitation; and (4) that its gravity is neglible at great distances.

When such a star collapses beneath $r = 2$M, it is believed that horizons form, that properties (1), (3), (4) persist, and, as the collapse settles down to a black hole, (2) is reestablished (Hawking, Ellis 1973; Misner, Thorne, Wheeler 1973). A sequence of results by Werner Israel, Brandon Carter, Stephen Hawking, and David Robinson leads to the conclusion that "a black hole has no hair": the collapse destroys the individual properties of the star leaving a final state *uniquely determined* by its mass and angular momentum (Hawking, Ellis 1973). Since the Kerr black hole covers these two parameters, it is left as the unique possibility.

The physical realism of the region $r \gg$ M in block I of the Kerr model makes it plausible that regions near, and perhaps within, the horizons are also realistic. Although the existence of black holes in our universe is generally accepted, it remains to be seen where and how an actual black hole departs from the Kerr model—as depart it must unless the universe is a stranger than we think.

MAXIMAL EXTENSIONS

In the preceding chapter the Kerr metric was introduced and its domain extended to Kerr-star spacetime K^*. This spacetime includes the axis of rotation and the horizons (when the latter exist) but not the ring singularity, where curvature becomes infinite.

Is K^* a maximally extended analytic spacetime? Yes, in the fast case, $a^2 > \text{M}^2$ but only then. One point that might arouse suspicion—other than experience with Schwarzschild spacetime—is the asymmetry in the construction of K^*, where *ingoing* principal null geodesics were preferred over outgoing. In this chapter we begin by repeating the construction but with the outgoers preferred. The resulting *star-Kerr spacetime* $^*K$ is built of the same Boyer–Lindquist blocks as K^*, but they are not assembled in the same way. Thus the two spacetimes $^*K$ and K^* can themselves be glued together, along a same-numbered block, to produce a spacetime that is an extension of both. Continuing this assembly process—with the principal null geodesics as warp and woof—leads quickly to a maximal extension in the extreme case $a^2 = \text{M}^2$. The slow case, as usual, is more ornate, and in its maximal extension, the horizons self-intersect.

With the maximal Kerr black holes well defined we can examine their global properties.

Kerr isometries are cconstructed, analogously to the spacetimes themselves, by assembling isometries of the Boyer–Lindquist blocks, and we find the isometry groups for the three rotation types: fast, extreme, and slow.

The topology of the maximal Kerr black holes—particularly in the slow case—is not simple. However, by retaining the ring singularity in each block III of M, we get a smooth manifold $\tilde{M}$ that is a 2-sphere bundle over the (north or south) axis A of M—indeed, a product manifold $A \times S^2$. This makes it possible to find the topology, in fact the homotopy type, of M itself.

In relativity, *chronology* [causality] refers to the relationship between events that can be joined by a timelike [nonspacelike] curve. Chapter 2 has already shown that in Boyer–Lindquist block III chronology is violated by the existence of *closed* timelike curves. In Section 3.10 we consider global chronology of Kerr black holes where, as usual, the slow case is the most interesting.

The three standard maximal Kerr spacetimes are by no means the only ones. In particular, there are "small" variants whose geometric conciseness is paid for by radical violations of chronology.

3.1   Star-Kerr Spacetime

In Boyer–Lindquist coordinates the principal null geodesics are given by

$$r' = \pm 1, \qquad \vartheta' = 0, \qquad \varphi' = a/\Delta, \qquad t' = (r^2 + a^2)/\Delta,$$

and the Kerr-star coordinate system was constructed by arranging for the ingoers, $r' = -1$, to be coordinate curves. To replace ingoers by outgoers, for which $r' = +1$, we again use the functions $A(r) = \int a/\Delta \, dr$ and $T(r) = \int (r^2 + a^2)/\Delta \, dr$. But now we define

$$^*\varphi = \varphi - A(r)$$

and

$$^*t = t - T(r)$$

with subtraction replacing the addition used for φ^*, t^*. Then in the proof of Lemma 2.5.1, the resulting sign changes show that $^*\xi = (r, \vartheta, {}^*\varphi, {}^*t)$ is a coordinate system on each Boyer–Lindquist block, axes omitted. We call $^*\xi$ the *star-Kerr coordinate system*.

Outgoing principal null geodesics, expressed in terms of these coordinates, still satisfy $r' = +1$ and $\vartheta' = 0$, and furthermore,

$$^*\varphi' = \varphi' - dA/dr \, r' = 0$$

and

$$^*t' = t' - dA/dr\, r' = 0$$

Thus the outgoers are now r-coordinate curves parametrized so that $r(s) = r_0 + s$.

Lemma 3.1.1 *The Kerr metric, expressed in terms of star-Kerr coordinates, has line-element*

$$ds^2 = g_{tt}d(^*t)^2 + 2g_{t\varphi}d(^*t)d(^*\varphi) + g_{\varphi\varphi}d(^*\varphi)^2 + \rho^2 d\vartheta^2$$
$$- 2d(^*t)\, dr + 2a\sin^2\vartheta d(^*\varphi)\, dr.$$

Here the components of g are the same as in Section 2.1. Comparison with the Kerr-star line-element (Section 2.5) shows that the only essential difference is the sign of the last two terms. And, just as for K^*, coordinate extension (Chapter 1) covers the horizons, and polar extension (Chapter 2) covers the axes, so the metric is extended to all of $\mathbf{R}^2 \times S^2$ except the ring singularity Σ.

Definition 3.1.2 Star-Kerr spacetime $^*K$ has
1. *Analytic manifold* $\mathbf{R}^2(^*t, r) \times S^2(\vartheta, {}^*\varphi) - \Sigma$,
2. *Metric tensor as described in Lemma 3.1.1, and*
3. *Time-orientation such that the null vectors* $^*\partial_r$ *are future-pointing.*

Here $^*\partial_r$ denotes the r-coordinate vector field of star-Kerr coordinates; its global existence shows that $^*K$ is time-orientable. As in the Kerr-star case, for each Boyer–Lindquist block B the natural imbedding of B in $^*K$ is denoted by $^*j$.

The radius function r has served in each of the three coordinate systems used so far: Boyer–Lindquist, Kerr-star, and star-Kerr. But its respective coordinate vector fields ∂_r, ∂_r^*, $^*\partial_r$, are different in each system. In Chapter 2 we found $\partial_r^* = \partial_r - \Delta^{-1}V$; now the corresponding calculation gives $^*\partial_r = \partial_r + \Delta^{-1}V$. The other three coordinate vector fields ∂_ϑ, ∂_φ, ∂_t of the three systems are the same.

The choice of time-orientation for $^*K$ is consistent with the convention that ∂_t be future-pointing for $r \gg 1$, since $<^*\partial_r, \partial_t> = -1$.

By construction the outgoing principal null geodesics on $^*K$ run unimpeded from $r = -\infty$ to $r = \infty$. Along the way they cut "outward" through the horizons of slow and extreme Kerr, and hence *leave the black hole*—supposedly a forbidden operation! To resolve this apparent contradiction, we must accept the fact that $^*K$ is not just a duplicate of K^*.

Following the analogous calculations for K^*, we find that the ingoers in $^*K$ can be expressed as the integral curves of the vector field $^*N = -(\Delta/2)^*\partial_r + V$.

Here, as usual, the principal null geodesics are parametrized as future-pointing. As before, if parametrization is neglected, there is a unique ingoer and a unique outgoer through each point of *K, axis included, and their tangent lines determine the principal planes Π on *K.

In Figure 3.1 the spacetime *K is pictured somewhat more abstractly than K^* was in Figure 2.7 of Chapter 2.

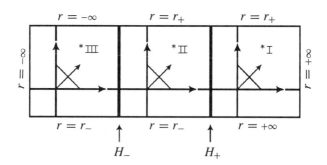

FIGURE 3.1.   Slow star-Kerr spacetime *K. Outgoers, drawn as horizontal lines, run from $r = -\infty$ to $r = +\infty$. Ingoers, drawn vertically, have limit values as indicated.

The spacetimes K^* and *K are composed of the same Boyer–Lindquist blocks assembled in two different ways. It will soon be important to relate a block in K^* with the same numbered block in *K. So for $B = \mathrm{I}, \mathrm{II}, \mathrm{III}$, let B^* and *B denote that block, imbedded by j^* and *j as open submanifolds of K^* and *K, respectively. Thus the isometry $\mu =$ *$j \circ (j^*)^{-1}$ is the natural way to identify B^* and *B.

Lemma 3.1.3 (1) *The isometry $\mu: B^* \to$ *B has coordinate description*

$$r = r, \qquad \vartheta = \vartheta, \qquad {}^*\varphi = \varphi^* - 2A(r), \qquad {}^*t = t^* - 2T(r),$$

where $A(r)$ and $T(r)$ are the functions used above (and in Chapter 2).
  (2) μ *preserves time-orientation for $B = \mathrm{I}, \mathrm{III}$, but reverses it for $B = \mathrm{II}$.*
  (3) μ *carries outgoers to outgoers, ingoers to ingoers, except that for $B = \mathrm{II}$ the image curves are parametrized as past-pointing.*

Proof. (1) Kerr and Boyer–Lindquist coordinates share the functions r, ϑ. By definition, $\varphi^* = \varphi + A(r)$ and *$\varphi = \varphi - A(r)$, hence *$\varphi = \varphi^* - 2A(r)$. The case of *$t$ and t^* is similar.

(2) It is easy to see that the differential map of μ carries ∂_t on B^* to ∂_t on $^*B$, and similarly for ∂_ϑ and ∂_φ. Since ∂_t is future-pointing for $|r|$ large, it follows that μ preserves time-orientation for I and III. The block II case will follow from assertion (3).

(3) We have found that the principal null geodesics are integral curves of vector fields as follows:

On K^*:   ingoers, $-\partial_r^*$,   outgoers, $\partial_r^* + (2/\Delta)V$,

On $^*K$:   ingoers, $-^*\partial_r + (2/\Delta)V$,   outgoers, $^*\partial_r$,

with these curves future-pointing *except* for outgoers in II* and ingoers in *II. The coordinate formulas for μ give

$$d\mu(\partial_r^*) = \Sigma \partial(x^i \circ \mu)/\partial r \partial_i = {}^*\partial_r - 2\,dA/dr\partial_\varphi - 2dT/dr\partial_t$$
$$= {}^*\partial_r - 2(a/\Delta)\partial_\varphi - 2[(r^2 + a^2)/\Delta]\partial_t = {}^*\partial_r - (2/\Delta)V.$$

This formula (with signs reversed) shows that μ carries ingoers to ingoers, except that the parametrizations are past-pointing in block *II. In particular, μ reverses time-orientation on II*—but, as we saw, preserves it on I* and III*. Since $d\mu$ preserves ∂_t and ∂_φ, it also preserves V. The corresponding result for outgoers follows from

$$d\mu(\partial_r^* + (2/\Delta)V) = {}^*\partial_r - (2/\Delta)V + (2/\Delta)V = {}^*\partial_r.$$

$\square$

The definition of μ shows that its Boyer–Lindquist coordinate expression is just the identity map. Thus preservation of time-orientation of block I is to be expected, since Kerr time-orientations derive from the natural choice for $r \gg 1$. Its reversal on block II—crucial to later developments—stems from the different way the same blocks are assembled in K^* and $^*K$ (see Figure 3.2).

Remark 3.1.4 In the rapidly rotating case, since there are no horizons, the imbeddings j^* and $^*j$ carry the single Boyer–Lindquist block B isometrically onto K^* and $^*K$, respectively. The unique spacetime M_f thus determined is called *maximal fast Kerr spacetime*.

In view of the different arrangement of blocks in K^* and $^*K$ when horizons exist it is somewhat unexpected to find that the two spacetimes are still isometric. However, comparing the line-element for $^*K$ in Lemma 3.1.1 with that of K^* in

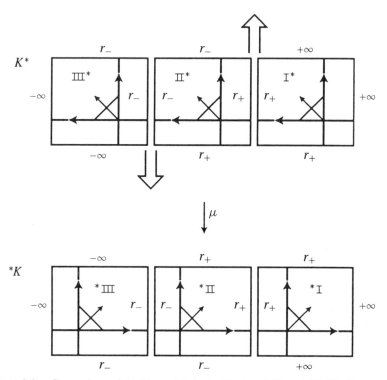

FIGURE 3.2.   Comparison of the Boyer–Lindquist blocks of K^* and $^*K$. The isometry μ explodes on horizons and, as the hollow arrows suggest, rotates blocks I^* and III^* by $-90°$ (preserving time-orientation), but rotates block II^* by $+90°$ (reversing time-orientation). Ingoers are always carried to ingoers, outgoers to outgoers—except that in block II^* the image curves are past-pointing.

Lemma 2.5.2 shows that, but for notation, the two differ only in the signs of their $d\varphi\,dr$ and $dt\,dr$ terms. Differentials are squared in the other terms, hence the coordinate map

$$r = r, \qquad \vartheta = \vartheta, \qquad {}^*\varphi = -\varphi^*, \qquad {}^*t = -t^*$$

is an isometry, and since this map is readily extendible over axes, it yields an isometry $\beta\colon K^* \to {}^*K$.

The Boyer–Lindquist formula for β is just $(r, \vartheta, \varphi, t) \to (r, \vartheta, -\varphi, -t)$, so β is the "backwards" isometry noted at the beginning of Chapter 2, which reverses time and the direction of rotation. Thus we can view K^* and $^*K$ as the same spacetime run oppositely.

3.2   Maximal Extreme Kerr Spacetime

For fast Kerr spacetime the models K^* and $^*K$ represent a single maximal space-time M_f. But horizons make matters more interesting. In this section we construct a maximal analytic extension in the extreme case, where a horizon $r = M$ sepa-rates the two Boyer–Lindquist blocks I and III in K^* and similarly in $^*K$. (The maximality assertions in this chapter are proved in Section 4.15).

A valuable guide to extending manifolds is to look for running room for geodesics. For example, the K^* and $^*K$ extensions of the Boyer–Lindquist blocks were constructed so that outgoing principal nulls could run freely in K^*, and ingoing ones in $^*K$. We want to combine these extensions in a single extension so that (with exceptions noted below) both ingoers and outgoers run from $r = \pm\infty$ to $r = \mp\infty$.

We say that a geodesic γ is *future [past] complete* if $|r| \to \infty$ along γ in the future [past] direction. Then γ is *complete* if both conditions hold. Since Kerr spacetime is approximately Minkowskian when $|r|$ is large, it is a good guess (proved in the next chapter) that a geodesic complete in this sense is, in fact, geodesically complete.

Obviously, principal nulls that hit the ring singularity are exceptional, as are those that lie in horizons, where r is constant. All other principal nulls will be called *ordinary*. The construction in this section is guided by the requirement that all ordinary principal null geodesics be complete. In the fast case (Remark 3.1.4) this goal was achieved effortlessly, since the model K^* of M_f shows that its ordinary ingoers are complete, and the model $^*K$ does the same for outgoers.

We begin the construction with K^* and work into the future (the past is entirely similar). The only ordinary principal nulls in K^* that are not future-complete are the outgoers in block III. But in $^*K$ *all* ordinary outgoers are complete. This suggests gluing $^*K$ onto K^* using the isometry $\mu: III^* \to {}^*III$ as the matching map (see Figures 3.3 and 3.4).

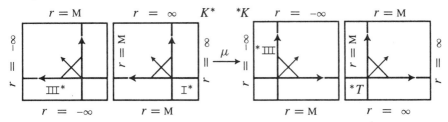

FIGURE 3.3.   Comparison of the Boyer–Lindquist blocks of extreme K^* and $^*K$. The isometry μ rotates both *I and *III by $-90°$, carrying ingoers to ingoers, outgoers to outgoers, and preserving time-orientation.

Lemma 3.2.1 *The Hausdorff condition (Definition 1.4.4) holds for the above construction of $K_1 = K^* \cup_\mu {}^*K$, where $\mu = \mu|\text{III}^*$.*

Proof. Suppose $\{p_n\}$ is a sequence in III^* that converges to a point $p \in K^* - \text{III}^*$. Then p is in the horizon $H: r = \text{M}$ of K^*, and $t^*(p_n)$ approaches a finite limit as $n \to \infty$. By Lemma 3.1.3 the sequence $\mu(p_n)$ has ${}^*t$ coordinate $t^*(p_n) - 2T(r)$. But $T(r) \to \infty$ as $n \to \infty$, so $\mu(p_n)$ is divergent as $n \to \infty$. $\square$

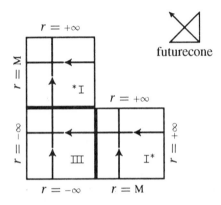

FIGURE 3.4.   The first stage K_1. The map μ requires that ${}^*K$ be rotated by $+90°$ to make ${}^*\text{III} = \text{III}^*$. Then the outgoers in III^* become complete in K_1.

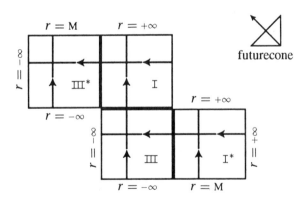

FIGURE 3.5.   The second stage, K_2. Gluing K^* to K_1 completes the ingoers of ${}^*\text{I}$. The heavy line segments represent three disjoint horizons $r = \text{M}$.

The results of such gluings are again Kerr manifolds as defined in Section 2.3. We supply the details of a typical proof; its theme is that objects on K^* and $^*K$ preserved by μ are well defined on $K^* \cup_\mu {}^*K$.

Lemma 3.2.2 $K_1 = K^* \cup_\mu {}^*K$ *is a Kerr spacetime.*

Proof. Since the matching map $\mu \colon \text{III}^* \to {}^*\text{III}$ is an isometry it follows that K_1 is at least a connected Lorentz 4-manifold. Since μ preserves time-orientation, K_1 is time-orientable, hence is a spacetime. Now we check the conditions in Definition 2.3.1 for it to be a Kerr spacetime:

Condition 1. Boyer–Lindquist blocks are dense, by construction, but we must exhibit *canonical* imbeddings of blocks I and III in K_1. As always, the gluing produces canonical imbeddings $^*i \colon K^* \to K_1$ and $i^* \colon {}^*K \to K_1$ such that $i^* = {}^*i \circ \mu$ (Section 1.4).

The definitions of K^* and $^*K$ by coordinate extension resulted in canonical imbeddings $j^* \colon B \to B^* \subset K^*$ and $^*j \colon B \to {}^*B \subset^* K$, for $B = $ I, III. Thus the standard block I is canonically imbedded in K_1 once by the composite map I $\to$ I$^* \subset K^* \to K_1$ and once by I $\to {}^*I \subset^* K \to K_1$ (see Figure 3.4.)

For block III, since the matching map μ is $^*j \circ (j^*)^{-1}$, it follows from $i^* = {}^*i \circ \mu$ that $i^* \circ j^* = {}^*i \circ {}^*j \colon \text{III} \to K_1$. This map j is the canonical imbedding of III in K_1 (see Figure 3.6).

Condition 2. Since μ preserves the function r and ϑ, the corresponding functions r and ϑ on K^* and $^*K$ agree on III $\subset K_1$, giving functions r and ϑ on K_1.

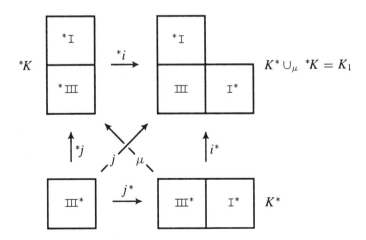

FIGURE 3.6. Canonical imbedding j of Boyer–Lindquist block III into K_1.

Condition 3. The equatorial isometry ϵ on Boyer–Lindquist blocks and on K^* and $^*K$ simply reverses the sign of ϑ leaving the other coordinate unchanged. It follows immediately that it commutes with the canonical imbedding mentioned earlier. Thus, by the Mapping Lemma 1.4.3, it defines an isometry $\epsilon: K_1 \to K_1$ that agrees on (canonically imbedded) blocks with the Boyer–Lindquist versions.

Condition 4. Similarly, since $d\mu$ preserves ∂_t and ∂_φ, the Killing vector fields $\tilde{\partial}\varphi$ and $\tilde{\partial}_t$ on K^* and $^*K$ agree on III, giving the required extensions on K_1. $\square$

The crucial objects preserved by μ are the principal null geodesics, so outgoers from the unique block III of K_1 are future complete (see Figure 3.4). As imbedded in K_1 (and in later stages including M_e itself) the spacetimes K^* and $^*K$ are called *Kerr patches*.

Now in K_1 the only ordinary principal nulls that are not future complete are the ingoers in block *I. Since all ordinary ingoers in K^* are complete, we shall glue K^* to K_1 by means of $\mu: I^* \to {}^*$I $\subset K_1$. Again the Hausdorff condition holds, and the properties of the isometry μ imply, as above, that the second stage $K_2 = K_1 \cup_\mu K^*$ is a Kerr spacetime (see Figure 3.5).

But in K_2 the outgoers of the new block III* in K_2 are future incomplete, so once again a K^* must be glued on to complete them. Evidently the process continues indefinitely into the future producing a Kerr spacetime in which all ordinary principal nulls are future complete.

Now consider the past. Back in block I of the original K^*, the outgoers are past incomplete. Gluing on a $^*K$ by means of $\mu: I^* \to {}^*$I cures this, but introduces past incomplete ingoers in the new block III*. So this process also continues indefinitely, as in the future construction. The result is *maximal extreme Kerr spacetime* M_e, in which all principal null geodesics that miss Σ are complete (see Figure 3.7).

Note that each horizon $H: r = $ M is in a unique Kerr patch, but every Boyer-Lindquist block is in exactly two Kerr patches, one K^* and one $^*K$.

The following alternative construction of M_e, which involves only a single "zipper" gluing, is convenient later on. Let M^* be the disjoint union of a sequence, $-\infty < i < +\infty$, of copies K^i of K^*, and let $\mathcal{U}^*$ be the union of all the blocks I^i and III^i in all K^i (so $\mathcal{U}^*$ is K^* minus horizons). Let $^*M$ and $^*\mathcal{U}$ be correspondingly defined for star-Kerr spacetime. Reading from Figure 3.7, define the matching map $\mu: \mathcal{U}^* \to {}^*\mathcal{U}$ by combining alternately the isometries $\mu: III^i \to $III^i and $\mu: I^i \to $I^{i+1} of Lemma 3.1.3. Then $M_e = M^* \cup_\mu {}^*M$ agrees with the inductive description above.

By construction, the ordinary principal null geodesics of M_e all run from $r = \pm\infty$ to $r = \mp\infty$ (which will imply completeness). Those that hit the ring singularity

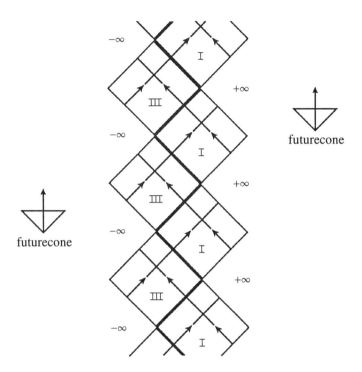

FIGURE 3.7.   Maximal extreme Kerr spacetime M_e (drawn here with 45° futurecones) is the union of a chain of alternating K^* and $^*K$ patches, the former rising I–III to the left, the latter rising III–I to the right. All its ordinary principal null geodesics are complete.

are obviously not complete, but we now show that those in the horizons of M_e are complete. The slow case will demonstrate that this issue is not a trivial one.

Lemma 3.2.3 *On the horizons $r = r_\pm$ of K^* and $^*K$, the canonical vector field V has $\nabla_V V = \frac{1}{2}(r_\pm - r_\mp)V$. Hence $\nabla_V V = 0$ in the extreme case, where $r_+ = r_- = M$.*

Proof. We first compute $\nabla_V V$ on a Boyer–Lindquist blocks I and III, where $\varepsilon = \text{sgn } \Delta = +1$, and then take the limit as $r \to r_\pm$.

In terms of the Boyer–Lindquist frame field, $V = \rho\sqrt{\Delta}E_0$. Since V is a linear combination of ∂_φ and ∂_t, while r and Δ involve only r and ϑ,

$$\nabla_V V = \rho\sqrt{\Delta}\nabla_{E_0}(\rho\sqrt{\Delta}E_0) = \Delta\rho^2\nabla_{E_0}E_0 = \Delta\rho^2[FE_1 - a^2SC\rho^{-3}E_2],$$

where the last equality uses Corollary 2.6.7.

As $r \to r_{\pm}$ and hence $\Delta \to 0$, the E_2 term also approaches 0 since $E_2 = \partial_{\vartheta}/\rho$ is well defined on horizons. To evaluate the remaining term, $\Delta\rho^2 F E_1$, we use Kerr-star coordinates, for which $E_1 = \rho^{-1}\sqrt{\Delta}\partial_r = \rho^{-1}\sqrt{\Delta}(\partial_r^* + \Delta^{-1}V)$. Then, taking limits as $r \to r_{\pm}$,

$$\nabla_V V = \lim \nabla_V V = \lim \Delta\rho^{-2}[\rho^2(r - M) - \Delta r](\partial_r^* + \Delta^{-1}V).$$

Here the factor Δ eliminates ∂_r^*, leaving

$$\nabla_V V = \lim \Delta\rho^{-2}[\rho^2(r - M) - \Delta r]\Delta^{-1}V$$
$$= \lim(r - M)V = (r_{\pm} - M)V = \tfrac{1}{2}(r_{\pm} - r_{\mp})V.$$

$\square$

Corollary 3.2.4 *In the extreme case, principal null geodesics in the horizons of K^* and $^*K$ are complete—and hence those of M_e are also.*

Proof. We know that every principal null in a horizon is an integral curve α of the canonical vector field V. By the lemma, $\alpha'' = \nabla_V V = 0$, so α is a geodesic. On the horizon, $V = (r_{\pm}^2 + a^2)\partial_t + a\partial_{\varphi}$, a *constant* linear combination of ∂_{φ} and ∂_t. Thus α is defined on the entire real line. $\square$

3.3   Extending Slow Kerr Spacetime

We now imitate, for slow Kerr spacetime, the extension method used for the extreme case in the preceding section. But the slow case is trickier, and has a quite different outcome.

Available for the construction are the Kerr-star and star-Kerr spacetimes: K^*, with complete *ingoers*, $^*K$, with complete *outgoers*. Any two like-numbered Boyer–Lindquist blocks in K^* and $^*K$ are related by the canonical isometry $\mu = {}^*j \circ (j^*)^{-1}: B^* \to {}^*B$, with properties as in Lemma 3.1.3.

Again we start with K^* and work at first into the future. The outgoers in blocks II^* and III^* of K^* are both future-incomplete. To cure this for II^*, we will glue $^*K$ onto it. Now the crucial novelty of the slow case appears; the gluing isometry $\mu: \mathrm{II}^* \to {}^*\mathrm{I}$ *reverses* time-orientation. But if the extended manifold is to be time-orientable (as required by the definition of spacetime) μ must *preserve* time-orientation. To achieve this, we have no choice but to reverse the time-orientation of the entire spacetime $^*K$ before the gluing. (The spacetime gotten by reversing the time-orientation of a spacetime N is denoted by N'.)

For III^* the isometry $\mu\colon \text{III}^* \to {}^*\text{III}$ preserves time-orientation, so no change in $^*K$ is needed when it is glued.

As in the extreme case, we drop the asterisk from the single block formed by gluing B^* and $^*B$, denoting this block simply by $B = \text{I}, \text{II}, \text{III}$ provided its time-orientation derives (via canonical isometry) from K^*. Thus, if $B = \text{I}$ or III, then $^*B \approx B^* \to B$, but $^*\text{II}' \approx \text{I}^* \to \text{II}$, so $^*\text{II} \approx \text{II}^{*\prime} \to \text{II}'$.

With these conventions, Figure 3.8 illustrates the spacetime K_1 produced by two gluings:

1. $^*K'$ onto K^* by means of $\mu\colon {}^*\text{II}' \approx \text{II}^*$, and
2. $^*K$ onto K^* by means of $\mu\colon {}^*\text{III} \approx \text{III}^*$.

As in the extreme case, the Hausdorff condition holds for the gluings, and K_1 is a Kerr spacetime. In particular, there is a canonical imbedding $j\colon B \to K$ for each Boyer–Lindquist block in K_1.

Since principal null geodesics are always future-pointing, their parametrizations must be reversed when time-orientations are reversed, so *ingoer* and *outgoer* are also reversed. For example, an outgoer γ in $^*K$ runs from $r = -\infty$ to $r = +\infty$, but in $^*K'$ its reversal $\bar{\gamma}$ runs from $r = +\infty$ to $r = -\infty$. Thus, as in Figure 3.8, $\bar{\gamma}$ serves to extend an incomplete ingoer of $\text{II}^* \subset K^*$.

In the newly added blocks $^*\text{II}$ and $^*\text{III}'$ of K_1, the outgoers are future-incomplete (these run horizontally, right to left, in Figure 3.8). This double defect can be

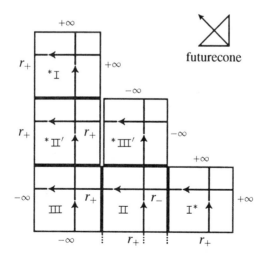

FIGURE 3.8.   The result K_1 of gluing $^*K'$ and $^*K$ (drawn vertically) onto the initial (horizontal) K^*. All the ordinary outgoers of K^* are now future-complete. Note the gap between $^*\text{II}'$ and $^*\text{III}'$.

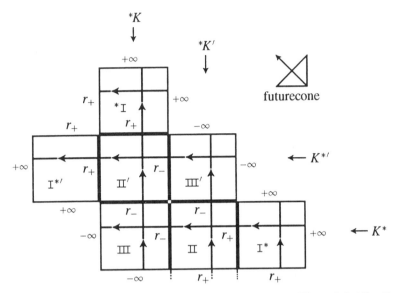

FIGURE 3.9.   The result K_2 of gluing $K^{*\prime}$ onto the spacetime K_1 of Figure 3.8. All ordinary outgoers in blocks II' and III' are now future-complete.

corrected—and the two blocks welded together—by gluing a single K^* to K_1. This time the orientation of K^* must be reversed, and $K^{*\prime}$ is glued onto K_1 using the map

$$\mu \colon \mathrm{I}^{*\prime} \cup \mathrm{II}^{*\prime} \to {}^*\mathrm{I}' \cup {}^*\mathrm{II},$$

which, as required, *preserves time-orientation*. The resulting spacetime K_2 is shown in Figure 3.9.

It should be clear how to continue this extension scheme into the future. Then we go back to the initial K^* and extend symmetrically into the past. The resulting spacetime M_0 is called the *basic extension* of slow Kerr spacetime (see Figure 3.10).

Looking at the pattern of M_0 in Figure 3.10 leads to two observations that guide future developments. First, instead of interpreting M_0 as a double sequence of overlapping vertical and horizontal three-block rectangles, it is simpler to view it as a single sequence in which two kinds of overlapping four-block domains alternate. These *Kruskal domains* $\mathcal{D}(r_\pm)$, shown in Figure 3.11, are open submanifolds of the Kerr spacetime M_0.

The second observation is that at the hollow dots at the center of $\mathcal{D}(r_\pm)$, four Kerr horizons seem to meet. But M_0 merely assembles K^* and $^*K$ patches; none of its points provide for such a meeting. A new set S is needed to connect the

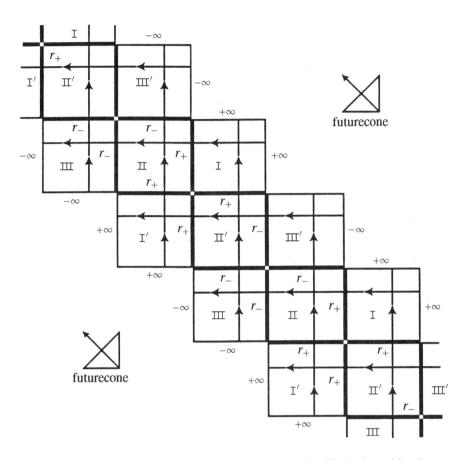

FIGURE 3.10.   The basic extension M_0 of slow Kerr spacetime.The horizontal levels are K^* spacetimes, the verticals are $^*K$. All ordinary principal nulls are complete. Note how time-orientations—and infinities—alternate, by contrast with the simpler extreme case, Figure 3.7.

four horizons in a natural way. Recall that a horizon in K^* or $^*K$ is a hypersurface diffeomorphic to $S^2 \times \mathbf{R}^1$—a 2-sphere enduring through time. At S, each opposite pair of these "short horizons" H will join to form a single "long horizon" $\mathbb{H}$. Then, two long horizons in $\mathcal{D}(r_\pm)$ will intersect at S, a 2-sphere called a *crossing sphere*.

As geometry this is natural enough, but physically the spacetime around crossing spheres will be very strange. We can imagine that an observer in $\mathcal{D}(r_\pm)$ could—if horizons were visible—see two horizon-spheres of the same radius, r_- or r_+,

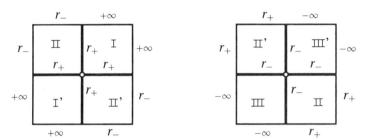

FIGURE 3.11. The Kruskal domains.

approach and pass through each other, analogously to the way an expanding 2-sphere in $\mathbf{R}^3$ passes through a concentric contracting 2-sphere.

Support for the creation of long horizons comes from the idea that all principal null geodesics that avoid ring singularities should be complete. In the extreme spacetime M_e this is true for the restphotons generating the short horizons of K^* and $^*K$. But in the slow case we will now see that these restphotons are not complete (the crossing spheres must make them so).

Restphotons in a Kerr horizon are (possibly reparametrized) integral curves of the canonical vector field V. Lemma 3.2.3 shows that in the extreme case they already have geodesic parametrization, but this too will fail in the slow case. On the horizons $r = r_\pm$, the vector field V is given by the same formula, $(r_\pm^2+a^2)\partial_t +a\partial_\varphi$, in both K^* and $^*K$ coordinates. We write $^*t^*$ in formulas that are valid for both t^* and $^*t$; similarly for $^*\varphi^*$. It suffices to consider just one integral curve α of V in each horizon, since the axial and time symmetries allow the initial values of its $^*t^*$ and $^*\varphi^*$ coordinates to be adjusted to any desired value. As *standard*, we pick $u \to \alpha(u)$ with simplest coordinates:

$$r = r_\pm, \quad \vartheta = \vartheta_0, \quad ^*\varphi^* = au, \quad ^*t^* = (r_\pm^2 + a^2)u.$$

To reparametrize α as a geodesic γ, write $\gamma(s) = \alpha(f(s))$. Then, by Lemma 3.2.3,

$$\gamma'' = f'^2\beta''(f) + f''\beta'(f) = f'^2\nabla_V V + f''V = \left[(r_\pm - r_\mp)/2f'^2 + f''\right]V.$$

Consequently, γ is a geodesic if and only if $f(s)$ is a solution of the differential equation $f'' + (r_\pm - r_\mp)/2f'^2 = 0$. By convention, principal nulls are always future-pointing, and V is future-pointing on the horizons of K^* and $^*K$, so there we require $f' > 0$. Thus, our standard null geodesics have

$$f(s) = k_\pm^{-1}\ln(k_\pm s), \quad \text{where } k_\pm = (r_\pm - r_\mp)/2,$$

with domain $\{s > 0\}$ for $r = r_+$, and $\{s < 0\}$ for $r = r_-$.

If the time-orientations of K^* and $^*K$ are reversed, then $f' < 0$ is required to make γ future-pointing, and the domains reverse. Thus, the four cases work out as follows.

Lemma 3.3.1 *The standard (future-pointing) principal null geodesic γ in the horizon $r = r_\pm$ of the slow Kerr spacetimes K^* and $^*K$ has coordinates*

$$r = r_\pm, \quad \vartheta = \vartheta_0, \quad {}^*\varphi^* = af(s), \quad {}^*t^* = (r_\pm^2 + a^2)f(s)$$

where $f(s)$ is
1. $k_+^{-1} \ln k_+ s$ *on $s > 0$; in $r = r_+$ for K^* and $^*K$.*
2. $k_+^{-1} \ln(-k_+ s)$ *on $s < 0$; in $r = r_+$ for $K^{*\prime}$ and $^*K'$.*
3. $k_-^{-1} \ln k_- s$ *on $s < 0$, in $r = r_-$ for K^* and $^*K$.*
4. $k_-^{-1} \ln(-k_- s)$ *on $s > 0$ in $r = r_-$ for $K^{*\prime}$ and $^*K'$.*

In all four cases, completeness obviously fails as $s \to 0$, and $\gamma(s)$ approaches the center of $\mathcal{D}(r_\pm)$ as $s \to 0$.

3.4   Building the Crossing Spheres

In this section we enlarge each Kruskal domain $\mathcal{D}_0(r_\pm)$ to an analytic manifold $\mathcal{D}(r_\pm)$ by constructing the spheres at which its long horizons intersect. In the next section the Kerr metric is extended over these spheres; then chaining domains $\mathcal{D}(r_+)$ and $\mathcal{D}(r_-)$ as suggested by Figure 3.10 will produce maximal slow Kerr spacetime.

The crossing spheres will be built by extending a coordinate system $\eta^\pm$ on $\mathcal{D}(r_\pm)-$ (axes) called "Kruskal-like" by its discoverers, R. H. Boyer and R. W. Lindquist (1967). Thus, we refer to *KBL coordinates*. (Boyer and Lindquist called the coordinates that now bear their names "Schwarzschild-like.") The globally defined radius function r, which has appeared in all earlier coordinate systems, cannot be a KBL coordinate function because every point p of a crossing sphere S will be a critical point of r. This is clear since the tangent space at each point p of S is spanned by vectors tangent to horizons, but r is constant on horizons, so $dr = 0$ at p.

What can replace r? Since each Boyer–Lindquist block is contained in two Kerr patches, one K^* and one $^*K$, a reasonable plan (traceable to Arthur Eddington in the 1920s) is to replace the Boyer–Lindquist coordinates r, t by the two Kerr time coordinates $t^*, {}^*t$. The coordinate t fails on all horizons, but t^* is at least extendible over K^* horizons (though it fails on those of ${}^*K$) and correspondingly for ${}^*t$. In addition to these semi-failures the values of $t^*, {}^* t$ do not distinguish between the different Boyer–Lindquist blocks of $\mathcal{D}(r_\pm)$. However, there is a simple correction whose parameters require some preparation.

Remarks 3.4.1

1. Since horizon radii r_+ and r_- are roots of $\Delta = r^2 - 2\mathrm{M}r + a^2$, they satisfy the identities $r_+ + r_- = 2\mathrm{M}$, $r_+r_- = a^2$, and $r_\pm^2 + a^2 = r_\pm(r_\pm + r_\mp) = 2\mathrm{M}r_\pm$. The last of these gives

$$\frac{r_-^2 + a^2}{r_+^2 + a^2} = \frac{r_-}{r_+}.$$

2. For slow Kerr spacetime the constants κ_+ and κ_- are defined to be

$$\kappa_\pm = \frac{r_\pm - r_\mp}{2(r_\pm^2 + a^2)}.$$

Thus, $\kappa_- < 0 < \kappa_+$, and, by the identity above,

$$2\kappa_+r_+ = \frac{r_+ - r_-}{r_+ + r_-} = -2\kappa_-r_-$$

and hence $\kappa_+/\kappa_- = -r_-/r_+$.   □

The time and angle functions $T(r)$ and $A(r)$ have been described so far merely as antiderivatives of $(r^2 + a^2)/\Delta$ and a/Δ, respectively. We now choose these two functions explicitly in the slow case.

Lemma 3.4.2

$$T(r) = r + \frac{1}{2\kappa_+} \ln |r - r_+| + \frac{1}{2\kappa_-} \ln |r - r_-|,$$

$$A(r) = \frac{a}{r_+ - r_-} \ln \left| \frac{r - r_+}{r - r_-} \right|.$$

For a proof it suffices to differentiate these functions. (The more elegant choice of A and T in Misner, Thorne, and Wheeler (1973) makes these functions zero at 0, M, and $2\mathrm{M}$, but leads to slightly more complicated formulas.) A and T separately are undefined at both r_- and r_+, but a combination of the two does better.

Corollary 3.4.3 *With notation as above,*

$$A(r) - \frac{a}{r_\pm^2 + a^2} T(r) = \left(\frac{-a}{r_\pm^2 + a^2} \right) [r + (r_\pm + r_\mp) \ln |r - r_\mp|.$$

The following immediate consequences will soon be needed.

$$A - aT/(r_+^2 + a^2) \text{ is an analytic function of } r \text{ for } r \neq r_-;$$
$$A - aT/(r_-^2 + a^2) \text{ is an analytic function of } r \text{ for } r \neq r_+$$

Corollary 3.4.4

$$\exp(2\kappa_\pm T(r)) = \exp(2\kappa_\pm r) |r - r_+|^{\kappa_\pm/\kappa_+} |r - r_-|^{\kappa_\pm/\kappa_-}.$$

Proof.

$$\exp(2\kappa_\pm T(r)) = \exp\left[\kappa_\pm(2r + \kappa_+^{-1} \ln |r - r_+| + \kappa_-^{-1} \ln |r - r_-|\right]$$
$$= \exp\left[2\kappa_\pm r + (\kappa_\pm/\kappa_+)\right] \ln |r - r_+| + (\kappa_\pm/\kappa_-) \ln |r - r_-|.$$

$\square$

Recall that in this formula, $\kappa_+/\kappa_- = -r_-/r_+$.

To eliminate the difficulties with $t^*, {}^*t$ mentioned above, it suffices to shrink a $t^*, {}^*t$ plane onto each of the four quadrants of a single plane, using variants of the map $(x, y) \to (e^x, e^y)$. (See Figure 3.12.)

Definition 3.4.5 *The real-valued functions U^+ and V^+ on the four Boyer–Lindquist blocks of $\mathcal{D}_0(r_+)$ are as follows:*

On I,	$U^+ = \exp(-\kappa_+^* t),$	$V^+ = \exp(\kappa_+ t^*)$
On II,	$U^+ = -\exp(-\kappa_+^* t),$	$V^+ = \exp(\kappa_+ t^*)$
On I′,	$U^+ = -\exp(-\kappa_+^* t),$	$V^+ = -\exp(\kappa_+ t^*)$
On II′,	$U^+ = \exp(-\kappa_+^* t),$	$V+ = -\exp(\kappa_+ t^*).$

For $\mathcal{D}_0(r_-)$, change all sub- and superscript plus signs to minus, and change block numbers thus: I → III′, II → II′, I′ → III, II′ → II.

Other sign patterns would still produce good coordinates; our choices make the usual quadrants of the $U^\pm$, $V^\pm$ plane consistent with Figure 3.11—and, for $\mathcal{D}(r_+)$, with Figure 3.12. The particular constants $\kappa_\pm$ prove their worth in Lemma 3.4.10. (Note that the signs preceding $\kappa_\pm$ make the product $U^+ V^+$ a function of r only.)

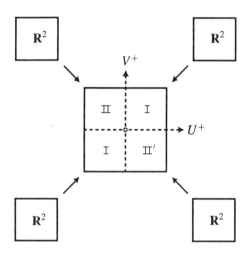

FIGURE 3.12. The $t^*,^*t$ plane mapping to each quadrant of the U^+V^+ plane.

$U^\pm$ and $V^\pm$ replace r and t, and the dependable colatitude function ϑ is kept. It remains to find a suitable longitude function $\varphi^\#$. The purpose of KBL coordinates is to construct crossing spheres at which the long horizons intersect. This led Boyer and Lindquist to determine $\varphi^\#$ so that *the principal null geodesics in the long horizons are KBL coordinate curves*. (Recall that this scheme of turning principal nulls into coordinate curves has already been used to construct Kerr-star and star-Kerr coordinates.) Two versions of $\varphi^\#$ are needed, φ^+ and φ^-, corresponding to the two kinds of Kruskal domain.

In both K^* and $^*K$ the principal nulls in the horizon $r = r_\pm$ are the integral curves of the canonical vector field V restricted to $r = r_\pm$, namely, $V_\pm = (r_\pm^2 + a^2)\partial_t + a\partial_\varphi$. For definiteness, take the K^* case and consider $\varphi^\pm$ as a function of Kerr-star coordinates r, ϑ, φ^*, t^*. Then $\varphi^\pm$ is constant on integral curves α of $V_\pm$ provided

$$0 = d(\varphi^\pm \circ \alpha)/ds = \alpha'[\varphi^\pm] = V_\pm[\varphi^\pm] = (r_\pm^2 + a^2)\partial\varphi^\pm/\partial t^* + a\partial\varphi^\pm/\partial\varphi^*.$$

Thus a natural choice for $\varphi^\pm$ is $\varphi^* - [a/(r_\pm^2 + a^2)]t^*$. The same argument on $^*K$ patches yields $\varphi^\pm = {}^*\varphi - [a/(r_\pm^2 + a^2)]^*t$. These two formulas are inconsistent on $K^* \cap {}^*K$, so we average them.

Definition 3.4.6 *On each Boyer–Lindquist block of* $\mathcal{D}(r_{\pm})$ *the KBL* longitude
functions $\varphi^{\pm}$ *are*

$$\varphi^{\pm} = \frac{1}{2}\left[\varphi^* + {}^*\varphi - \frac{a}{r_{\pm}^2 + a^2}(t^* + {}^*t)\right].$$

We now prove the global validity of KBL coordinates.

Proposition 3.4.7

A. *The functions* $U^{\pm}$, $V^{\pm}$, ϑ, $\varphi^{\pm}$ *on the Boyer–Lindquist blocks of* $\mathcal{D}_0(r_{\pm})-$ *(axes),*
   have unique analytic extensions over $\mathcal{D}_0(r_{\pm})-$ *(axes).*
B. *There they form a coordinate system* $\eta^{\pm}$.
C. *Furthermore, the mapping* $\eta^{\pm}$ *has a unique analytic extension to a diffeomor-*
   phism of $\mathcal{D}_0(r_{\pm})$ *onto* $(\mathbf{R}^2 - (0, 0)) \times S^2$ *(with ring singularity deleted in the*
   r_- *case).*

Proof. For definiteness consider the r_+ case.

Proof of assertion A. $\mathcal{D}_0(r_+)$ is covered by parts of four Kerr patches as indicated
in Figure 3.13. It suffices to show—working solely with off-axis points—that for
each Kerr patch, the functions U^+, V^+, ϑ, φ^+ can be analytically extended from
two Boyer–Lindquist blocks in $K \cap \mathcal{D}_0(r_+)$ over the short horizon that joins them.

Consider first blocks I and II of $K^* \subset \mathcal{D}_0(r_+)$. Naturally, we express the
functions in terms of the Kerr-star coordinates of K^*. Since only I and II are
involved, the inequality $r > r_-$ holds. To extend the four functions in this case we
first prove convenient formulas for them.

(1) We assert that $U^+ = (r - r_+)(r - r_-)^{-r_-/r_+} \exp[\kappa_+(2r - t^*)]$.
    To see this, note that on I $\cup$ II, the double definition of U^+ can be written
    simply as $\mathrm{sgn}(r - r_+) \exp(-\kappa_+^* t)$. Then substituting ${}^*t = t^* - 2T(r)$ and using
    Corollary 3.4.4, we find

$$U^+ = \mathrm{sgn}(r - r_+) \exp(-\kappa_+ t^*) \exp(2\kappa_+ T(r))$$
$$= \mathrm{sgn}(r - r_+) \exp(-\kappa_+ t^*) \exp(2\kappa_+ r)|r - r_+| \, |r - r_-|^{-r_-/r_+}$$
$$= (r - r_+)(r - r_-)^{-r_-/r_+} \exp[\kappa_+(2r - t^*)].$$

Here the sgn eliminates one absolute value, and $r > r_-$ the other.
(2) By definition, $V^+ = \exp(\kappa_+ t^*)$ on I $\cup$ II.
(3) ϑ is unchanged.

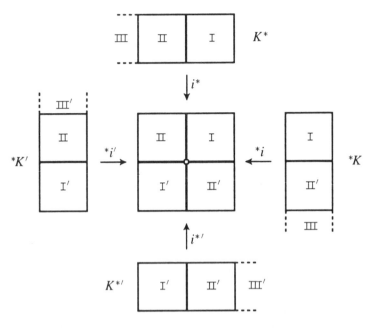

FIGURE 3.13. $\mathcal{D}_0(r_+)$, with its four partial Kerr coordinate patches. (For the pattern of primes see Figure 3.10.)

(4) By definition, $\varphi^+ = (1/2)[\varphi^* + {}^*\varphi - a(t^* + {}^*t)/(r_+^2 + a^2)]$. Substituting ${}^*\varphi = \varphi^* - 2A(r)$ and ${}^*t = t^* - 2T(r)$ gives

$$\varphi^+ = \varphi^* - \frac{a}{r_+^2 + a^2}\, t^* - \left(A(r) - \frac{a}{r_+^2 + a^2}\, T(r)\right)$$

Obviously, the functions U^+, V^+, and ϑ can be analytically extended over the horizon to the off-axis region $r > r_-$ in K^*. The same is true for φ^+, since—as noted after Corollary 3.4.3—the final summand in the expression for φ^+ is analytic on $r > r_-$.

Now consider blocks I' and II in the Kerr patch ${}^*K'$. There, using star-Kerr coordinates, we find formulas analogous to (1) through (4) above.

(1') $U^+ = -\exp(-\kappa_+ {}^*t)$ on $\mathrm{I}' \cup \mathrm{II}$.

(2') $V^+ = -\mathrm{sgn}(r - r_+)\exp(\kappa_+\, t^*) = -\mathrm{sgn}(r - r_+)\exp(\kappa_+({}^*t + 2T(r)))$
$\quad = -\mathrm{sgn}(r - r_+)\exp(\kappa_+\, {}^*t)\exp(2\kappa_+ T(r))$
$\quad = -\mathrm{sgn}(r - r_+)\exp(\kappa_+\, {}^*t)\exp(2\kappa_+ r)|r - r_+|(r - r_-)^{-r_-/r_+}$
$\quad = -(r - r_+)(r - r_-)^{-r_-/r_+}\exp[\kappa_+(2r + {}^*t)]$.

(3′) ϑ is unchanged.

(4′) Substituting $\varphi^* = {}^*\varphi + 2A(r)$ and $t^* = {}^*t + 2T(r)$ in the definition of φ^+ gives

$$\varphi^+ = {}^*\varphi - \frac{a}{r_+^2 + a^2} \, {}^*t - \left(A(r) - \frac{a}{r_+^2 + a^2} \, T(r)\right).$$

As before these functions are all analytically extendible over the off-axis region $r > r_-$ in ${}^*K$'.

The other two Kerr coordinate patches involve only similar sign changes; thus the extended functions constitute an analytic mapping $\eta^+ = (U^+, V^+, \varphi^+, \vartheta)$ from $\mathcal{D}_0(r_+) -$ (axes) into $R^2 \times S^2$.

Proof of assertion B. To show that η^+ is actually a coordinate system, note that Definition 3.4.5 expresses η^+, on each block, as the map $\tau = (t^*, {}^*t, \vartheta, \varphi^+)$ followed by a variant of the diffeomorphism $(x, y) \to (e^x, e^y)$. But on any Boyer–Lindquist block, we assert that τ is a coordinate system. In fact, it is easy to check that it is one-one, and its Jacobian determinant, relative to, say, Kerr-star coordinates, is $2(r^2 + a^2)/\Delta \neq 0$. Thus, η^+ is a coordinate system on individual blocks. Sign choices make the images of the blocks disjoint, so η^+ is one-one on $\mathcal{D}_0(r_+)$. The only remaining question is whether the Jacobian of η^+ is nonzero on horizons. Since the four Kerr coordinate systems differ only by minor sign changes, the K^* case should suffice. We need only show that the Jacobian determinant

$$J = \partial(U^+, V^+, \varphi, \vartheta)/\partial(r, \varphi^*, t^*, \vartheta)$$

is never zero on the horizon $r = r_+$ of K^*. Evidently, ϑ can be dropped. Furthermore, the only partial derivative of V^+ that is not identically zero is $\partial V^+/\partial t^* = \kappa_+ \exp(\kappa_+ t^*)$, which is never zero. Thus, we need only examine $\partial(U^+, \varphi^+)/\partial(r, \varphi^*)$. But $\partial U^+/\partial \varphi^* = 0$ and $\partial \varphi^+/\partial \varphi^* = 1$. Consequently, the Jacobian J and $\partial U^+/\partial r$ are nonzero at exactly the same points.

Now it suffices to check that $\partial U^+/\partial r$ is positive when $r = r_+$. Write U^+ from the earlier formula (1) as $(r - r_+) f(r)$, where

$$f(r) = (r - r_-)^{-r_-/r_+} \exp(2\kappa_+ r).$$

Then $dU^+/dr = f(r) + (r - r_+)f'(r)$. For $r = r_+$ this reduces to $f(r)$, which is certainly positive. (This argument actually shows that the Jacobian J is positive for all $r > r_-$, since it is positive at r_+ and nonvanishing elsewhere.)

Proof of assertion C. Before extending the KBL coordinate system over the axes, we show that the coordinates U^+, V^+ map $\mathcal{D}_0(r_+) -$ (axes) onto a plane $\mathbf{R}^2$ with origin $(0, 0)$ deleted.

On Boyer–Lindquist blocks the map (U^+, V^+) is $(t^*, {}^*t)$ followed by exponential diffeomorphisms that carry $t^*, {}^*t$ planes onto the four U^+, V^+ quadrants. By Remark 3.4.8, the open quadrants of $\mathbf{R}^2$ are entirely covered. Now consider the short horizons. It follows from the formula (1) for U^+ in the proof of assertion A that, in the Kerr patch K^*, the horizon $r = r_+$ is given by $U^+ = 0$. The other quadrants are similar; thus the image of (U^+, V^+) is $\mathbf{R}^2 - (0, 0)$.

It is easy to see that for fixed U^+, V^+ every nonpolar point of S^2 is covered. Hence the coordinate system η^+ carries $\mathcal{D}_0(r_+) -$ (axes) diffeomorphically onto all of $\mathbf{R}^2 \times (S^2 - (\text{poles}) = (\mathbf{R}^2 \times S^2) -$ (axes) *except* the crossing sphere $(0, 0) \times S^2$.

It suffices to extend η^+ on each of the four Kerr patches separately; there is no consistency problem because the patches intersect on Boyer–Lindquist blocks, where η^+ is uniquely determined by its off-axis values.

Consider, for example, the K^* patch. As in Figure 3.13 let i^* be the canonical imbedding of the region $r > r_-$ of K^* into $\mathcal{D}(r_+)$. As before, let R_α be rotation of S^2 through angle α around its polar axis. Then the extension of η^+ over the axes sends each point $i^*(r, t^*, y)$ to $\left(U^+, V^+, R_{\alpha(r,t^*)}(y)\right)$, where

$$\alpha(r, t^*) = -\frac{a}{r_+^2 + a^2}\, t^* - \left(A(r) - \frac{a}{r_+^2 + a^2}\, T(r)\right).$$

This analytic map is an extension of η^+ since if $y \in S^2$ is nonpolar the rotation preserves its colatitude ϑ and increases its longitude φ^* by $\alpha(r, t^*)$, giving, by formula (4) above, just φ^+.

The other Kerr patches give analogous results. Evidently the image of the entire extension is $(\mathbf{R}^2 - (0, 0)) \times S^2$. This completes the proof of Proposition 3.4.7.   □

Remark 3.4.8   Each Boyer–Lindquist block is mapped by $(t^*, {}^*t)$ onto the entire plane $\mathbf{R}^2$. This asserts that the equations

$$a = t^* = t + T(r), \qquad b = {}^*t = t - T(r)$$

can always be solved for t and r when r is restricted, in turn, to each of the three components of $\mathbf{R} - \{r_-, r_+\}$. Adding the equations produces $t = (a + b)/2$. Subtracting them gives $a - b = 2T(r)$. But the explicit formula for $T(r)$ in Lemma 3.4.2 shows that $T(r)$ maps each interval of $\mathbf{R} - \{r_-, r_+\}$ onto $\mathbf{R}$.

Thus, a coordinate extension (Section 1.4) produces the *full Kruskal domain* $\mathcal{D}(r_+) \approx \mathbf{R}^2 \times S^2$, in which the crossing sphere $(0, 0) \times S^2$ is the set $U^+ = V^+ = 0$. The r_- case is strictly analogous, except that since the ring singularity is in block III, $\mathcal{D}(r_-)$ is diffeomorphic to $\mathbf{R}^2 \times S^2 - \Sigma'$, where Σ' is the KBL image of the ring singularity $r = 0$, $\vartheta = \pi/2$.

Although the function r is no longer a coordinate function it is still well behaved, and the functions $r - r_\pm$ and $U^\pm V^\pm$ are closely related.

Lemma 3.4.9 (1) *The radius function r is a well defined analytic function on $\mathcal{D}(r_\pm)$.*

(2) *There exists a nonvanishing analytic function $G_\pm$ on $\mathcal{D}(r_\pm)$, depending solely on r, such that*

$$G_\pm(r) = \frac{r - r_\pm}{U^\pm V^\pm}$$

Proof. Comparing the definitions of $U^\pm$ and $V^\pm$ with the quadrant arrangements in Figure 3.11 shows:

$$\text{On } \text{I} \cup \text{I}' \subset \mathcal{D}(r_+): \quad U^+ V^+ > 0 \text{ and } r - r_+ > 0.$$
$$\text{On } \text{III} \cup \text{III}' \subset \mathcal{D}(r_-): \quad U^- V^- > 0 \text{ and } r - r_- < 0.$$

Thus, on all Boyer–Lindquist blocks, sgn $U^\pm V^\pm = \pm$sgn $(r - r_\pm)$.
Hence Definition 3.4.5 gives

$$U^\pm V^\pm = \pm\text{sgn}(r - r_\pm)\exp[\kappa_\pm(t^* - {}^*t)] = \pm\text{sgn}(r - r_\pm)\exp[2\kappa_\pm T(r)].$$

By Corollary 3.4.4 this becomes

$$U^\pm V^\pm = \pm\text{sgn}(r - r_\pm)\exp[2\kappa_\pm r + (\kappa_\pm/\kappa_+)\ln|r - r_+| + (\kappa_\pm/\kappa_-)\ln|r - r_-|]$$
$$= \pm\text{sgn}(r - r_\pm)\exp(2\kappa_\pm r)|r - r_+|^{\kappa_\pm/\kappa_+}|r - r_-|^{\kappa_\pm/\kappa_-}.$$

Separating the two cases produces

$$U^+ V^+ = \text{sgn}(r - r_+)\exp(2\kappa_+ r)|r - r_+|\,|r - r_-|^{\kappa_+/\kappa_-}$$
$$= (r - r_+)\exp(2\kappa_+ r)|r - r_-|^{\kappa_+/\kappa_-},$$
$$U^- V^- = -\text{sgn}(r - r_-)\exp(2\kappa_- r)|r - r_+|^{\kappa_-/\kappa_+}|r - r_-|$$
$$= -(r - r_-)\exp(2\kappa_- r)|r - r_+|^{\kappa_-/\kappa_+}.$$

We can now prove assertions (1) and (2) in the statement of the lemma.

(1) Take the r_+ case. Write the first equation above as $U^+ V^+ = F(r) = (r - r_+)f(r)$. Then $dF/dr = f(r) + (r - r_+)df/dr$. Thus, at $r = r_+$, $dF/dr = f(r_+)$, which is clearly nonzero. Hence, $r = F^{-1}(U^+ V^+)$ is well defined and analytic

for r near r_+. This suffices since we already know that r is well defined and analytic except possibly at the crossing sphere. (Actually, F^{-1} exists globally.)

(2) It follows at once from the above equations for $U^+ V^+$ and $U^- V^-$ that the function $G_\pm(r) = \pm \exp(-2\kappa_\pm r)|r - r_\mp|^p$, where $p = -\kappa_\pm/\kappa_\mp = r_\mp/r_\pm$, has the required properties.   □

The horizons in $\mathcal{D}(r_\pm)$ can be seen to live up to expectations. The two hypersurfaces $U^\pm = 0$ and $V^\pm = 0$ are the long horizons, each consisting of a pair of short horizons joined at the crossing sphere.

Finally let us check that KBL coordinate extension actually does merge pairs of (incomplete) restphotons into (complete) coordinate curves. For the r_+ case, the discussion preceding Lemma 3.3.1 reduces the question to just four standard principal null geodesics.

Lemma 3.4.10   *The two standard restphotons in the long horizon $U^+ = 0$ of $\mathcal{D}(r_+)$ extend through the crossing sphere to form the KBL coordinate curve*

$$U^+ = 0, \quad V^+ = s, \quad \vartheta = const, \quad \varphi^+ = const, \quad \text{for all } s \in \mathbf{R}.$$

Proof.   Consider first the horizon $H_+ : r = r_+$ in $K^* \subset \mathcal{D}(r_+)$. H_+ is described in KBL coordinates as $U^+ = 0$, $V^+ > 0$. The standard principal null geodesic γ is H_+ is case (1) of Lemma 3.3.1. Substituting the geodesic parametrization from that lemma into the definition of KBL coordinates leaves ϑ constant. For V^+, since $k_+ = (r_+ - r_-)/2 = (r^2 + a^2)\kappa_+$, the simplification is radical:

$$V^+ = \exp(\kappa_+ t^*) = \exp[\kappa_+ (r_+^2 + a^2)k_+^{-1} \ln s] = \exp \ln s = s.$$

Earlier we found

$$\varphi^+ = {}^*\varphi - \frac{a}{r_+^2 + a^2} \, {}^*t - \left(A(r) - \frac{a}{r_+^2 + a^2} \, T(r)\right).$$

But φ^+ was defined so that the first two terms cancel (substitute $\varphi^* = af(s)$ and $t^* = (r_+^2 + a^2)f(s)$ to check this). The remaining term is analytic for $r > r_-$; so for $r = r_+$ it is just some finite constant.

Since it is defined only for $s > 0$, this null geodesic γ in H_+ is past-incomplete, but as $s \to 0$ it converges to the point $\gamma(0)$ in the crossing sphere $U^+ = V^+ = 0$ with $\varphi^+ = \varphi^+(0)$, $\vartheta = \vartheta_0$. There it meets the geodesic of case (2) in Lemma 3.3.1, whose KBL coordinates are given by the same formulas but defined on $s < 0$. Thus, γ is a coordinate curve running the length of the long horizon $U^+ = 0$.   □

3.5   Maximal Slow Kerr Spacetime

In the previous section we extended the Kruskal domain $\mathcal{D}_0(r_+)$ over its crossing sphere to obtain the analytic manifold $\mathcal{D}(r_+) \approx \mathbf{R}^2 \times S^2$. The r_- case is strictly analogous once ring singularities are deleted. Now we extend the metric tensor of $\mathcal{D}_0(r_\pm)$ over the crossing sphere to make $\mathcal{D}(r_\pm)$ a Kerr spacetime. Chaining these domains in the pattern of Figure 3.10 will then give the sought-for maximal slow Kerr spacetime.

Our plan is to compute the Kerr line-element ds^2 in terms of KBL coordinates $U^\pm$, $V^\pm$, ϑ, $\varphi^\pm$ and verify that it is analytically extendible over all of $\mathcal{D}(r_\pm)$— and remains nondegenerate there. We find the line-element formula by changing coordinates in the Boyer–Lindquist version.

There is no explicit formula for the radius function r in terms of $U^\pm$, $V^\pm$, but $T(r)$ suffices.

Proposition 3.5.1   *Boyer–Lindquist coordinates and KBL cooordinates in $\mathcal{D}(r_\pm)$ are related as follows: The function ϑ is the same for both, and*

$$|U^\pm| = \exp\!\big(\kappa_\pm(T(r) - t)\big), \qquad |V^\pm| = \exp\!\big(\kappa_\pm(T(r) + t)\big),$$
$$\varphi^\pm = \varphi - at/(r_\pm^2 + a^2).$$

Reciprocally,

$$T(r) = \frac{1}{2\kappa_\pm}\ln|U^\pm V^\pm|, \qquad t = \frac{1}{2\kappa_\pm}\ln|V^\pm/U^\pm|,$$

and

$$\varphi = \varphi^\pm + \frac{a}{r_+^2 + a^2}\,t = \varphi^\pm + \frac{a}{r_\pm - r_\mp}\ln|V^\pm/U^\pm|.$$

Proof.   The first three equations appear promptly when the Boyer–Lindquist expressions for t^*, $^*t$ and φ^*, $^*\varphi$ are substituted into Definition 3.4.5 for $U^\pm$, $V^\pm$ and Definition 3.4.6 for $\varphi^\pm$. After taking logarithms, we readily solve the first two equations for $T(r)$ and t. Then substituting for t gives φ.   □

The primary purpose of the KBL line-element is to extend the Kerr metric over the crossing spheres. Accordingly, its computation is a game in which we try to rid denominators of the threatening terms, notably $r - r_\pm$, that are zero on $\mathcal{D}(r_\pm)$. To do so, the nonthreatening function $G_\pm(r)$ from Lemma 3.4.9 will prove useful.

Corollary 3.5.2 *The differentials of the Boyer–Lindquist coordinate functions are expressed in terms of KBL coordinates as*

$$dr = \frac{(r - r_\pm)G_\pm}{2\kappa_\pm(r^2 + a^2)}\,(U^\pm dV^\pm + V^\pm dU^\pm),$$

$$dt = \frac{G_\pm}{2\kappa_\pm(r - r_\pm)}\,(U^\pm dV^\pm - V^\pm dU^\pm),$$

$$d\varphi = d\varphi^\pm + \frac{aG_\pm}{(r - r_\pm)(r_\pm - r_\mp)}\,(U^\pm dV^\pm - V^\pm dU^\pm),$$

and, of course, $d\vartheta = d\vartheta$.

Proof. Since the differential of $\ln|f|$ is just df/f the differentials of t and φ are easily found, using the function $G_\pm$ from Lemma 3.4.9. For dr, note first that $dT = (dT/dr)dr = \Delta^{-1}(r^2 + a^2)\,dr$; hence $dr = \Delta(r^2 + a^2)^{-1}dT$. But the formula in Lemma 3.5.1 produces

$$dt = \frac{1}{2\kappa_\pm U^\pm V^\pm}\,d(U^\pm V^\pm) = \frac{G_\pm}{2\kappa_\pm(r - r_\pm)}\,(U^\pm dV^\pm + V^\pm dU^\pm).$$

$\square$

We are now ready to find the KBL line-element. The two cases $r_\pm$ are essentially the same, so for clarity consider the r_+ case. The start is the Boyer–Lindquist line-element in the orthogonal version in Corollary 2.6.3, where ds^2 is

$$\underset{(0)}{-\rho^{-2}\Delta[dt - aS^2 d\varphi]^2} + \underset{(1)}{\Delta^{-1}\rho^2 dr^2} + \underset{(2)}{\rho^2 d\vartheta^2} + \underset{(3)}{\rho^{-2}S^2\big[(r^2 + a^2)d\varphi - a\,dt\big]^2}.$$

Term (2), at least, presents no problems. To change the others to KBL coordinates, we first find the two square-bracketed one-forms. By the preceding corollary

$$(r^2 + a^2)\,d\varphi - a\,dt = (r^2 + a^2)\,d\varphi^+ + \frac{a(r + r_+)G_+}{r_+ - r_-}\,(U^+ dV^+ - V^+ U^+),$$

$$dt - aS^2 d\varphi = \frac{\rho_+{}^2 G_+}{(r - r_+)(r_+ - r_-)}\,(U^+ dV^+ - V^+ dU^+) - aS^2 d\varphi^+,$$

where $\rho_+{}^2 = r_+{}^2 + a^2C^2$. Clearly the first of these forms is extendible over $\mathcal{D}(r_+)$. Substituting it in term (3) of the line-element expresses (3) in KBL coordinates and shows that it is analytically extendible to all of $\mathcal{D}(r_+)$.

For term (1), squaring dr from Corollary 3.5.2 gives

$$\frac{\rho^2(r - r_-)G_+{}^2}{4\kappa_+{}^2(r^2 + a^2)^2(r - r_+)}\,(U^+ dV^+ + V^+ dU^+)^2.$$

The $r - r_+$ in the denominator is still a problem. Finally, for term (0), substitution from above gives

$$-\frac{1}{\rho^2}\left[\frac{(r-r_-)\rho_+^{\,4}G_+^{\,2}}{(r-r_+)(r_+-r_-)^2}(U^+dV^+ - V^+dU^+)^2\right.$$
$$\left.-\frac{2(r-r_-)\rho_+^{\,2}G_+aS^2}{r_+-r_-}(U^+dV^+ - V^+dU^+)\,d\varphi^+ + \Delta a^2S^2d\varphi^{+2}\right]$$

The last two summands are clearly extendible, and will appear in the final form of ds^2. There remain the first summand and the term (l); their sum is

$$-\frac{(r-r_-)\rho_+^{\,4}G_+^{\,2}}{\rho^2(r-r_+)(r_+-r_-)^2}(U^+dV^+ - V^+dU^+)^2$$
$$+\frac{(r-r_-)\rho^2G_+^{\,2}}{4\kappa_+^{\,2}(r^2+a^2)^2(r-r_+)}(U^+dV^+ + V^+dU^+)^2.$$

To simplify these two summands, replace $(r_+ - r_-)^2$ in the first by $4\kappa_+^{\,2}(r_+^{\,2}+a^2)^2$ and adjust the powers of ρ in the second so that a common factor can be extracted as follows:

$$\frac{G_+^{\,2}(r-r_-)}{4\kappa_+^{\,2}\rho^2(r-r_+)}\left[\frac{\rho^4}{(r^2+a^2)^2}(U^+dV^+ + V^+dU^+)^2\right.$$
$$\left.-\frac{\rho_+^{\,4}}{(r_+^{\,2}+a^2)^2}(U^+dV^+ - V^+dU^+)^2\right]$$

Now U^+ and V^+ (which are zero on $r = r_+$) must manage to eliminate the factor $r - r_+$ in the denominator. After the two dU^+, dV^+ forms are squared, let

$$A = \text{coefficient of } dU^+dV^+,$$
$$B = \text{coefficient of } V^{+2}dU^{+2} + U^{+2}dV^{+2}.$$

Then
$$A = \frac{G_+^{\,2}(r-r_-)}{4\kappa_+^{\,2}\rho^2(r-r_+)}2U^+V^+\left[\frac{\rho^4}{(r^2+a^2)^2}+\frac{\rho_+^{\,4}}{(r_+^{\,2}+a^2)^2}\right]$$
Since $U^+V^+/(r-r_+) = 1/G_+$, this becomes

$$A = \frac{G_+(r-r_-)}{2\kappa_+^{\,2}\rho^2}\left[\frac{\rho^4}{(r^2+a^2)^2}+\frac{\rho_+^{\,4}}{(r_+^{\,2}+a^2)^2}\right]$$

The other coefficient is

$$B = \frac{G_+^{\,2}a^2S^2}{4\kappa_+^{\,2}\rho^2}\frac{(r-r_-)(r+r_+)}{(r^2+a^2)(r_+^{\,2}+a^2)}\left[\frac{\rho^2}{r^2+a^2}+\frac{\rho_+^{\,2}}{r_+^{\,2}+a^2}\right].$$

A preliminary collection of all these contributions is in order. On the Boyer–Lindquist blocks of $\mathcal{D}(r_+)$ the Kerr line-element ds^2 is given in KBL coordinates as

$$
\frac{G_+^{\,2} a^2 S^2}{4\kappa_+^{\,2} \rho^2} \frac{(r - r_-)(r + r_+)}{(r^2 + a^2)(r_+^{\,2} + a^2)} \left[\frac{\rho^2}{r^2 + a^2} + \frac{\rho_+^{\,2}}{r_+^{\,2} + a^2} \right] (U^{+2} dV^{+2} + V^{+2} dU^{+2})
$$

$$
+ \frac{G_+(r - r_-)}{2\kappa_+^{\,2} \rho^2} \left[\frac{\rho^4}{(r^2 + a^2)^2} + \frac{\rho_+^{\,4}}{(r_+^{\,2} + a^2)^2} \right] dU^+ dV^+ + \rho^2 d\vartheta^2
$$

$$
- \frac{aS^2}{\rho^2} \left[\Delta a S^2 d\varphi^+ + \frac{G_+(r - r_-)}{\kappa_+} \frac{\rho_+^{\,2}}{r_+^{\,2} + a^2} (U^+ dV^+ - V^+ dU^+) \right] d\varphi^+
$$

$$
+ \frac{S^2}{\rho^2} \left[(r^2 + a^2) d\varphi^+ - \frac{aG_+}{2\kappa_+} \frac{r + r_+}{r_+^{\,2} + a^2} (U^+ dV^+ - V^+ dU^+) \right]^2 .
$$

For $\mathcal{D}(r_-)$ the same formula holds with + and − reversed in all sub- and super-scripts.

This formula can be improved by expanding its last two lines. Both these lines contribute to the coefficient of $(U^+ dV^+ - V^+ dU^+)\, d\varphi^+$, which is

$$
- \frac{G_+ a^2 S^2}{\kappa_+ \rho^2 (r_+^{\,2} + a^2)} \left[\rho_+^{\,2}(r - r_-) + (r^2 + a^2)(r + r_+) \right].
$$

In the last line only, $(U^+ dV^+ - V^+ dU^+)^2$ has coefficient

$$
\frac{G_+^{\,2} a^2 S^2}{4\kappa_+^{\,2} \rho^2} \frac{(r + r_+)^2}{(r_+^{\,2} + a^2)} .
$$

The remaining terms from these two lines give

$$
\frac{S^2}{\rho^2} \left[-a^2 S^2 \Delta + (r^2 + a^2)^2 \right] d\varphi^{+2} .
$$

Substituting $\Delta = r^2 - 2Mr + a^2$ here shows that the term in square bracket is nothing but the Boyer–Lindquist metric component $g_{\varphi\varphi}$. These changes give the final result.

Proposition 3.5.3 *The Kerr line-element ds^2 on the Boyer–Lindquist blocks of $\mathcal{D}(r_+)$ is given in KBL coordinates as*

$$
\frac{G_+^{\,2} a^2 S^2}{4\kappa_+^{\,2} \rho^2} \frac{(r - r_-)(r + r_+)}{(r^2 + a^2)(r_+^{\,2} + a^2)} \left[\frac{\rho^2}{r^2 + a^2} + \frac{\rho_+^{\,2}}{r_+^{\,2} + a^2} \right] (U^{+2} dV^{+2} + V^{+2} dU^{+2})
$$

$$+\frac{G_+(r-r_-)}{2\kappa_+^2\rho^2}\left[\frac{\rho^4}{(r^2+a^2)^2}+\frac{\rho_+^4}{(r_+^2+a^2)}\right]dU^+dV^+$$

$$+\frac{G_+^2a^2S^2}{4\kappa_+^2\rho^2}\frac{(r+r_+)^2}{(r_+^2+a^2)}(U^+dV^+-V^+dU^+)^2$$

$$-\frac{G_+aS^2}{\kappa_+\rho^2(r_+^2+a^2)}\left[\rho_+^2(r-r_-)+(r^2+a^2)(r+r_+)\right](U^+dV^+-V^+dU^+)d\varphi^{+2}$$

$$+\rho^2d\vartheta^2+g_{\varphi\varphi}d\varphi^{+2}.$$

For $\mathcal{D}(r_-)$ the same formula holds with the signs $+$ and $-$ reversed in all sub- and superscripts.

At first sight this line-element may seem hopelessly complicated. But it is the formulas for its coefficients that are complicated; the *form* of ds^2—its dependence on the coordinate differentials—is quite reasonable.

Corollary 3.5.4 *The Kerr metric on Boyer–Lindquist blocks of $\mathcal{D}(r_\pm)$ has a unique analytic extension to all of $\mathcal{D}(r_\pm)$.*

Proof. Inspection of the KBL line-element shows at once that it is analytically extendible over $\mathcal{D}(r_+)-$ (axes). Then, as before, Lemma 2.2.2 supplies the analytic extension of ds^2 over the axes.

We must check nondegeneracy on the new territory, the part of $\mathcal{D}(r_+)$ not in any Boyer–Lindquist block. The density of blocks implies that ds^2 must agree with the (nondegenerate) Kerr metric already established on $\mathcal{D}_0(r_+)$, including axes. Thus, we need only examine ds^2 at crossing spheres. Setting $U^+ = 0$, $V^+ = 0$ (and hence $r = r_+$) in Proposition 3.5.5 reduces it to

$$ds^2|_0 = \frac{G_+\rho_+^2(r_+-r_-)}{\kappa_+^2(r_+^2+a^2)^2}dU^+dV^+ + \rho_+^2d\vartheta^2 + g_{\varphi\varphi}d\varphi^{+2},$$

where $g_{\varphi\varphi}$ is also evaluated at $U^+ = V^+ = 0$, that is, at $r = r_+$. The fact that the three coefficients are all nonzero suffices to prove nondegeneracy—except at the poles, where a final application of Lemma 2.2.2 completes the proof.   □

Setting $dU^+ = 0$, $dV^+ = 0$ in the formula above for $ds^2|_0$ gives the metric of the crossing sphere $S(r_+)$ as a submanifold of M_s. In particular, $S(r_+)$ is Riemannian. Furthermore, it is totally geodesic since it is the intersection of the two (totally geodesic) long horizons. The crossing spheres are certainly one of the most remarkable features of Kerr geometry. In spite of their critical location in the slow Kerr black hole they have not been easy to find, perhaps because we started in the distant regions of block I.

Before extending the metric on $\mathcal{D}(r_+)$ to all of M_s we relate $\mathcal{D}(r_+)$ to Schwarzschild spacetime.

Remark 3.5.5 (Schwarzschild Comparison.) By setting $a = 0$ in the previous work we can recover the Kruskal maximal extension of Schwarzschild spacetime. The initial extension of the Kerr metric from $r > 0$ to all values of r is impossible in the Schwarzschild case since its curvature formula $k = 48M^2/r^6$ (Section 2.7) shows that $r = 0$ is a curvature singularity. Thus only the Kruskal domain $\mathcal{D}(r_+)$ is relevant. Its Schwarschild limit as $a \to 0$, hence $r_- \to 0$ and $r_+ \to 2M$, is shown in Figure 3.14.

To find the metric on $\mathcal{D}$ we set $a = 0$ in the formulas of Proposition 3.5.3. This immediately eliminates all but three terms, of the form

$$d\sigma^2 = ds^2|_{a=0} = FdU^+ dV^+ + r^2 d\vartheta^2 + g_{\varphi\varphi} d(\varphi+)^2.$$

For $a = 0$, the square bracketed term in the expression for F becomes just 2, and Remark 3.4.1 shows that $\kappa_+ = 1/(4M)$ and $\kappa_+/\kappa_- = -r_-/r_+ = 0$. Hence $F = 16M^2 G_+/r$. Then Lemma 3.4.9 and its proof give $G_+ = (r - 2M)/(U^+ V^+) = \exp(-r/2M)$. Thus, $F = 16M^2 \exp(-r/2M)/r$. For $a = 0$, $g_{\varphi\varphi} = r^2 S^2$, and Proposition 3.5.1 shows that $\varphi^+ = \varphi$. Writing $u = U^+$, $v = V^+$, we find

$$d\sigma^2 = 16M^2 \exp(-r/2M)/r \, du \, dv + r^2(d\vartheta^2 + S^2 d\varphi^2),$$

one form of the Kruskal metric (Hawking, Ellis, 1973).

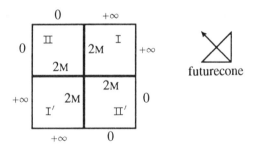

FIGURE 3.14.   The Kruskal maximal extension of Schwarzschild spacetime. Two long horizons $r = 2M$ intersect in a single crossing sphere. Any particle that falls from the exterior, I, through the horizon is destroyed at the singularity $r = 0$. Block II' is the distant past of I.

Returning to the construction of the maximal spacetime M_s; except for the relabeling of quadrants—and the ring singularity in each block III—the $\mathcal{D}(r_-)$ case is the same as $\mathcal{D}(r_+)$. Guided by Figure 3.10 we now chain the spacetimes $\mathcal{D}(r_\pm)$ together to attain M_s. As in the alternative description of the extreme case, a single gluing suffices, as follows:

Let M_+ be the disjoint union of a sequence, $-\infty < i < +\infty$, of copies $\mathcal{D}(r_+)_i$ of the spacetime $\mathcal{D}(r_+)$. M_- is defined analogously. Let $\mathcal{U}_\pm$ be the union of all Boyer–Lindquist blocks II and II$'$ in $M_\pm$. For each index i there is a canonical isometry μ_i from block II$_i$ of M_+ to block II$_i$ of M_- (the identity map is its expression in terms of Boyer–Lindquist coordinates). Similarly, μ_i' is the canonical isometry from II$_i'$ of M_+ to II$_{(i-1)}'$ of M_-. These constitute a single isometry $\mu : \mathcal{U}_+ \to \mathcal{U}_-$. Then *maximal slow Kerr spacetime* M_s is the spacetime $M_s = M_+ \cup_\mu M_-$ obtained by gluing M_+ and M_-, using μ as the matching map (see Figure 3.15.)

Clearly M_s is a spacetime; to show it is a Kerr spacetime, we need a preliminary result.

Lemma 3.5.6 *The Boyer–Lindquist coordinate vector fields are expressed (off axis) in terms of KBL cooordinate vector fields on $\mathcal{D}(r_+)$ by*

$$\partial_t = \kappa_+\big[-U^+\partial_{U^+} + V^+\partial_{V^+}\big] - \frac{a}{r_+^2 + a^2}\,\partial_{\varphi^+}$$

$$\partial_t = \kappa_+\frac{r^2 + a^2}{\Delta}\big[U^+\partial_{U^+} + V^+\partial_{V^+}\big],$$

$$\partial_\vartheta = \partial_\vartheta,\quad \partial_\varphi = \partial_{\varphi^+}.$$

(On the axis, set $\partial_{\varphi^+} = 0$.)

Proof. According to the Basis Theorem (Section 1.1), the coefficients of ∂_t, for example, are $\partial U^+/\partial t$, $\partial V^+/\partial t$, $\partial \vartheta/\partial t$, $\partial \varphi^+/\partial t$. Using Definition 3.4.5, these are readily found to be $U^+(-\kappa_+)$, $V^+(\kappa_+)$, 0, $-a/(r^2 + a^2)$, respectively. This gives ∂_t, and the other cases are similar.   □

Corollary 3.5.7 *M_s is a Kerr spacetime.*

Proof. As in the corresponding proof for M_e, most of the requirements in Definition 2.3.1 are easily met. Canonically imbedded Boyer–Lindquist blocks are dense by construction. The matching maps μ are identity maps in Boyer–Lindquist coordinates, so r and ϑ are preserved, and μ commutes with the equatorial isometry ϵ. Thus r, ϑ, and ϵ are globally defined on M_s. Also $d\mu$ preserves ∂_φ and ∂_t; hence, in particular, ∂_{φ^+} is a extension of ∂_φ. The formula for ∂_t in the preceding lemma

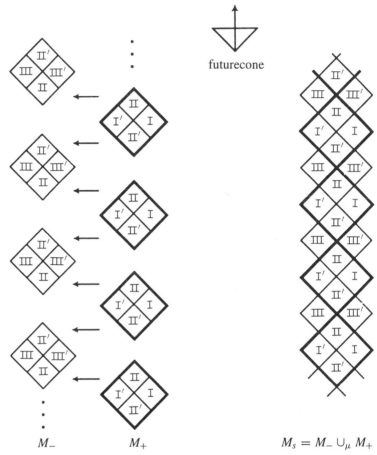

FIGURE 3.15. $M_+ = \cup\{\mathcal{D}(r_+)_i\}$, and $M_- = \cup\{\mathcal{D}(r_-)_i\}$, are glued to produce maximal slow Kerr spacetime M_s. The matching map μ consists of copies of the natural isometry $\mu\colon \mathrm{II} \to \mathrm{II}$.

provides an extension $\tilde{\partial}_t$ over all of $\mathcal{D}(r_+)$, axes included. By continuity, $\tilde{\partial}_t$ is still a Killing vector field.   □

Although the simplest definition of M_s is in terms of the domains $\mathcal{D}(r_\pm)$, the pattern of Kerr patches in M_s remains important. As Figure 3.10 illustrates, every Boyer–Lindquist block is in exactly two Kerr patches (one K^* and one $^*K$), and every short horizon is in exactly one (either K^* or $^*K$). Only crossing spheres do not meet a Kerr patch.

The Killing vector field $\tilde{\partial}_t$ is never zero on either M_f or M_e. However, the formula for ∂_t in Lemma 3.5.6 shows that in M_s it vanishes exactly at the poles of crossing spheres. Thus, there is no hope of factoring out a t-axis from M_s.

Remark 3.5.8 The vector fields $\tilde{\partial}_\varphi$ and $\tilde{\partial}_t$ are complete, that is, their integral curves are defined on the entire real line. This is clear for $\tilde{\partial}_\varphi$ since its integral curves are periodic (or constant). For $\tilde{\partial}_t$ the formula in Lemma 3.5.6 can readily be integrated explicitly. Off the axes, we find the complete curves:

$$U^+(s) = U_0^+ e^{-\kappa_+ s}, \; V^+(s) = V_0^+ e^{\kappa_+ s}, \; \varphi^+(s) = \varphi_0^+ - as/(r^2+a^2), \; \vartheta(s) = \vartheta_0.$$

The vector field $\tilde{\partial}_t$ is tangent to the axis, and there its integral curves are given by $U^+(s)$ and $V^+(s)$ as above. (Note the constant integral curves in the axis at $U_0^+ = V_0^+ = 0$, that is, at the poles of $S(r_+)$.)

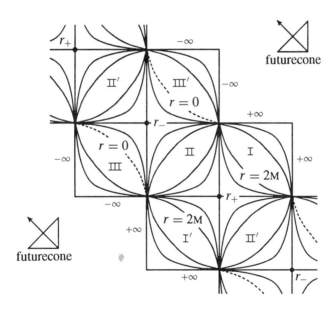

FIGURE 3.16. Level hypersurfaces of the radius function r on slow Kerr spacetime. By definition $r > r_+$ on I, $r_- < r < r_+$ on II, $r < r_-$ on III, with $0 < r_- < r_+ < 2M$. The only critical points of r are at the crossing spheres.

Although the function r cannot be a coordinate function at points of the crossing spheres, it retains its dominant role in Kerr geometry, for example, in studying geodesic orbits (Chapter 4). Figure 3.16 gives a schematic representation of the level hypersurfaces of r.

3.6   Bundle Structure of Kerr Spacetime

The provisional definition of Kerr spacetime in Chapter 2 merely described some basic features needed for the construction of the maximal Kerr spacetimes $M = M_f$, M_e, M_s. These (and variants as in Section 3.9) are the actual Kerr spacetimes, and now that we have them, we can consider what they are like. The key to finding their global structure is to begin by "leaving the ring singularities in," thereby producing, as we shall see, a very simple smooth manifold $\widehat{M}$.

In any Kerr spacetime the axis A consists of two connected components A_N and A_S that are isometric under the equatorial isometry ϵ. To avoid choices it is convenient to regard them intrinsically as a single *abstract axis* $\mathcal{A}$, identified by the isometry ϵ. Then, $\mathcal{A}$ is a connected Lorentz 2-manifold.

For example, the abstract axis of K^* appeared already in Section 2.5 as $\mathbf{R}^2(r, t^*)$, with line-element $ds^2 = g_{tt}dt^{*2} + dr\,dt^*$. For $\mathcal{A}(^*K)$, replace $(+)$ by $(-)$. The abstract axis of $\mathcal{D}(r_+)$ is also $\mathbf{R}^2$, with its natural coordinates now denoted by U^+, V^+ and line-element derived from Proposition 3.5.3.

The abstract axis of a maximal Kerr spacetime M can be defined—independently of M—by the same pattern of gluing used to construct M. This is not surprising, since the matching maps μ used in the gluing all preserve $C = \cos\vartheta$, and hence preserve axes. Furthermore, such μ all commute with the equatorial isometry ϵ, by which the two axis components are identified. Thus, restricting μ to axes gives the corresponding matching map m for the abstract axis. Of course, M_f requires no gluing; its abstract axis is just $\mathcal{A}(K^*) \approx \mathcal{A}(^*K) \approx \mathbf{R}^2$.

Recall that the ring singularity was excluded from block Ⅲ only because the Kerr metric fails at $r = 0$, $\vartheta = \pi/2$; the manifold structure remains valid there. Failing to remove Σ from $K^* \subset \mathbf{R}^2 \times S^2$ and $^*K \subset \mathbf{R}^2 \times S^2$ leaves $\mathbf{R}^2 \times S^2$ in both cases. Again, the matching maps used in the construction of M_e and M_s preserve r and ϑ, hence preserve the set Σ. Thus, M is enlarged to a smooth manifold $\widehat{M} = \widehat{M}_f$, $\widehat{M}_e$, $\widehat{M}_s$ in which ring singularities (one in each block Ⅲ) remain—though of course without metric tensors. $\widehat{M}$ is called the *smooth enlargement* of M. Evidently, M is a dense open submanifold of $\widehat{M}$.

We show first that $\widehat{M}$ can be expressed as a 2-sphere bundle over the abstract axis $\mathcal{A}(M)$ of M. Intuitively, the projection map π of this bundle sends each point

$p \in M$ to the single point $\pi(p) \in \mathcal{A}(M)$ formed by the identification of the two points at which the full axis meets the sphere $r = r(p)$ in the t-slice $t = t(p)$ (see Figure 2.2). Formally, the bundle structure derives from the following variant of a standard construction method (Steenrod 1951). Fix the following notation:

1. G is a Lie group operating effectively on a smooth manifold F (Section 1.1).
2. $M = B \times F$ and $M' = B' \times F$ are smooth product manifolds, with π and π' their natural projections onto B and B', respectively.

We use the gluing machinery from Section 1.4.

Proposition 3.6.1 *With previously established notation, let* (B, B', U, U', m) *and* $(M, M', U \times F, U' \times F, \mu)$ *be Hausdorff gluing data such that there is a smooth map* $\gamma \colon U \to G$ *for which*

$$\mu(b, y) = (m(b), \gamma(b)y) \quad \text{for all } b \in U, \ y \in F.$$

Then there exists a unique fiber bundle with projection

$$\hat{\pi} \colon M \cup_\mu M' \to B \cup_m B',$$

with group G operating on fiber F as above, and whose bundle coordinate systems are the natural injections $j \colon M \to M \cup_\mu M'$ and $j' \colon M' \to M \cup_\mu M'$.

Proof. The projection $\hat{\pi}$ is produced from π and π' by the Mapping Lemma (Section 1.4) since $m \circ \pi = \pi' \circ \mu$ on $U \times F$. Then, if $i \colon B \to B \cup_m B'$ and $i' \colon B' \to B \cup_m B'$ are the natural injections, the lemma asserts that $\hat{\pi} \circ j = i \circ \pi$ and $\hat{\pi} \circ j' = i' \circ \pi'$.

As natural injections, j and j' are diffeomorphisms onto their respective images. Explicitly, the preceding commutativities imply

$$j \colon M = B \times F \approx \hat{\pi}^{-1}(i(B)), \quad j' \colon M' = B' \times F \approx \hat{\pi}^{-1}(i'(B')),$$

and show that $\hat{\pi}$ is uniquely determined.

It remains to verify the overlap condition for bundle coordinate systems. Since $ib = i'(m(b))$, we must show that

$$\big((j')^{-1} \circ j)\big)(b, y) = (m(b), \gamma(b)y) \quad \text{for all } b \in U, \ y \in F.$$

By gluing generalities, $j = j' \circ \mu$ on $U \times F$, and, using the hypothesis on μ,

$$j(b, y) = j'(\mu(b, y)) = j'(m(b), \gamma(b)y).$$

Applying $(j')^{-1}$ to both sides gives the desired result. $\square$

In applications of this result to Kerr spacetimes the fiber will always be the 2-sphere S^2, and the group G will be the circle group S^1 acting on S^2 by rotation R_α through angle $\alpha \epsilon S^1$ about the z-axis. For example, let us "enlarge" the first gluing in the construction M_e, namely, $K_1 = K^* \cup_\mu^* K$, shown in Figure 3.4. As noted above, the enlargement of both K^* and $^*K$ is $\mathbf{R}^2 \times S^2$, with $\mathbf{R}^2$ considered as $\mathcal{A}(K^*)$ and $\mathcal{A}(^*K)$, respectively. According to Lemma 3.1.3, the coordinate expression of the matching map μ: $\amalg^* \to {}^*\amalg$ is

$$r = r, \quad \vartheta = \vartheta, \quad {}^*\varphi = \varphi^* - 2A(r), \quad {}^*t = t^* - 2T(r).$$

Expressed in terms of rotations R_α, and with t^* written merely as t, this becomes

$$\mu(r, t, y) = (r, t - 2T(r), R_{-2A(r)}(y)) \quad \text{for all } (r, t) \epsilon \mathbf{R}^2, y \epsilon S^2,$$

We emphasize that μ is a smooth (indeed, analytic) map, well defined on both axes and ring singularity. The corresponding matching map m for the abstract axes of K^* and $^*K$ is just the restriction of μ to $A_N \approx A_S$, given explicitly by $m(r, t) = (r, t - 2T(r))$. Now μ has the form required by the proposition, so $\hat{K}_1$ is a well-defined two-sphere bundle over $\mathcal{A}(K_1)$.

In the enlargement $\mathbf{R}^2 \times S^2$ of K^* and $^*K$, the ring singularity consists of all points $(0, t, y)$ for which y is on the equator $\vartheta = \pi/2$ of S^2. Axial rotation preserves this equator, showing again that μ preserves ring singularities; hence Σ is a well-defined, two-dimensional submanifold of $\hat{K}_1$.

In the zipper definition of M_e as $M^* \cup_\mu {}^*M$ described in Section 3.2, the gluing map μ is the union of infinitely many disjoint gluings, half using μ: $\amalg^* \to {}^*\amalg$ as for K_1 above, and half analogously using μ: $\mathrm{I}^* \to {}^*\mathrm{I}$. Since M^* and $^*M$ are product manifolds, the proposition applies, presenting M_e as a 2-sphere bundle over $\mathcal{A}(M_e)$.

The same plan works for slow Kerr spacetime M_s, with K^* and $^*K$ replaced by $\mathcal{D}(r_+)$ and $\mathcal{D}(r_-)$, respectively. The single-gluing definition of M_s, illustrated in Figure 3.15, also satisfies the hypotheses of Proposition 3.6.1 giving this result.

Corollary 3.6.2 *The smooth enlargement $\widehat{M}_s$ of slow Kerr spacetime is a 2-sphere bundle over the abstract axis $\mathcal{A}(M_s)$, with bundle coordinate systems given by the canonical imbeddings $j_\pm: \mathcal{D}(r_\pm) \to M_s$. The group of the bundle is the circle group S^1 acting as axial rotations of $S^2 \subset \mathbf{R}^3$.*

This bundle structure clarifies many earlier figures that give *schematic* descriptions of Kerr spacetimes (for example, Figure 3.7 for M_e and 3.10 for M_s). These figures for M can now be recognized as *literal* descriptions of the abstract axis $\mathcal{A}(M)$ of M. The full spacetime M can now be pictured (Figure 3.17) as locating a 2-sphere over each point of $\mathcal{A}(M)$ but with ring singularities removed by excising the equator $\vartheta = \pi/2$ from all spheres over points $r = 0$ in $\mathcal{A}(M)$.

Our goal now is to show that $\widehat{M}$ is, in fact, a product manifold.

Lemma 3.6.3 *The abstract axes of M_f, M_e, and M_s are diffeomorphic to $\mathbf{R}^2$.*

Proof. Take $\mathcal{A} = \mathcal{A}(M_e)$. The abstract axes of K^* and $^*K$ are diffeomorphic to $\mathbf{R}^2$, and in the first step of the inductive construction of M_e, they were glued along open sets diffeomorphic to $\mathbf{R}^2$ to produce $\mathcal{A}(K_1)$ (see Figure 3.4). By the Seifert–van Kampen theorem (Massey 1987), $\mathcal{A}(K_1)$ is simply connected. This argument can be repeated—alternately past and future—to show that $\mathcal{A}(K(n))$ is simply connected, where $K(n)$ is the result of n future-directed and n past-directed gluings.

The sets $\mathcal{A}(K(n))$ for all $n \geq 1$ form an increasing open covering of $\mathcal{A} = \mathcal{A}(M_e)$. It follows that $\mathcal{A}$ is simply connected because any continuous loop in M_e is contained in some $\mathcal{A}(K(n))$, hence is homotopic to a constant. Then the Riemann mapping theorem asserts that $\mathcal{A}$ is diffeomorphic to either $\mathbf{R}^2$ or S^2. Since $\mathcal{A}$ is not compact it is not S^2.

The proof in the slow case replaces with Kerr patches by Kruskal domains. □

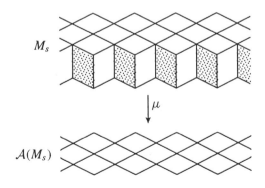

FIGURE 3.17.   Slow Kerr spacetime M_s as a fiber bundle over its abstract axis $\mathcal{A}(M_s)$. The fiber 2-spheres are rendered as intervals.

A general fact about fiber bundles $\pi\colon M \to B$ (Corollary 11.6 in Steenrod 1951) asserts that if the base B is contractible, then the bundle is *trivial*, that is, there is a fiber-preserving diffeomorphism $M \approx B \times F$. By the preceding lemma the Kerr axes $\mathcal{A} \approx \mathbf{R}^2$ are contractible, hence $\widehat{M}_f$, $\widehat{M}_e$, and $\widehat{M}_s$ are diffeomorphic to $\mathbf{R}^2 \times S^2$. To draw some consequences of this fact we recall that a smooth manifold N is *parallelizable* provided that its tangent bundle is trivial, $TN \approx N \times \mathbf{R}^n$, or equivalently, that N admits a globally defined frame field. For example, $\mathbf{R}^n$ is parallelizable but even-dimensional spheres are not.

Corollary 3.6.4   *The maximal Kerr spacetimes M_f, M_e, M_s are parallelizable, hence orientable.*

Proof.   Let M be any of the three manifolds. Orientability is a consequence of parallelizability but is clear *a priori* since M is an open subset of the orientable manifold $\mathbf{R}^2 \times S^2$. (We did not include orientability in the definition of spacetime.) An orientable spacetime, being time-orientable by definition, is also space-orientable.

To prove parallelizability, note that the map $\mathbf{R}^1 \times S^2 \to \mathbf{R}^3 - 0$ given by $(t, v) \to e^t v$ (scalar multiplication) is a diffeomorphism. Also, an open submanifold of a parallelizable manifold is clearly parallelizable. Then, writing "$<$" for "is an open submanifold of," we have $M < \mathbf{R}^4$ since

$$M < \mathbf{R}^2 \times S^2 = \mathbf{R}^1 \times (\mathbf{R}^1 \times S^2) \approx \mathbf{R}^1 \times (\mathbf{R}^3 - 0) < \mathbf{R}^4.$$

$\square$

Now we examine the consequences of the bundle structure of $\widehat{M}$ on some submanifolds of M.

THE FIBER SPHERES

Because the projection $\pi\colon M \to \mathcal{A}(M)$ is invariantly defined, so are the *fiber spheres* $\pi^{-1}(b)$, for $b \in \mathcal{A}(M) \approx A_N \approx A_S$. On $\pi^{-1}(b)$ the line-element of M restricts—poles excepted—to

$$ds^2 = \rho^2 d\vartheta^2 + g_{\varphi\varphi} d\varphi^2, \quad \text{where, as usual, } g_{\varphi\varphi} = (r^2 + a^2 + 2\mathrm{M}ra^2 S^2/\rho^2)S^2.$$

At the poles, Lemma 2.2.2 shows that $\pi^{-1}(b)$ is spacelike, so the induced metric is nondegenerate on the whole sphere only if $g_{\varphi\varphi} > 0$. Thus, away from the time machine (Section 2.4) the fiber spheres are Riemannian surfaces, and, for $|r|$ large, it is natural to picture them as spatial spheres at constant distance from the black hole.

Since the intrinsic geometry of the fiber sphere $\pi^{-1}(b)$ involves only the r-coordinate of b, we write it as $S^2\langle r \rangle$ to distinguish it from an ordinary round sphere $S_r = \{x : |x| = r\}$ in $\mathbf{R}^3$. Unlike S_r the geometry of $S^2\langle r \rangle$ depends on colatitude ϑ, with rotations $\varphi \to \varphi + c$ as isometries (For Schwarzschild spacetime, $a = 0$ implies $g_{\varphi\varphi} = r^2 S^2$, hence $S^2\langle r \rangle \approx S_r$.)

The area of $S^2\langle r \rangle$, given by $\iint (g_{\vartheta\vartheta} g_{\varphi\varphi})^{1/2} d\vartheta \, d\varphi$, is an elementary integral, and the Gaussian curvature of $S^2\langle r \rangle$ can be computed using Corollary 1.8.10. We consider only the most interesting case: the *horizon spheres* $S^2\langle r_\pm \rangle$, spacelike slices of the horizons. In a given horizon, all these spheres are isometric under the flow of the restphotons, which thus maintain the same relative positions as time passes. On $S^2\langle r_\pm \rangle$,

$$g_{\varphi\varphi} = (r_\pm^2 + a^2)^2 S^2 / (r_\pm^2 + a^2 C^2)$$

and

$$g_{\vartheta\vartheta} = r_\pm^2 + a^2 C^2.$$

Thus, $(g_{\vartheta\vartheta} g_{\varphi\varphi})^{1/2} = (r_\pm^2 + a^2)S$; so the horizon spheres have area $4\pi(r_\pm^2 + a^2)$, larger by a rotation term than the area $4\pi r_\pm^2$ of the ordinary sphere $S_{r_\pm}$. This area plays a key role in black hole dynamics (Chapter 9 in Hawking, Ellis 1973).

The circumference $C_\pm$ of $S^2\langle r_\pm \rangle$ measured on its equator $\vartheta = \pi/2$ is independent of the rate of rotation, hence agrees with the Schwarzschild case:

$$C_\pm = 2\pi (g_{\varphi\varphi})^{1/2} = 2\pi (r_\pm^2 + a^2)/r_\pm = 4\pi M$$

since $r_\pm^2 + a^2 = 2Mr_\pm$, on horizons.

The Gaussian curvature of $S^2\langle r_\pm \rangle$ is nontrivial.

$$K_\pm(\vartheta) = (r_\pm^2 + a^2)(r_\pm^2 - 3a^2 C^2)/(r_\pm^2 + a^2 C^2)^3.$$

Its maximum, at the equator, is $1/r_\pm^2$, the constant curvature of $S_{r_\pm}$. The Gauss–Bonnet theorem asserts that both $S^2\langle r_\pm \rangle$ and $S_{r_\pm}$ have total curvature $\iint K \, dA = 4\pi$, so the larger area of $S^2\langle r_\pm \rangle$ balances its smaller curvature.

By contrast, recall from Section 2.3 that the central sphere $S^2\langle 0 \rangle$ in the throat $r = 0$ actually consists of two flat disks of radius a, so $S^2\langle 0 \rangle$ has area $2\pi a^2$. (Given that $S^2\langle 0 \rangle$ is flat, the Gauss–Bonnet theorem shows that it cannot be an entire sphere and hence must "contain" singularities.)

In slow Kerr spacetime M_s the crossing spheres $S(r_\pm)$ are horizon spheres and share their intrinsic properties. But *extrinsically*, that is, in their relation to the geometry of M_s, they have two unique properties.

Lemma 3.6.5   *The crossing spheres $S\langle r_\pm\rangle$ in slow Kerr spacetime M_s are* (1) *the only integral manifolds of $\Pi^\perp$, the distribution orthogonal to the principal planes Π, and* (2) *the only fiber spheres that are totally geodesic.*

Proof. (1) Take $S = S\langle r_+\rangle$ for definiteness. The tangent space $T_p(S)$ is spanned by ∂_ϑ and ∂_{φ^+} (except at the poles). A formula in proof of Corollary 3.5.4 shows that on S these two vectors are orthogonal to ∂_{U^+} and ∂_{V^+}. Hence $T_p(S)^\perp$ is spanned by ∂_{U^+} and ∂_{V^+}. But by construction the restphotons in the long horizons of $\mathcal{D}(r_+)$ are just the KBL U^+- and V^+-coordinate curves (see Corollary 3.4.10). Since restphotons are principal null geodesics, $\mathrm{span}\{\partial_{U^+}, \partial_{V^+}\} = \Pi$, hence $T_p(S) = \Pi^\perp$. However, as noted following Lemma 2.5.7, $\Pi^\perp$ has no integral surfaces in K^* (nor in $^*K$).

(2) Crossing spheres are totally geodesic since they are intersections of (totally geodesic) horizons. Let $S^2\langle r_\pm\rangle$ be any other horizon sphere, that is, one in K^* or $^*K$. A necessary condition for it to be totally geodesic is that, on it, covariant derivatives involving only ∂_ϑ and ∂_φ are tangent to it. In particular the Christoffel symbol $\Gamma^r_{\vartheta\varphi}$ must vanish on $S^2\langle r_\pm\rangle$. The classical formula for the Christoffel symbols in Section 1.3 gives $\Gamma^r_{\vartheta\varphi} = 2Ma^3 r\rho^{-4}CS^3$ for Boyer–Lindquist coordinates—and this formula remains valid for K^* and $^*K$ coordinates since their coordinate vector fields ∂_ϑ, ∂_φ, ∂_t are the same as the Boyer–Lindquist ones. Thus, if $S^2\langle r\rangle$ is totally geodesic, then $r = 0$. But a simple computation in the proof of Lemma 4.15.7 shows that central spheres $S^2\langle 0\rangle$, are not totally geodesic. $\square$

THE EQUATORIAL PLANE

The equator Eq of M_e or M_s could, like the axes, be constructed by gluing the equators of K^*, $^*K$ or $\mathcal{D}(r_\pm)$ in the same patterns as for M itself. But Eq can be derived from the bundle structure of $\widehat{M}$ simply by replacing each fiber sphere $S^2\langle b\rangle$ in M by its equatorial circle. The result, $Eq\,\widehat{}$ (with ring singularities retained), is diffeomorphic to $\mathbf{R}^2 \times S^1$. The fiber is now the group S^1, acting as usual on S^1, hence $Eq\,\widehat{}$ is the *associated principal bundle* of the bundle of $\widehat{M}$.

The bundle structure makes it easy to describe the connected components of $Eq = Eq\,\widehat{} - \Sigma$. Removing the ring singularity removes the entire fiber over the set $r = 0$ in the abstract axis, thereby disconnecting Eq. Thus, viewing Figure 3.10

as a description of $\mathcal{A}(M_s)$, it is clear that the $r > 0$ side of Σ is connected, but the $r < 0$ side consists of infinitely many components, each isometric to the equator in the $r < 0$ side of Boyer–Lindquist block Ⅲ.

POLAR PLANES

No geometric feature of slow Kerr spacetime profits more from the construction of the crossing spheres than do the polar planes. In the nearly Minkowskian regions $|r| \gg 1$ of blocks I and Ⅲ, we can expect that polar planes, $\varphi = \varphi_0$, $t = t_0$, resemble ordinary Euclidean planes through the axis of rotation. But what happens for $|r|$ small is not clear. Figure 2.2 seems to suggest that a single polar plane cuts through all three Boyer–Lindquist blocks and that the separate pieces meet at the horizons. But when horizons were constructed in K^* and $^*K$ there were no such meetings; the polar planes diverged to infinity, and this is also what happens in M_e.

But in maximal slow Kerr spacetime the meetings at last take place—at the crossing spheres. Consider first the domain $\mathcal{D}(r_+)$. There the Boyer–Lindquist equations $\varphi = \varphi_0$, $t = t_0$ for a polar plane transform (using Proposition 3.5.1) to KBL equations $V^+ = \pm k_0 U^+$, $\varphi^+ = \varphi^+_0$, where $k_0 = \exp(2\kappa_+ t_0) > 0$. With positive sign, the first equation shows that the polar planes $\varphi = \varphi_0$, $t = t_0$ in blocks I and I′ join at the crossing sphere $U^+ = V^+ = 0$ to form a single "large" polar plane P_{I}. With negative sign, the polar planes $\varphi = \varphi_0$, $t = t_0$ in Ⅱ and Ⅱ′ are similarly joined. Thus, polar planes from different numbered blocks do not meet: like meets like.

These larger polar planes admit global coordinates U^+, ϑ, with ϑ now a circular coordinate (Section 1.1), and they have well defined metrics. Analogously, in $\mathcal{D}(r_-)$, pairs of polar planes in Ⅲ and Ⅲ′ join at $S\langle r \rangle$ to form a single $P_{\text{Ⅲ}}$, and pairs from Ⅱ and Ⅱ′ are again joined.

We know from Chapter 2 that the polar planes in I and Ⅲ are spacelike and those in Ⅱ are timelike. For the timelike planes the enlargement is not finished; each double polar plane in $\mathcal{D}(r_+)$ overlaps smoothly with one in $\mathcal{D}(r_-)$, and a whole chain of these forms a single polar plane $P_{\text{Ⅱ}}$ that runs through every block Ⅱ of M_s and meets every crossing sphere.

In the next section we find that the polar planes P_{I}, $P_{\text{Ⅱ}}$, $P_{\text{Ⅲ}}$ are fixed-point sets of isometries, hence they are closed, totally geodesic 2-dimensional submanifolds of M_s.

The bundle structure of $\widehat{M}$ expresses each polar plane as a circle bundle over a curve $C \approx \mathbf{R}^1$ in the abstract axis of M (see Figure 3.18). Thus, $P = P(\varphi_0, t_0)$ is

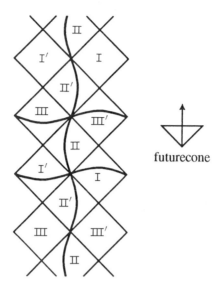

futurecone

FIGURE 3.18.   Polar planes represented by their projections in the axis of slow Kerr spacetime. The long timelike planes P_{II} run through every block II and every crossing sphere. Spacelike planes meet only two blocks and a single crossing sphere: $S(r_+)$ for P_{I}, and $S(r_-)$ for P_{III}. There is a polar plane through every nonhorizon point of M_s.

diffeomorphic to $\mathbf{R}^1 \times S^1$, where the circles $r \times S^1$, are full circles of longitude, all given by the same constant $\pm\varphi_0$.

All polar planes share two basic formulas. According to Remark 2.7.9, their Gaussian curvature is $K = K(r, \vartheta) = -\mathrm{M}r\rho^{-6}(r^2 - 3a^2C^2)$, and the circumference of the circle $r \times S^1$ is given by the elliptic integral

$$L(r) = \int\limits_0^{2\pi} \sqrt{g_{\vartheta\vartheta}}\,d\vartheta = 4 \int\limits_0^{\pi/2} \sqrt{r^2 + a^2\cos^2\vartheta}\,d\vartheta.$$

Nevertheless, the three kinds of polar plane differ sharply by causal character. The spacelike types P_{I} and P_{III} represent a spatial surface at an instant of time. When $|r|$ is large, both have $K \approx 0$ and $L(r) \approx 2\pi r$, confirming their resemblence to Euclidean planes, with $|r|$ and ϑ as ordinary polar coordinates. But for smaller $|r|$, curvature asserts itself, and we can picture P_{I} as two "parallel" near-planes—one in I, the other in I'—joined by a neck passing through $S(r_+)$. P_{III} is analogous. In each, four geodesics are picked out, by intersection with the equator at $\vartheta = \pm\pi/2$ and with the north and south axes at $\vartheta = 0, \pi$. P_{III} is somewhat more relativistic than P_I, since to go from $r = -\infty$ to $S(r_-)$ and back it must pass twice through $r = 0$, where its equatorial geodesics are severed by the ring singularity.

Timelike polar planes $P_{II} \approx C \times S^1$ are very different. Each represents a full circle of longitude that endures forever, making a grand time-tour of the black hole, as indicated in Figure 3.18. As time passes, the radius of the circle S^1 oscillates between a minimum circumference $L(r_-)$ for S^1 in $S\langle r_- \rangle$ and a maximum $L(r_+)$ for S^1 in $S\langle r_+ \rangle$. Like the other types, each P_{II} has four distinguished geodesics given by intersections with the axis and equator; they are timelike and run its entire length.

3.7 Isometries of Boyer–Lindquist Blocks

The goal of this section is to determine the isometry groups of Boyer–Lindquist blocks. This is the essential step in determining the isometry groups of the Kerr spacetimes M_e and M_s, because, just as the spacetimes are assembled from Boyer–Lindquist blocks, so their isometries are assembled from those of the blocks. (Of course, fast Kerr spacetime M_f is a single block.) Also, since isometries and Killing vector fields are each uniquely determined by their behavior in a single neighborhood, information about one block in a Kerr spacetime can yield global information.

For an arbitrary semi-Riemannian manifold M the set of all isometries $M \to M$ forms a Lie group $I(M)$ whose identity component $I_0(M)$ is a closed normal subgroup. $I_0(M)$ is generated by the stages ψ_s of the flows of the complete Killing vector fields on M (see Helgason 1978). The quotient group $I(M)/I_0(M)$ is a discrete group whose (left-coset) elements are the connected components of $I(M)$. $I(M)/I_0(M)$ could be called the group of *large isometries* of M, since isometries in $I(M) - I_0(M)$ are not deformable, in $I(M)$, back to the identity.

We will see that for Kerr spacetimes the isometry group $I(M)$ varies with rotation type and topological structure but that $I_0(M)$ is always the same.

On every Kerr spacetime there are, by definition, Killing vector fields $\tilde{\partial}_t$ and $\tilde{\partial}_\varphi$ that are extensions of the Boyer–Lindquist coordinate vector fields ∂_φ and ∂_t. As noted earlier, on the maximal Kerr spacetimes M the vector fields $\tilde{\partial}_t$ and $\tilde{\partial}_\varphi$ are complete; consequently, their flow isometries are globally defined. These consist of the *t-translations* τ_c generated by $\tilde{\partial}_t$, and the *φ-rotations* ρ_α generated by $\tilde{\partial}_\varphi$. (In Boyer–Lindquist and Kerr coordinates, τ_c adds c to the t-coordinate, and, except on the axis, ρ_α adds α to the circular φ-coordinate.) These isometries commute and hence generate an abelian group $K = \{\tau_c \rho_\alpha : c \epsilon \mathbf{R}^1, \alpha \epsilon S^1\}$. We call $\tau_c \rho_\alpha$ a

Killing isometry and K the *Killing group*. Evidently the map $(c, \alpha) \to \tau_c \rho_\alpha$ is a group isomorphism $\mathbf{R} \times S^1 \approx K$, where $\mathbf{R}$ is the additive group of real numbers and $S^1 = \mathbf{R}^1/(2\pi\mathbf{R})$. Every Killing isometry $\psi \epsilon K$ is in $I_0(M)$, since $\psi = \tau_c \rho_\alpha$ is the stage $s = 1$ of the flow of the Killing vector field $c\tilde{\partial}_t + \alpha\tilde{\partial}_\varphi$. This proves $K \subset I_0(M)$—and Corollary 3.7.5 will show in fact that $K = I_0(M)$.

Because the Killing vector fields $\tilde{\partial}_t$, $\tilde{\partial}_\varphi$ on M are extensions of ∂_t, ∂_φ on B (the latter also complete), the isometries $\{\tau_c \rho_\alpha : c\epsilon\mathbf{R}, \alpha\epsilon S^1\}$ of M are extensions of those of B. Thus, $K(M) \approx \{\tau_c \rho_\alpha : c\epsilon\mathbf{R}, \alpha\epsilon S^1\} \approx K(B)$. Since these extensions are so natural, we write simply K for both.

The study of Kerr isometries is based on a fundamental result from Chapter 5: *the functions r and $C^2 = \cos^2 \vartheta$, and the principal planes Π, are isometric invariants.* Since the Boyer–Lindquist metric components depend only on r and C^2, they too are preserved by isometries. This justifies the natural expectation that Kerr isometries preserve the axis, equatorial plane, and horizons.

Although $\cos^2 \vartheta$ is invariant, the function ϑ itself need not be, as the equatorial isometry ϵ illustrates. However, if an isometry ϕ does not preserve ϑ, then, since ϕ preserves $\cos^2 \vartheta$, the isometry $\epsilon\phi$ preserves ϑ.

Recall that the principal null directions at a point p are the null lines in the (timelike) principal plane Π_p at p. Consequently, an isometry preserves principal planes, $d\phi(\Pi_p) = \Pi_{\phi(p)}$, if and only if it preserves principal null directions, that is, carries the null directions at p to those at $\phi(p)$.

As in Chapter 2, we say that tangent vectors in $\text{span}\{V, W\} = \text{span}\{\partial_t, \partial_\varphi\}$ are *vertical* and those in the orthogonal subspace $\text{span}\{\partial_r, \partial_\vartheta\}$ are *horizontal*. Figure 2.3 illustrates the relation between these and $\Pi = \text{span}\{\partial_r, V\}$, $\Pi^\perp = \text{span}\{\partial_\vartheta, W\}$.

Proposition 3.7.1   *If $\phi: B \to B$ is an isometry of a Boyer–Lindquist block, then*

$$d\phi(\partial_r) = \partial_r, \quad d\phi(\partial_\vartheta) = \pm\partial_\vartheta, \quad d\phi(\partial_\varphi) = \varepsilon\partial_\varphi, \quad d\phi(\partial_t) = \varepsilon\partial_t,$$

where $\varepsilon = \pm 1$ is independent of the sign $\pm$ for ∂_ϑ.

Proof.   Because $r \circ \phi = r$ and $\vartheta \circ \phi$ is either ϑ or $\pi - \vartheta$, the general formula $d\phi(\partial_i) = \Sigma \left[\partial(x^j \circ \phi)/\partial x^i\right] \partial_j$ shows that

$$d\phi(\partial_r) = \partial_r + X, \qquad d\phi(\partial_\vartheta) = \pm\partial_\vartheta + Y,$$

where X and Y depend only ∂_φ and ∂_t, hence are vertical vector fields. Since $d\phi$ preserves scalar products and $X, Y \perp \partial_r, \partial_\vartheta$, it follows that $<X, X> = <Y, Y> = 0$.

Thus, X and Y are either 0 or null. Since $d\phi$ preserves principal planes Π, and hence also $\Pi^\perp$, we can write

$$d\phi(\partial_r) = A\partial_r + BV, \qquad d\phi(\partial_\vartheta) = F\partial_\vartheta + GW,$$

for functions A, B, F, G. Comparison with the preceding equations then yields

$$\partial_r + X = f\partial_r + FV, \qquad \pm\partial_\vartheta + Y = g\partial_\vartheta + GW.$$

In each of these equations the horizontal vectors must be equal and likewise the vertical, hence

$$f = 1, \qquad g = \pm 1, \qquad X = FV, \qquad Y = GW.$$

Since $<V, V> = -\Delta\rho^2 \neq 0$ on B, while $<X, X> = 0$ we conclude that $F = 0$. Similarly, $<W, W> = \rho^2 S^2 \neq 0$ at off-axis points, so $G = 0$. Thus, $d\phi(\partial_r) = \partial_r$ and $d\phi(\partial_\vartheta) = \partial_\vartheta$. These vector field equations imply

$$d\phi(\partial_r|_p) = \partial_r|_{\phi(p)}, \qquad d\phi(\partial_\vartheta|_p) = \partial_\vartheta|_{\phi(p)} \quad \text{for all } p.$$

This shows that $d\phi$ carries horizontal tangent planes $H = \text{span}\{\partial_r, \partial_\vartheta\}$ to horizontal planes; hence it carries vertical planes $H^\perp$ to vertical planes. Thus, ϕ carries integral surfaces of $H^\perp$ to integral surfaces of $H^\perp$. These surfaces are just the Killing orbits F (Section 2.3). Intrinsically, each F is a flat cylinder on which ∂_φ and ∂_t are parallel vector fields. In the intrinsic geometry of F the only closed geodesics are the integral curves of ∂_φ. The restriction $\phi|F : F \to F'$ is an isometry, so closed geodesics are carried to closed geodesics. Thus, $d\phi(\partial_\varphi|_p) = h(p)\partial_\varphi|_{\phi(p)}$ for all $p \epsilon F$. Since ϕ preserves Boyer–Lindquist metric components, taking scalar products gives $g_{\varphi\varphi}(p) = h^2(p)g_{\varphi\varphi}(\phi p) = h^2(p)g_{\varphi\varphi}(p)$; hence $h^2(p) = 1$ and $d\phi(\partial_\varphi) = \varepsilon\partial_\varphi$, where $\varepsilon = \pm 1$.

It remains to show that $d\phi(\partial_t) = \varepsilon\partial_t$. For definiteness, take $\varepsilon = +1$, so $d\phi(\partial_\varphi) = \partial_\varphi$. There are two possibilities for $d\phi(\partial_t)$: ∂_t and $Z = -\partial_t + 2(g_{\varphi t}/g_{\varphi\varphi})\partial_\varphi$, since Z is the only other vector such that $<Z, Z> = <\partial_t, \partial_t>$ and $<Z, \partial_\varphi> = <\partial_t, \partial_\varphi>$.

To rule out the latter possibility, note first that the canonical vector field W lies in both $H^\perp$ and $\Pi^\perp$, and these planes are preserved by $d\phi$, so $d\phi(W)$ and W are collinear. In fact, since $<W, W>$ depends only on r and C^2 it is also invariant under ϕ, so $d\phi(W) = \pm W$. But if $d\phi(\partial_t) = Z$ as above, then

$$d\phi(W) = d\phi(\partial_\varphi + aS^2\partial_t) = \partial_\varphi + aS^2Z = [1 + 2aS^2g_{\varphi t}/g_{\varphi\varphi}]\partial_\varphi - aS^2\partial_t.$$

The coefficient of ∂_φ must be -1, but using the metric identity (m1) we compute it to be

$$[g_{\varphi\varphi} + 2aS^2 g_{\varphi t}]/g_{\varphi\varphi} = [(r^2 + a^2)S^2 + aS^2 g_{\varphi t}]/g_{\varphi\varphi}$$
$$= S^2[r^2 + a^2 - 2Mra^2 S^2/\rho^2]/g_{\varphi\varphi},$$

If the sign of $r^2 + a^2$ here were reversed the result would be -1, but as things stand, the possibility $d\phi(\partial_t) = Z$ is ruled out.   □

Consider the *sign function* sg that assigns to each isometry $\phi \epsilon I(M)$ signs $(\pm 1, \pm 1)$, where the first is the ∂_ϑ sign in Proposition 3.7.1, the second the common sign ε of ∂_φ and ∂_t. For example, the equatorial isometry ϵ has sg$(\epsilon) = (-1, +1)$, and the backwards isometry β, which sends $(r, \vartheta, \varphi, t)$ to $(r, \vartheta, -\varphi, -t)$, has sg$(\beta) = (+1, -1)$.

The chain rule $d(\phi_1 \circ \phi_2) = d\phi_1 \circ d\phi_2$ implies that sg is a homomorphism from $I(M)$ to the four element multiplicative group $\{(\pm 1, \pm 1)\}$. (This group is the well-known *four-group*, written in additive notation as $Z_2 \times Z_2$.) Furthermore, since sg is a continuous function into a discrete space, it is constant on each connected component of $I(M)$. In particular, since the identity map has sg $= (+1, +1)$, so does every element of $I_0(M)$.

For a Boyer–Lindquist block B we now have the isometries $K = \{\rho_\alpha \tau_c\}$, the equatorial isometry ϵ, and the backwards isometry β. These are enough.

Corollary 3.7.2   *The isometry group $I(B)$ of a Boyer–Lindquist block B (including the case $B = M_f$, fast Kerr spacetime) is generated by ϵ, β, and Killing isometries K.*

Proof.   Let ϕ be an isometry of B. It suffices to show that composing ϕ with suitable choices of ϵ, β, and isometries in K produces the identity map of B. Using ϵ if necessary, we can assume that ϕ preserves ϑ as well as r. Hence sg$(\phi) = (+1, \pm 1)$. If the latter sign is negative, then following ϕ by β will make it positive. We have reached

$$d\phi(\partial_r) = \partial_r, \quad d\phi(\partial_\vartheta) = \partial_\vartheta, \quad d\phi(\partial_\varphi) = \partial_\varphi, \quad d\phi(\partial_t) = \partial_t.$$

Let p be an off-axis point of B. The point $\phi(p)$ has the same r and ϑ coordinates as p, hence by following ϕ by a suitable Killing isometry $\tau_c \rho_\alpha$ (if necessary) we can arrange that $\phi(p) = p$. The equations above then assert that $d\phi_p$ is the identity map of $T_p(M)$. An isometry of a connected manifold is uniquely determined by its differential map at a single point, hence $\phi = id$.   □

We now consider the algebraic relations among the isometries of B. The isometries ϵ and β are idempotent, $\epsilon^2 = id = \beta^2$, and they commute, $\epsilon\beta = \beta\epsilon$. Thus, the subgroup $Q = \{id, \epsilon, \beta, \epsilon\beta\}$ of $I(B)$ they generate is a four-group. Indeed, sg is an isomorphism of Q onto $\{(\pm 1, \pm 1)\}$.

Also ϵ commutes with the isometries $\tau_c\rho_\alpha$ since the only Boyer–Lindquist coordinate changed by ϵ is ϑ. Thus, ϵ commutes with every isometry of B.

However, β does not generally commute with these Killing isometries. If $\psi = \tau_c\rho_\alpha$, then $\psi^{-1} = \tau_{-c}\rho_{-\alpha}$, and

$$(\beta\psi\beta)(r, \vartheta, \varphi, t) = \beta\psi(r, \vartheta, -\varphi, -t) = \beta(r, \vartheta, -\varphi + \alpha, -t + c)$$
$$= (r, \vartheta, \varphi - \alpha, t - c) = \psi^{-1}(r, \vartheta, \varphi, t).$$

Thus $\beta\psi\beta = \psi^{-1}$ for all $\psi\epsilon K$.

Proposition 3.7.3 *In the isometry group $I(B)$ of a Boyer–Lindquist block B, the subgroup $I_0(B)$ is $K = \{\tau_c\rho_\alpha\} \approx S^1 \times \mathbf{R}^1$, and $I(B)$ is a semi-direct product of K and the four-group $Q = \{id, \epsilon, \beta, \epsilon\beta\}$. Explicitly,*

(1) K is a normal subgroup of $I(B)$,   (2) $I(B) = K \cdot Q$,   and (3) $K \cap Q = \{id\}$.

Furthermore, the only failure of commutativity among ϵ, β, and $\psi\epsilon K$ is $\beta\psi = \psi^{-1}\beta$.

Proof. Consider first the numbered assertions.

(1) This is clear from the algebraic relations.

(2) By Corollary 3.7.2 every isometry ϕ of B can be expressed as a word in ϵ, β, and Killing isometries. The relations just established show that this word can be reduced to $\phi = \psi q$ for some $\psi\epsilon K$ and $q\epsilon Q$.

(3) Every Killing isometry has sg $= (+1, +1)$ since the differential maps of τ_c and ρ_α preserve the Boyer–Lindquist coordinate vector fields. But sg$(\epsilon) = (-1, +1)$ and sg$(\beta) = (+1, -1)$, hence sg$(\epsilon\beta) = (-1, -1)$. Thus, if ϕ is in both K and Q, then ϕ can only be id.

To show that $I_0(B) = K$ we need only prove $I_0(B) \subset K$ since the reverse inclusion is known. Let $\phi\epsilon I_0(B)$, so sg$(\phi) = (+1, +1)$. By (2), $\phi = \psi q$ with $\psi\epsilon K$ and $q\epsilon Q$; hence

$$(+1, +1) = \text{sg}(\phi) = \text{sg}(\psi q) = sg(\psi)\text{sg}(q) = \text{sg}(q).$$

Hence $q = id$, so $\phi\epsilon K$. (This gives another proof of (1), since $I_0(M)$ is always a normal subgroup of $I(M)$.) $\square$

This result supplies a complete algebraic description of $I(B)$, showing that it is completely determined by Q, K, and automorphisms $\{C_q : q \in Q\}$ of K, where $C_q(\psi) = q\psi q^{-1}$. Since $I(B) = KQ$, this is clear from the identity

$$(\psi q)(\psi' q') = \psi q \psi' q^{-1} qq' = \psi C_q(\psi')qq'.$$

If every C_q were the identity automorphism, $I(B)$ would be the direct product $Q \times K$, since the elements of Q would commute with those of K. Here $C_\epsilon = id$, but $C_\beta(\psi) = C_\beta(\psi) = \psi^{-1}$. For example, suppose $\psi' \neq (\psi')^{-1}$ (which is always the case unless $\psi = \rho_\pi$); then

$$\psi \beta \psi' \epsilon = \psi(\psi')^{-1}\beta\epsilon \neq \psi\psi'\beta\epsilon.$$

The preceding proposition also shows that the homomorphism $q \to qK$ is an isomorphism of Q onto $I(B)/K$. Thus, for the group of connected components of $I(B)$ we have

$$I(B)/I_0(B) = I(B)/K = \{K, \epsilon K, \beta K, \epsilon\beta K\} \approx Q$$

Thus, every large (i.e., non-Killing) isometry of B differs from an element of Q only by a Killing isometry.

In the next section we need to know which isometries of B preserve orientation or time-orientation. This depends only on components of the isometry group since the choice is unaffected by a continuous deformation. Accordingly, every isometry in K preserves both orientation or time-orientation since the identity map obviously does, and every isometry in ϵK preserves time-orientation but reverses orientation since ϵ does. The isometry β preserves orientation on all blocks since it reverses the signs of two coordinates, but its effect on time-orientation is irregular. On blocks I or III, it *reverses* time-orientation since ∂_t is timelike for $|r|$ large and $d\beta(\partial_t) = -\partial_t$. But on block II where ∂_r is timelike, β preserves time-orientation, since $d\beta(\partial_r) = \partial_r$. Consequently, $\epsilon\beta$ reverses orientation and has the same effect on time-orientation as does β.

The results above for a single Boyer–Lindquist block have two important consequences for Kerr spacetimes. The first deals with Killing vector fields not assumed to be complete.

Corollary 3.7.4 *Every Killing vector field X on a Kerr spacetime M is a constant linear combination of $\tilde{\partial}_t$ and $\tilde{\partial}_\varphi$.*

Proof. Let p be an off-axis point in a Boyer–Lindquist block B of M, and let $\{\psi_s\}$ be a local flow of X defined on a neighborhood $\mathcal{U}$ of p (Section 1.1). The proof of Proposition 3.7.1 is entirely local, so its conclusion holds for each (locally defined) isometry ψ_s. In fact, since ψ_s varies continously to $\psi_0 = id$, the signs $\pm$ and ε are both positive.

Thus, in the proof of Corollary 3.7.2 the isometries ϵ and β are not needed, and we conclude that each ψ_s has the form $\rho_{f(s)}\tau_{g(s)}$ for continuous functions f and g on some interval $(-\varepsilon, \varepsilon)$. The map $S \to \psi_s$ has the homomorphism properties $\psi_0 = id$ and $\psi_{s+t} = \Psi_s \circ \psi_t$ for all $s, t, s + t$ in $(-\varepsilon, \varepsilon)$; hence the corresponding properties hold for f and g. It follows that $f(s) = cs$, $g(s) = ds$ for constants c, d.

In coordinate terms $(\rho_{cs}\tau_{ds})(r, \vartheta, \varphi, t) = (r, \vartheta, \varphi + cs, t + ds)$. Thus, at any point p the velocity of the curve $S \to \rho_{cs}\tau_{ds}(p)$ is $(c\partial_\varphi + d\partial_t)|_p$. This means that $\{\rho_{cs}\tau_{ds}\}$ is the flow of the vector field $c\partial_\varphi + d\partial_t$. Since a vector field is uniquely determined by its flow, $X = c\partial_\varphi + d\partial_t$ on the neighborhood $\mathcal{U}$. Because these are Killing vector fields, the equality holds globally. □

Corollary 3.7.5 *For any Kerr spacetime M on which the vector fields $\tilde{\partial}_\varphi$ and $\tilde{\partial}_t$ are complete, (e.g., M_e and M_s) the identity component $I_0(M)$ of $I(M)$ is the Killing group $K = \{\tau_c \rho_\alpha\}$.*

Proof. We saw above that $K \subset I_0(M)$. For the reverse inclusion, suppose ϕ is in $I_0(M_e)$; thus ϕ can be deformed through isometries back to the identity map. Fix a Boyer–Lindquist block B in M_e. Because isometries leave each Boyer–Lindquist block B invariant, $\phi|B$ is deformed to id through isometries $B \to B$, that is, $\phi|B$ is in $I_0(B) = K$. Since ϕ agrees with some Killing isometry $\rho_\alpha\tau_c$ on B, it does so also on M. Thus, $I_0(M) \subset K$. □

3.8 Isometries of M_e and M_s

On both extreme and slow Kerr spacetime, the preceding section has shown that we have at least the equatorial isometry ϵ and the group K of Killing isometries $\rho_\alpha\tau_c$. The backwards isometry of Section 3.1 is useful in constructing more isometries. When it is restricted to a single Boyer–Lindquist block B we denote it by $\beta_0: B \to B$. Recall from Section 3.1 that β_0 can be extended to an isometry $\beta: K^* \to {}^*K$, reversing the Kerr-star and star-Kerr spacetime. However, when horizons exist, β_0 cannot be extended over a horizon to produce an isometry $K^* \to K^*$ (or ${}^*K \to$

$^*K$). In fact, the original Boyer–Lindquist formula $\beta_0(r, \vartheta, \varphi, t) = (r, \vartheta, -\varphi, -t)$ gives

$$\varphi^* \to -\varphi^* + 2A(r), \qquad t \to -t + 2T(r),$$

but $A(r)$ and $T(r)$ become infinite on the horizons.

Consider first the extreme Kerr spacetime M_e. As in Section 3.7, we first find enough isometries to generate the isometry group $I(M_e)$, then organize its algebra.

Recall from Section 3.2 the zipper construction of M_e as $M^* \cup_\mu {}^*M$, where $M^* = \cup\{K^i\}$, $^*M = \cup\{^iK\}$, and the single matching map μ is built from the isometries of Lemma 3.1.3. This makes it easy to construct the intuitively natural isometry λ of M_e that moves it one step into the future. Referring to the Mapping Lemma (Section 1.4), let $\lambda^*\colon M^* \to M^*$ send each K^i to K^{i+1} by what is, but for relabeling, the identity map; similarly for $^*\lambda$. The commutativity condition in the lemma is satisfied trivially, and the desired isometry is $\lambda = \lambda^* \cup {}^*\lambda$.

Next, we use $\beta\colon K^* \to {}^*K$ to construct an isometry $\beta_e\colon M_e \to M_e$ that can be vizualized in Figure 3.7 as reflection in a horizontal line. Evidently the gluing operation satisfies the commutativity, $A \cup_\mu B \approx B \cup_{\mu-1} A$. Thus, for M_e as above, we define

$$\beta_e\colon M^* \cup_\mu {}^*M \to {}^*M \cup_{(\mu^{-1})} M^*$$

to be $\beta^* \cup {}^*\beta$, where, for all i, β^* is $\beta\colon K^i \to {}^{-i}K$ and $^*\beta$ is $\beta^{-1}\colon {}^iK \to K^{-i}$. But for indexing, the commutativity condition of the Mapping Lemma is $\mu^{-1} \circ \beta = \beta^{-1} \circ \mu$ on a single block B. This holds since in terms of Boyer–Lindquist coordinates on B, μ is the identity map and $\beta_0 = \beta_0{}^{-1}$. It should be safe now to drop the subscript e from β_e.

The isometries above suffice to generate the isometry group of M_e.

Proposition 3.8.1 *The isometry group $I(M_e)$ of extreme Kerr spacetime M_e is generated by ϵ, β, λ, and the Killing isometries $K = \{\tau_c \rho_\alpha\}$.*

Proof. We must show that an arbitrary isometry ϕ of M_e can be expressed as a word in the specified isometries. Choose a Boyer–Lindquist block B of M_e. As noted earlier, ϕ carries B to a block B' of the same type, so some power λ^n carries B' back to B. The restriction $\lambda^n \phi | B$ is a self-isometry of B, and hence by Proposition 3.7.3, it can be expressed as ψq_0, where $\psi \epsilon K$ and q_0 is in $Q = \{id, \epsilon, \beta_0, \epsilon\beta_0\}$. These isometries all have unique extensions to M_e, so $\lambda^n \phi = \psi q_0$, that is, $\phi = \lambda^{-n} \psi q_0$. $\qquad\square$

Let Q_e be the group generated by ϵ, λ, and β. Evidently the equatorial isometry ϵ commutes with λ; also, ϵ commutes with β on a single Boyer–Lindquist block, hence everywhere. It follows that Q_e is a direct sum of $\{id, \epsilon\}$ and the group D

generated by β and λ. The isometry λ generates an infinite cyclic group, and β, like ϵ, is idempotent.

However, λ and β do not commute. Applying $\beta\lambda\beta$ to the set K^0, gives $K^0 \rightarrow {}^0K \rightarrow {}^1K \rightarrow K^{-1}$. This map agrees with λ^{-1}, so $\beta\lambda\beta = \lambda^{-1}$ holds globally. Thus, D is an infinite dihedral group.

To show that Q_e is the large isometry group $I(M_e)/I_0(M_e)$ of M_e, we consider the effect of its isometries on orientation and time-orientation. Evidently λ preserves both. For the others, their effect on M_e is already determined by their effect on a single Boyer–Lindquist block (see Section 3.7). Writing $+1$ for preservation and -1 for reversal, we have

isometry	orientation	time-orientation
ϵ	-1	$+1$
β	$+1$	-1
λ, ψ	$+1$	$+1$

where $\psi \epsilon K$. Again the chain rule $d(\phi_1\phi_2) = d\phi_1 \circ d\phi_2$ shows that the function that assigns these *orientation signs* $(\pm 1, \pm 1)$ to each isometry is a homomorphism.

Proposition 3.8.2 *The isometry group $I(M_e)$ of extreme Kerr spacetime M_e is the semi-direct product of $I_0(M_e) = K \approx \mathbf{R}^1 \times S^1$ and the group Q_e generated by ϵ, β, and λ. Explicitly,*

(1) K is a normal subgroup of $I(M_e)$, (2) $I(M_e) = K \cdot Q_e$, and (3) $K \cap Q_e = \{id\}$.

Proof. Recall that Lemma 3.7.5 implies $K = I_0(M_e)$.

(1) $I_0(M_e)$ is a closed normal subgroup of $I(M_e)$.

(2) $I(M_e) = K \cdot Q_e$. The proof of Proposition 3.8.1 showed that any isometry ϕ of M_e can be expressed as $\phi = \lambda^{-n}\psi q_0$, with $q_0 \epsilon Q$ and $\psi \epsilon K$. Since λ and ψ commute, $\phi = \psi q$, where $q = \lambda^{-n} q_0 \epsilon Q_e$.

(3) $K \cap Q_e = \{id\}$. Suppose that $\psi \epsilon K$ is also in Q_e. Then $\psi = \lambda^n \epsilon^a \beta^b$, where a and b are 0 or 1, and n is an integer. Of these isometries, only ϵ reverses orientation, so $a = 0$. Only β reverses time-orientation, so $\psi = \lambda^n$. However, no power of λ except $\lambda^0 = id$ is in K since Killing isometries carry each Boyer–Lindquist block to itself. $\square$

To complete the algebraic description of $I(M_e)$ it remains only to record the automorphisms $C_q(\psi) = q\psi q^{-1}$ of K, where $q \epsilon Q$. Since $q \to C_q$ is a homomorphism, it suffices to take q successively to be ϵ, β, λ. Of these only C_β is nontrivial, with $C_\beta(\psi) = \beta\psi\beta = \psi^{-1}$.

The fixed point set Fix(ϕ) of an isometry ϕ is a closed, totally geodesic submanifold (Section 1.7). In M_e, Fix(ϵ) is the equator Eq, and λ has no fixed points. The only Boyer–Lindquist block invariant under the backwards isometry β is the selected I block, I^0, of K^0. The Boyer–Lindquist coordinate expression of $\beta|\mathrm{I}^0$ is $(r, \vartheta, \varphi, t) \to (r, \vartheta, -\varphi, -t)$, so Fix($\beta$) = $\{(r, \vartheta, 0, 0) : r > \mathrm{M}, 0 \le \vartheta \le \pi\}$, which is just the polar plane $P(0, 0)$ in I^0. The other polar planes P_I of M_e are congruent to $P(0, 0)$ under isometries of M_e, hence are also fixed point sets of an isometry. Corresponding results for polar planes P_III in III blocks follow by replacing β by the isometry $\lambda\beta$. We conclude that all the polar planes in M_e are closed, totally geodesic surfaces.

Now we turn to the isometries of slow Kerr spacetime M_s. As before, the Killing isometry group K and the equatorial isometry ϵ are present. To find more isometries of M_s, we first consider those of the Kruskal domains, starting with $\mathcal{D}(r_+)$.

There are four natural mappings of the U^+V^+ plane that (as isometries must) preserve the function r, namely, those sending (U^+, V^+) to

$$(U^+, V^+) \quad (-U^+, -V^+), \quad (V^+, U^+), \quad \text{or} \quad (-V^+, -U^+).$$

Looking for isometries, we extend these over $\mathcal{D}(r_+)$ by leaving ϑ and φ^+ unchanged. Then the large formula for ds^2 in Proposition 3.5.3 shows at once that the first two mappings are isometries. The last two fail, but only because the one-form $U^+dV^+ - V^+dU^+$ changes sign. This is readily corrected by also changing the sign of $d\varphi^+$, that is, by extending the above maps by $\vartheta \to \vartheta, \varphi^+ \to -\varphi^+$.

Strictly speaking, these four isometries are defined only on $\mathcal{D}(r_+)$-(axes), but all have obvious extensions to diffeomorphisms of $\mathcal{D}(r_+)$, and by continuity the extensions are isometries. The first one is just the identity map; so we look at the other three.

1. The isometry γ given by $(U^+, V^+, \vartheta, \varphi^+) \to (-U^+, -V^+, \vartheta, \varphi^+)$ expresses the radial symmetry of $\mathcal{D}(r_+)$ about the crossing sphere $S(r_+)$.

2. Proposition 3.5.1 shows that reversing U^+ and V^+ amounts to changing the sign of t. Since $\varphi^+ = \varphi$, it follows that the isometry $(U^+, V^+, \vartheta, \varphi^+) \to (V^+, U^+, \vartheta, -\varphi^+)$ agrees with the backwards isometry β_0 on block I and block I'—thus leaving them invariant—but switches blocks II and II'. We call this isometry the *backwards isometry* β_+ of $\mathcal{D}(r_+)$. (From Figure 3.13 we can

see how β_+ agrees with β: $K^* \to {}^*K$ for $\mathrm{I} \cup \mathrm{II} \to \mathrm{I} \cup \mathrm{II}'$ and correspondingly for $\mathrm{I}' \cup \mathrm{II}' \to \mathrm{I}' \cup \mathrm{II}$.)

3. The isometry $(U^+, V^+, \vartheta, \varphi^+) \to (-V^+, -U^+, \vartheta, -\varphi^+)$, denoted by σ, also derives from β_0, but their sign differences cause σ and β_+ to permute the blocks of $\mathcal{D}(r_+)$ differently. Thus, σ agrees with β_0 on block II and block II', but switches I and I'. On block II, where r measures time, the effect of β_0 is to reverse space, so we call σ the *space-reversal* isometry of $\mathcal{D}(r_+)$. (Again Figure 3.13 shows how σ agrees with β: $K^* \to {}^*K'$ for $\mathrm{I} \cup \mathrm{II} \to \mathrm{I}' \cup \mathrm{II}$ and correspondingly for $\mathrm{I}' \cup \mathrm{II}' \to \mathrm{I} \cup \mathrm{II}'$.)

Evidently these isometries constitute a four-group $\{id, \beta_+, \sigma, \gamma\}$. Its geometry is summarized in Figure 3.19.

Writing $+1$, as before, for preservation of orientation, we find

isometry	orientation	time-orientation
ϵ	-1	$+1$
β_+, γ	$+1$	-1
σ	$+1$	$+1$

These signs derive readily from effects on ∂_{U^+}, ∂_{V^+}, or from the results in the preceding section, since β_+ and σ are extensions of β_0 from blocks I and II, respectively.

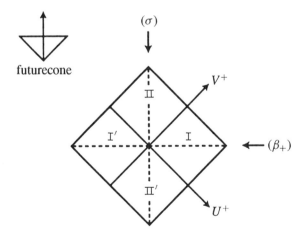

FIGURE 3.19.   Isometries of the Kruskal domain $\mathcal{D}(r_+)$. β_+ combines a reflection of the U^+, V^+ plane about the dashed line (β_+): $V^+ = U^+$ with the map $\varphi \to -\varphi$ of $S^2(\vartheta, \varphi)$. Analogously, σ reflects in (σ): $V^+ = -U^+$. The radial symmetry $\gamma = \beta\sigma = \sigma\beta$ sends (U^+, V^+) to $(-U^+, -V^+)$ leaving $S^2(\vartheta, \varphi)$ unchanged.

These isometries suffice to generate the isometry group of $\mathcal{D}(r_+)$.

Lemma 3.8.3   *The isometry group $I(\mathcal{D}(r_+))$ of $\mathcal{D}(r_+)$ is generated by ϵ, any two of $\{\beta_+, \sigma, \gamma\}$, and the Killing isometries K.*

Proof. Let ϕ be an isometry of $\mathcal{D}(r_+)$. Using ϵ, if necessary, we can assume that ϕ preserves ϑ. We know that ϕ permutes the Boyer–Lindquist blocks of $\mathcal{D}(r_+)$. Consider block $\mathtt{I}$; it can be sent only to $\mathtt{I}$ or $\mathtt{I}'$.

If $\phi(\mathtt{I}) = \mathtt{I}$, then since ϑ is also preserved, Corollary 3.7.2 implies that $\phi \mid \mathtt{I}: \mathtt{I} \to \mathtt{I}$ must be either ψ and $\psi\beta_0$ for some $\psi \epsilon K$. These isometries have unique extensions ψ and $\psi\beta_+$ to all of $\mathcal{D}(r_+)$, hence ϕ is either ψ or $\psi\beta_+$.

If $\phi(\mathtt{I}) = \mathtt{I}'$, then $\sigma\phi(\mathtt{I}) = \mathtt{I}$, so by the preceding argument, $\sigma\phi$ is either ψ or $\psi\beta_+$. Since $\sigma = \sigma^{-1}$, ϕ is either $\sigma\psi$ or $\sigma\psi\beta^+$.   □

Corresponding results hold for the domains $\mathcal{D}(r_-)$; only the numbering changes. Accordingly, β_- agrees with β_0 on $\mathtt{III}$ and on $\mathtt{III}'$ but switches $\mathtt{II}$ and $\mathtt{II}'$, and σ agrees with β_0 on $\mathtt{II}$ and $\mathtt{II}'$ but switches $\mathtt{III}$ and on $\mathtt{III}'$.

To find the isometry group of M_s we begin as usual with K and ϵ, and look for sufficiently many large isometries to generate the entire group $I(M_s)$. As with M_e, one such is the isometry λ that moves each $\mathcal{D}(r_\pm)_i$ to the next future one, $\mathcal{D}(r_\pm)_{i+1}$ (see Figure 3.15). Isometries of $\mathcal{D}(r_\pm)$ supply more.

Lemma 3.8.4   *The isometries β_+, σ, γ on the Kruskal domain $\mathcal{D}(r_+)_0$ of M_s extend to isometries β, σ, γ of M_s. All three are idempotent.*

Proof. We apply the Mapping Lemma to the definition of M_s (illustrated in Figure 3.15). The simplest case is σ, where using $\sigma: \mathcal{D}(r_\pm)_i \to \mathcal{D}(r_\pm)_i$ for every index i produces the required extension. The consistency condition in the lemma is trivial since the matching isometry in the definition of M_s consists of copies of the identity map of $\mathtt{II}$.

For β, start with the isometry β_+ on $\mathcal{D}(r_+)_0$, and then (see Figure 3.15) use $\beta_+: \mathcal{D}(r_+)_i \to \mathcal{D}(r_+)_{-i}$ and $\beta_-: \mathcal{D}(r_-)_i \to \mathcal{D}(r_-)_{-i-1}$ for all i.

These extensions then extend $\gamma \mid \mathcal{D}(r_+)_0$ to $\gamma = \beta\sigma = \sigma\beta$ on M_s.   □

The backwards isometry β now runs the entire spacetime M_s backwards, and γ can be imagined in Figure 3.15 as radial symmetry (or a 180° rotation) of M_s about the crossing sphere $S(r_+)$ in $\mathcal{D}(r_+)_0$.

Both β and γ depend on the choice of $\mathcal{D}(r_+)_0$, but it will soon appear that algebra in $I(M_s)$ gives corresponding isometries for any $\mathcal{D}(r_\pm)$. By contrast, σ already spatially reverses every Boyer–Lindquist domain of M_s.

Lemma 3.8.5 *The isometry group $I(M_s)$ is generated by ϵ, λ, any two of $\{\beta, \sigma, \gamma\}$, and the Killing isometries K.*

Proof. An arbitrary isometry ϕ of M_s must carry $\mathcal{D}(r_+)_0$ to some $\mathcal{D}(r_+)_n$. Thus, $\lambda^{-n}\phi$ is a self-isometry of $\mathcal{D}(r_+)_0$. Employing ϵ as usual, we can assume that ϕ preserves ϑ. Then the proof of Lemma 3.8.3 shows that $\lambda^{-n}\phi \, | \mathcal{D}(r_+)_0$ has the form ψq_+, where $q_+ \epsilon \{id, \beta_+, \sigma, \gamma\}$. These isometries are all extendible to M_s; hence $\phi = \lambda^i \psi \tilde{q}_+$. □

The effect of these isometries on orientation and time-orientation is readily found. Clearly λ preserves both orientation and time-orientation. The behavior of the others is the same as on any Kruskal domain that they leave invariant. With sign conventions as before, we find

isometry	orientation	time-orientation
λ, ψ, σ	$+1$	$+1$
ϵ	-1	$+1$
β, γ	$+1$	-1

Now consider the commutativity relations among these isometries. The equatorial isometry ϵ is central, that is, commutes with everything. Since the KBL coordinate expressions for λ are identity maps, λ commutes with any isometry that leaves each $\mathcal{D}(r_\pm)$ invariant, notably, $\psi \epsilon K$ and σ. Also, we know that β, σ, and γ commute.

Lemma 3.8.6 *Among the isometries ϵ, λ, β, σ, and $\psi \epsilon K$ of M_s the only failures of commutativity are $\beta \psi \beta = \psi^{-1}$, $\sigma \psi \sigma = \psi^{-1}$, and $\beta \lambda \beta = \lambda^{-1}$.*

Proof. Since β is an extension of β_0 on block I of $\mathcal{D}(r_+)_0$, the relation $\beta \psi \beta = \psi^{-1}$ follows from the corresponding formula for a single block. The same argument applies to σ since it is an extension of β_0 on block II of $\mathcal{D}(r_+)_0$.

The isometry λ relates differently to β and σ. The latter leaves each Kruskal domains invariant, hence commutes with λ. However, β sends $\mathcal{D}(r_+)_i$ to $\mathcal{D}(r_+)_{-i}$, so

$$(\beta\lambda\beta)\mathcal{D}(r_+)_0 = \beta\lambda\mathcal{D}(r_+)_0 = \beta\mathcal{D}(r_+)_1 = \mathcal{D}(r_+)_{-1} = \lambda^{-1}\mathcal{D}(r_+)_0.$$

The isometries $\beta\lambda\beta$ and λ^{-1} agree pointwise on $\mathcal{D}(r_+)_0$—hence everywhere— since λ preserves KBL coordinates and $\beta^2 = id$. We have seen that the other pairs commute. □

Thus, as before, λ and β generate an infinite dihedral group D.

Let Q_s be the group generated by ϵ, λ, and any two of β, σ, γ. (Q_s will turn out to be the large isometry group of M_s.)

Lemma 3.8.7 *The group Q_s is the direct product of the four-group $\{id, \epsilon, \sigma, \epsilon\sigma\}$ and the infinite dihedral group D generated by λ and β.*

Proof. Let $F = \{id, \epsilon, \sigma, \epsilon\sigma\}$; then evidently, $Q_s = F \cdot D$. The relations above show that σ commutes with every element of Q_s, and so does ϵ. It remains to show that $F \cap D = \{id\}$, so suppose $\phi \epsilon F \cap D$. Among the generators $\epsilon, \sigma, \lambda, \beta$ only ϵ is orientation-reversing. Thus, $\phi \epsilon F$ implies that ϕ is either σ or id.

Assume $\phi = \sigma$. Then $\phi \epsilon D$ implies $\sigma = \beta^a \lambda^n$, where $a = 0$ or 1. Since σ and λ preserve time-orientation and β does not, $a = 0$. But $\sigma = \lambda^n$ is impossible since σ leaves Boyer–Lindquist domains invariant but λ^n does not when $n \neq 0$. Thus, $\phi = id$. □

We can now describe the isometry group of M_s.

Theorem 3.8.8 *The isometry group $I(M_s)$ of slow Kerr spacetime M_s is the semi-direct product of $I_0(M_s) = K = \{\tau_c \rho_\alpha\} \approx S^1 \times \mathbf{R}^1$ and the discrete group Q_s defined above. Explicitly,*

(1) *K is a normal subgroup,*   (2) *$I(M_s) = K \cdot Q_s$,*   (3) *$I_0(M_s) \cap Q_s = \{id\}$.*

(For the failures of commutativity between $\psi \epsilon K$ and the generators $\epsilon, \lambda, \beta, \sigma$ of Q_s, see Lemma 3.8.6.)

Proof. Lemma 3.7.5 shows that $K = I_0(M_s)$.

(1) As before, this is a general fact about $I_0(M_s)$.

(2) $I(M_s) = K \cdot Q_s$. The proof of Lemma 3.8.5 shows that any isometry ϕ of M_s can be written as $\phi = \lambda^n \psi \beta \tilde{q}_+ = \lambda^n \psi \epsilon^a \beta^b \sigma^c$, where n is some integer and a, b, c are either 0 or 1. Since $\lambda^n \psi = \psi^{-1} \lambda^n$, this becomes $\phi = \psi^{-1} q$, with $q = \lambda^n \epsilon^a \beta^b \sigma^c$ in Q_s.

(3) $I_0(M_s) \cap Q_s = \{id\}$. Suppose ϕ is contained in both $I_0(M_s)$ and Q_s. By Lemma 3.8.5, we can write $\phi = \epsilon^a \sigma^b \beta^c \lambda^n$, where n is some integer and each of a, b, c is either 0 or 1. Since ϵ is the only one of these isometries that reverses orientation, $a = 0$. Now $\phi = \sigma^b \beta^c \lambda^n$. And since β is the only one of these four that reverses time-orientation, $\phi = \sigma^b \lambda^n$. But λ is the only one of these that does not leave Kruskal domains invariant, so $\phi = \sigma^c$. Here $c = 1$ is impossible since ϕ leaves Boyer–Lindquist blocks invariant, but σ does not. Hence $\phi = id$. □

Multiplication in $I(M_s)$ is now characterized by the automorphisms C_q of K that for each $q \epsilon Q_s$ send ψ to $q\psi q^{-1}$. (Recall the identity $\psi q \psi' q' = \psi C_q(\psi')qq'$.) Because $q \to C_q$ is itself a homomorphism, we need only find C_q for the generators $\lambda, \epsilon, \beta, \sigma$. Of these, ϵ is central and λ commutes with ψ, so the only nontrivial cases among the generators are $C_\beta(\psi) = \psi^{-1}$ and $C_\sigma(\psi) = \psi^{-1}$.

Finally, we show that in M_s all large polar planes of the same kind (i.e., P_{I}, P_{II}, or P_{III}) are congruent under an isometry of M_s. It follows that each is the fixed point set of an isometry, thus verifying that polar planes are closed, totally geodesic surfaces.

The backwards isometry β leaves only $\mathcal{D}(r_+)_0$ invariant; there it fixes the points $U^+ = V^+$, $\varphi^+ = 0$. Thus, $\text{Fix}(\beta)$ is just the spacelike polar plane $P_{\text{I}}(0,0)$ in $\mathcal{D}(r_+)_0$. (We can then regard β as reflection of M_s in this plane.) The isometry $\rho_{-\alpha}\tau_{-c}$ carries $P_{\text{I}}(\alpha, c)$ to $P_{\text{I}}(0,0)$, hence every plane P_{I} in M_s is congruent to $P_{\text{I}}(0,0)$ under a suitable $\lambda^n \psi$. Thus, P_{I} is the fixed point set of the isometry $(\lambda^n \psi)^{-1} \beta (\lambda^n \psi) = \beta \lambda^{2n} \psi^2$.

The isometry $\lambda\beta$ fixes the polar plane $P_{\text{III}}(0,0)$ in $\mathcal{D}(r_-)_0$, and again every P_{III} is congruent to $P_{\text{III}}(0,0)$ and hence P_{III} is the fixed point set of an isometry.

The space-reversing isometry σ fixes the points of the timelike polar plane $P_{\text{II}}(0,0)$ in every domain $\mathcal{D}(r_\pm)$ in M_s. Thus, $\text{Fix}(\sigma)$ is the entire polar plane $P_{\text{II}}(0,0)$ running through all II blocks. Every other plane P_{II} in M_s is congruent to $P_{\text{II}}(0,0)$ under a Killing isometry ψ, so $P_{\text{II}} = \text{Fix}(\psi^{-1}\sigma\psi)$.

Remark 3.8.9 (Schwarzschild Comparison.) By construction, Schwarzschild spacetime is spherically symmetric, so the orthogonal group O(3) acts (isometrically) on it, and the same is true for its maximal analytic extension Kruskal spacetime Kr. By contrast, the axial symmetry of Kerr spacetime yields only the circle group $S^1 = \{\rho_\varphi\}$. Thus, $I(Kr)$ is larger than the (two-dimensional) Kerr isometry groups. In fact, Remark 3.5.5 expresses Kr as a warped product $Q \times_r S^2$, where the metric of $Q \approx \mathbf{R}^2$ has the form $F(r)\,du\,dv$. Since r is an isometric invariant, the isometry group of Q is the same as that of Minkowski 2-space, namely, the Lorentz group $O(1, 1)$, which is one-dimensional with four components. Then $I(Kr) \approx O(1, 1) \times O(3)$, which is four-dimensional. □

3.9   Topology of Kerr Spacetime

In this section we examine the topological properties of the maximal Kerr spacetimes M_f, M_e, and M_s and some noteworthy variants.

A basic tool is the following: Let A be a subspace of a topological space X; then a continuous map $\rho: X \to A$ such that $\rho(a) = a$ for all $a \epsilon A$ is called a *retraction* of X onto A. A is a *strong deformation retract* of X provided there exists a retraction $\rho: X \to A \subset X$ that is homotopic to the identity map *id* of X, with points of A remaining fixed during the homotopy. For example, the unit sphere $S^n \subset \mathbf{R}^{n+1}$ is a strong deformation retract of $\mathbf{R}^{n+1} - 0$ under the homotopy $h_s(x) = (1 - s)x + sx/|x|$, which, as s traverses [0,1], moves each nonzero point x of $\mathbf{R}^{n+1}$ radially toward S^n, leaving points of S^n fixed.

If a space X_1 is a strong deformation retract of a space X_2, then in particular, X_1 and X_2 are *homotopically equivalent* (or, *have the same homotopy type*), and hence have the same homotopy and homology properties. (For topological terms, see, for example, Massey 1987; Greenberg, Harper 1981). Our goal is to find simple spaces that are homotopically equivalent to various Kerr spacetimes.

Boyer–Lindquist blocks I and II are diffeomorphic to $\mathbf{R}^2 \times S^2$, and since $\mathbf{R}^n$ can be radially shrunk down to a single point, they are homotopically equivalent to a 2-sphere and, in particular, are simply connected. Block III is not as simple since the ring singularity has been removed from it.

Lemma 3.9.1   (1) *Fast Kerr spacetime* M_f, *the Kerr spacetimes* K^* *and* $*K$ *(all three rotation types), and a Boyer–Lindquist block* III *in* M_e *or* M_s *are all diffeomorphic.* (2) *Each is homotopically equivalent to a space* X_f *consisting of two 2-spheres joined at the north poles and south poles (see Figure 3.20).*

Proof.   (1) For example, consider M_f and III $\subset M_s$. By definition, $M_f = \mathbf{R}^2(r, t) \times S^2 - \Sigma$, where Σ is the ring singularity. Let $f: \mathbf{R} \to (-\infty, r_-)$ be a smooth (or analytic) function such that $f' > 0$, $f(0) = 0$, $\lim_{r \to -\infty} f(r) = -\infty$, and $\lim_{r \to +\infty} f(r) = r_-$. Then the mapping $\phi: M_f \to$ III given by $\phi(r, t, y) = (f(r), t, y)$ is a diffeomorphism.

(2) Consider III $\subset M_s$. (This proof is modelled on Figure 2.2 so that homotopy results can interpreted directly in III.) We first show that a t-slice of III, say $\{t = 0\}$, is homotopically equivalent to III itself. Now

$$\text{III} = [(-\infty, r_-) \times \mathbf{R}(t) \times S^2] - \Sigma = \mathbf{R}(t) \times \{t = 0\}$$

Thus, as s traverses [0, 1], the deformation $h_s(t, r, y) = ((1 - s)t, r, y)$ for $0 \leq s \leq 1$ squeezes the t-line $\mathbf{R}(t)$ down to the point $t = 0$, showing that $\{t = 0\}$ is a strong deformation retract of III. Explicitly, $\{t = 0\} = [(-\infty, r_-) \times S^2] - \Sigma_0$, where Σ_0 is the equator of $0 \times S^2$.

The exponential rendering of the t-slice in Figure 2.2 is realized formally by the diffeomorphism $\psi(r, y) = e^r y$ carrying $\{t = 0\}$ onto $B - (0 \cup C)$, where B is a ball of radius $e^r 1$ in $\mathbf{R}^3$ and $C \approx \Sigma_0$ is the unit circle in the xy-plane of $\mathbf{R}^3$.

$B - (0 \cup C)$ has a strong deformation retract X—shown on the left in Figure 3.20—consisting of two 2-spheres S_1 and S_2 with line segments joining their north and their south poles. The required deformation onto X sends points outside S_2 radially inward to S_2, and points inside S_1 radially outward to S_1. Points between S_1 and S_2 follow curves that initially move outward from C, then curve onto X. Topologically, X_f in Figure 3.20 is the same as X.   □

Comparing Figures 3.20 and 2.2 should make it clear that when X is pictured in block Ⅲ, the 2-spheres are $r = r_1$ and $r = r_2$ in $\{t = 0\}$, where $r_1 < 0 < r_2 < r_-$, and the segments joining the poles lie in the axis.

Since M_f, Ⅲ, and X_f are homotopically equivalent, it follows that the fundamental group of M_f is infinite cyclic, $\pi_1(M_f) \approx Z$, and its nonvanishing integer homology groups are $H_0 \approx H_1 \approx Z$, $H_2 \approx Z \times Z$. Hence the Euler–Poincaré number of M_f is 2.

This interpretation of homotopy in terms of Ⅲ makes it clear that the generators of both cyclic groups π_1 and H_1 derive from a simple closed curve that links the ring singularity Σ. The generating 2-cycles in H_2 "surround" $r = -\infty$ and $r = +\infty$ and are separated by Σ.

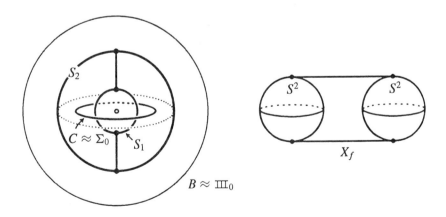

FIGURE 3.20.   Both fast Kerr spacetime M_f and Boyer–Lindquist block Ⅲ in M_s are homotopically equivalent to a t-slice Ⅲ$_0$ of Ⅲ and hence to a space X_f consisting of two 2-spheres linked at their poles.

The homotopy structure of M_s, with its infinitely many ring singularities, is more complicated, but as we saw in Section 3.6, restoration of its ring singularities gives a manifold $\widehat{M}_s$ diffeomorphic to $\mathbf{R}^2 \times S^2$. In preparation for defining the deformation of M_s onto a simpler space, we describe $M_s \subset \mathbf{R}^2 \times S^2$ as follows.

Let n and s be the north and south poles of S^2, and replace $\mathbf{R}^2$ above by an infinite strip $\mathbf{R}^1 \times (-1, +1)$. Then $\widehat{M} \approx \mathbf{R}^1 \times (-1, +1) \times S^2$ consists of the north axis $A_N = \mathbf{R}^1 \times (-1, +1) \times n$, the south axis $A_S = \mathbf{R}^1 \times (-1, +1) \times s$, and for every pair of corresponding points (x, n) and (x, s), a 2-sphere S_x with these points as its poles. The subspace M_s of $\mathbf{R}^1 \times (-1, +1) \times S^2$ is gotten by removing ring singularities, that is, by removing the equators of those spheres S_x for which x corresponds to an axis point at which $r = 0$.

Figure 3.21 shows how the infinitely many curves with $r = 0$ separate the components $C_n (n = \pm 1, \pm 2, \ldots)$ of the set $\{r < 0\}$ from the rest of the axis $\mathbf{R}^1 \times (-1, +1)$.

We can now describe a simpler subspace X that turns out to be a deformation retract of M_s. In each component C_n choose a point c_n, and pick a single point c_0 in $\{r > 0\}$. Then X is the union of A_N, A_S, the spheres S_{c_n} for all n, and S_{c_0}. Thus, X is an infinite sequence of 2-spheres clamped between two flat strips.

The deformation of M_s onto X will follow a continuous family of *flow lines*, as shown in Figure 3.21 (top), such that

1. In each component C_n of $\{r < 0\}$ the lines run radially inward to the selected point c_n from the C_n boundary points—both $r = 0$ and, in the limit, $r = -\infty$.

2. Outside these components, the flow lines run to the point c_0 from each point of $\{r = 0\}$ and from $r = +\infty$. (If the axis is remodeled as a disk with center c_0 these curves would just be radial lines.)

During the deformation, points will move so that their axial projections follow the flow lines. Furthermore, the spherical coordinate φ remains constant during the deformation. Thus, to describe the deformation it remains only to tell how the ϑ coordinate varies. This is suggested in the lower half of Figure 3.21, a rectangular strip with four subrectangles:

A. Points in C_n move from $r = -\infty$ into X, ending in A_N, A_S or the sphere S_{C_n}. Note that points in X remain fixed.

B. Points in C_n move away from the ring singularity, again ending at A_N, A_S or S_{c_n}. Note that points in $r = 0$ remain in the same sphere, moving due north or south away from the ring singularity to the poles $n \epsilon A_N$ and $s \epsilon A_S$.

C. This case is analogous to that of B but points move from $r = 0$ out through $\{r > 0\}$ to A_N, A_S, or the sphere S_{c_0}.

D. This case is like A, but points move from $r = +\infty$ to c_0.

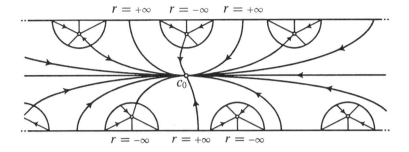

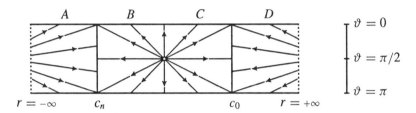

FIGURE 3.21.   Deformation of M_s onto a subspace X. (Top) Flow lines in the axis: from $r = 0$ and $r = -\infty$ in toward the points c_n, from $r = 0$ and $r = +\infty$ toward the central point c_0. (Bottom) Character of the deformation on a single meridian of longitude, $0 \le \vartheta \le \pi$.

Thus X is a deformation retract of M_s, so the two are homotopically equivalent. Finally, each of the axes A_N and A_S is shrunk to a straight line, giving an infinite version X_s of the model space X_f for M_f.

Proposition 3.9.2   *Maximal slow Kerr spacetime M_s has the homotopy type of a subspace X_s of $\mathbf{R}^3$ consisting of an infinite sequence of 2-spheres whose north poles are all linked (as in Figure 3.22) by a line N and south poles are linked by a line S.*

The lines N and S could be further collapsed to single points, but then the topology of the space would not be obvious.

The model X_s shows that the fundamental group $\pi_1(M_s)$ is a free group F^∞ with an countably infinite set of generators, and the nonzero homology groups (with integer coefficients Z) are $H_0 \approx Z$, and $H_1 \approx H_2 \approx Z^\infty$ (free abelian group).

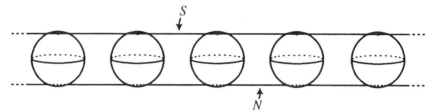

FIGURE 3.22.   Slow Kerr spacetime M_s is homotopically equivalent to this space X_s consisting of infinitely many 2-spheres linked at their poles.

The extreme Kerr spacetime M_e has the same homotopy type since it is diffeomorphic to M_s.

Using the elements of covering space theory (Greenberg, Harper 1981; Massey 1987; O'Neill 1983), many other maximal Kerr spacetimes can be found that either cover or are covered by the standard models. (Here the covering map is required to be a local isometry, that is, preserve the Kerr metric.) For any manifold its simply connected covering manifold is always the largest and in a sense its simplest covering manifold. However, in Kerr geometry, the simply connected spacetime $\tilde{M}_s$ covering M_s is not too valuable; it lacks the clear relation to the ring singularities in M_s expressed by $\widehat{M}_s$ and, as we see in the next section, still contains violations of chronology.

Alternatively, there are many spacetimes that are covered by standard Kerr models. To construct these we use the notion of *orbit manifold* M / Γ. Here Γ is a subgroup of the isometry group $I(M)$ that is *properly discontinuous*—roughly speaking, each point has a neighborhood $\mathcal{U}$ such that the sets $\phi(\mathcal{U})$ for all $\phi \epsilon \Gamma$ are well separated. (In particular, no $\phi \neq id$ has a fixed point.) Then M / Γ is constructed by identifying points p, q of M if $\phi(p) = q$ for some $\phi \epsilon \Gamma$. Finally the natural map $\pi : M \to M / \Gamma$ is a locally isometric covering map. (See pages 187, 188, 214 in O'Neill 1983).

For the maximal Kerr models, there are many such orbit manifolds. For example, on M_f or M_e we could take Γ to be the infinite cycle group generated by the isometry τ_c, $c \neq 0$. This would produce eternal recurrence of time (in the regions where $\tilde{\partial}_t$ is timelike). The scheme fails for M_s since $\tilde{\partial}_t$ has zeros, and hence its flow isometries τ_c have fixed points.

A more natural construction, applicable to M_e and M_s, derives from the isometry λ that moves the spacetimes one step into the future. Again the group $(\lambda) = \{\lambda^n : n \epsilon Z\}$ generated by λ is properly discontinuous.

Definition 3.9.3   *The orbit manifold* $S_e = M_e / (\lambda)$ *is called* small extreme Kerr spacetime S_e. *Similarly,* $S_s = M_s / (\lambda)$ *is* small slow Kerr spacetime.

Geometric repetition in M_e is eliminated in the elegant spacetime S_e, which possesses unique Boyer–Lindquist blocks Ⅰ and Ⅲ, a single ring singularity, and two short horizons (see Figure 3.23).

It is dependable rule that any property of M preserved by the isometries of Γ descends to M/Γ. Accordingly, S_e is both orientable and time-orientable since λ preserves these properties. However, the closed timelike curve τ shown in Figure 3.23 is a violation of chronology in S_e not present in M_e so the latter is considered preferable on physical grounds.

The small slow spacetime S_s is no less elegant. As Figure 3.24 shows, it has exactly one of everything, namely,

- Six Boyer–Lindquist blocks: one of each type, in each time-orientation,

- Eight short horizons, determined by the three double choices: r_+ or r_-, K^* or $^*K$, natural or reversed time-orientation.

- Two crossing spheres: $S(r_+)$ and $S(r_-)$.

These are arranged in two Boyer–Lindquist domains: one $\mathcal{D}(r_+)$ and one $\mathcal{D}(r_-)$.

The homotopy type of S_s can be found by modifying the construction used for M_s. Topologically, M_s was two strips with 2-spheres clamped between them at every pair of corresponding points—their equators deleted over the curves $r = 0$. But for S_s the strips $\mathbf{R}^1 \times (-1, +1)$ that topologically model the axis now become bands $S^1 \times (-1, +1)$. Pick points c, c' in the two regions $r < 0$ of Ⅲ and Ⅲ$'$ respectively. A deformation like that for M_s (Figure 3.21) moves points over flow lines (as suggested in Figure 3.25) to a subspace of S_s in which full spheres remain over the two points c, c' and the entire circle S^1. (The flow lines cannot be collected at anything smaller).

A further homotopy gives a simple model.

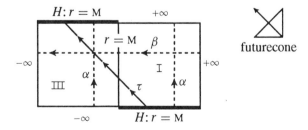

FIGURE 3.23. Small extreme Kerr spacetime S_e. The two horizons denoted by H are identified by the isometry λ; consequently, α is a single principal null geodesic, and the timelike curve τ is closed.

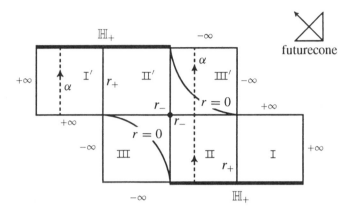

FIGURE 3.24.   Small slow Kerr spacetime S_s. The two long horizons denoted by $\mathbb{H}_+$ are identified in the obvious way so that α is a single principal null geodesic. $\mathcal{D}(r_-)$ is evident in the center of the figure; $\mathcal{D}(r_+)$ is produced by the identifications.

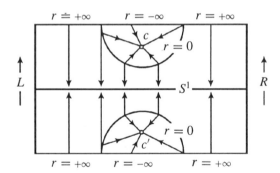

FIGURE 3.25.   Deformation of S_s. The left (L) and right (R) sides of the rectangle are identified, turning the central line into a circle S^1. The flow lines (some indicated) guide the deformation.

Lemma 3.9.4 *Small slow Kerr spacetime S_s is homotopically equivalent to the product space $S^1 \times S^2$ and two 2-spheres whose north and south poles are all identified with a single point of $S^1 \times S^2$.*

Thus, for example, the fundamental group of S_s is a free group on three generators. For S_e the model is the same except that only one 2-sphere is attached.

3.10   Kerr Chronology

We consider some fundamentals of causality theory—that is, the influence of causal character on geometry—and apply them in the Kerr case. For brevity we emphasize *chronology*, roughly speaking, time's orderly sequential progression (or its failure) as demonstrated in a spacetime by the timelike curves.

For any spacetime M a natural question is: For which pairs of points p, q is there a timelike future-pointing curve in M from p to q? Relativistically, can a observer starting at event p reach event q? In Newtonian physics, where there is no speed limit, the answer is yes if and only if q occurs later than p, as measured by the universal Newtonian clock. In relativity each spacetime presents a separate problem.

The *chronological future* of p in a spacetime M is the set $I^+(p, M)$ of all $q \in M$ for which there is a future-pointing (piecewise smooth) timelike curve from p to q in M. We write simply $I^+(p)$ when the particular manifold M is not at issue. If $q \in I^+(p)$, then evidently $I^+(q) \subset I^+(p)$. (Allowing piecewise smooth curves is only a minor convenience since the corners of a broken timelike curve can always be smoothed away.) It can be shown that $I^+(p)$ is always an open set of M (Hawking, Ellis 1973).

For the corresponding *chronological past* $I^-(p, M)$ of p in M *future-pointing* is replaced by *past-pointing* in the previous definition.

In Minkowski space $\mathbf{R}_1^4$, the chronological future $I^+(p)$ of any event p is just an ordinary solid cone with vertex (downward) at p; thus $I^+(0)$ consists of the points (x, y, z, t) such that $x^2 + y^2 + z^2 < t^2$ and $t > 0$.

It is seldom easy to determine where timelike curves go, but the following notion often helps to show where they do not go.

Definition 3.10.1   *A subset A of a spacetime M is* achronal *provided no timelike curve in M meets A more than once.*

For example, though the term was not used, it was shown in Chapter 2 that the horizons H in K^* (and $^*K$) are achronal. This followed easily from the fact that H is null and separates K^* (i.e., $K^* - H$ is not connected). The argument is quite general.

Corollary 3.10.2   *If a closed connected nontimelike hypersurface S separates a spacetime M, then S is achronal.*

Proof. By Corollary 1.7.8, $M-S$ has two components, C^- and C^+ whose common boundary is S, and any future-pointing timelike curve that meets S must cross S transversally from C^- to C^+. Thus, if such a curve α meets S once, say at $\alpha(s_0) \epsilon S$, it then enters C^+. To leave C^+, α must meet S—an impossibility, since that would produce a crossing from C^+ to C^-. □

Long horizons in M_s have the same general properties as do the short horizons in K^*.

Lemma 3.10.3 *The long horizons $\mathbb{H}$ in M_s are achronal, and each separates M_s into two components, a past side C^- and a future side C^+.*

Proof. Since these horizons are connected null hypersurfaces it suffices to show they separate M_s and are closed.

In a Kruskal domain $\mathcal{D}_0(r_+)$, consider $\mathbb{H}_+$. (The other three cases are analogous.) $\mathbb{H}_+$ is the set $U^+ = 0$ in $\mathcal{D}_0(r_+)$, so it separates $\mathcal{D}_0(r_+)$ into $U^+ > 0$ and $U^+ < 0$. The definition of M_s as a doubly infinite chain of open, connected domains $\mathcal{D}(r_\pm)$ shows that

$$M_s - \mathbb{H}_+ = C^+ \cup C^-,$$

where, as Figure 3.15 makes clear,

1. C^+ is the union of $U^+ < 0$ and a chain of overlapping domains $\mathcal{D}(r_\pm)$ starting with a $\mathcal{D}(r_-)$ that meets $U^+ < 0$ only in block II of $\mathcal{D}_0(r_+)$, and
2. C^- is the union of $U^+ > 0$ and a similar chain starting with a $\mathcal{D}(r_-)$ that meets $U^+ > 0$ only in block II' of $\mathcal{D}_0(r_+)$.

Thus, the sets C^+ and C^- are open, connected, and disjoint. Hence they are the connected components of $M_s - \mathbb{H}_+$, and the hypersurface $\mathbb{H}_+$ is closed. □

The character of horizons as one-way membranes for nonspacelike curves will have decisive effects on chronology.

Obviously, any subset of an achronal set is achronal. If A is achronal in M it may not be achronal in a larger spacetime $\tilde{M}$ since there will be more timelike curves to contend with. However, if a set A is achronal in a Boyer–Lindquist block B, it remains achronal in M_s, since a curve cannot leave B without crossing a long horizon, through which no timelike curve can return. Thus, for example, the level hypersurfaces $t = $ const in Boyer–Lindquist block I are achronal in block I (since $dt/ds > 0$ on future-pointing timelike curves in I). Hence they are achronal in M_s. The same is true for the hypersurfaces $r = $ const in block II. But achronal set can never meet block III, since through each point of III there passes a closed timelike curve.

Timelike *geodesics* starting at a point p of a spacetime cannot always cannot reach every point of $I^+(p)$. Nevertheless, the following fact is useful: If $\alpha: I \to M$ is a [future-pointing] null geodesic from p to q, then given any neighborhood $\mathcal{U}$ of q there is a [future-pointing] timelike geodesic from p to some point in $\mathcal{U}$. Since there are a timelike tangent vectors arbitrarily near every null vector, this follows from the continuous dependence of solutions of differential equations on initial conditions or, briefly, on the exponential map.

Lemma 3.10.4   *Let p, q be points in the spacetime K^*. If $q \in \mathrm{III}$, there exists a future-pointing timelike curve from p to q.*

Proof. We saw in Section 2.4 that III is vicious; that is, any ordered pair of points can be joined by a future-pointing timelike curve. Thus, it suffices to find a future-pointing timelike curve β starting at p and ending somewhere in III. The ingoing principal null geodesic through p enters III (even if it hits Σ). Then a timelike approximation as above gives the required curve β. $\square$

If the preceding lemma is expressed as $[K^* \to \mathrm{III}]$, then reversing time-orientations gives $[\mathrm{III}' \to K^{*\prime}]$. In $^*K$ there is an outgoing principal geodesic through every point that, in the past direction, reaches III'. Thus, $[\mathrm{III}' \to {}^*K]$, that is, from any point in III' there is a future-pointing timelike curve in $^*K$ to any given point in $^*K$ (first travel in III' to avoid Σ). Again, reversing time-orientations gives $[^*K' \to \mathrm{III}]$.

Consider the future set $I^+(p)$ of a point p in slow Kerr spacetime M_s. We cannot expect an explicit, rigorous description of $I^+(p)$ near the point p itself. For example, if p is in block I, then along timelike future-pointing curves that run outward, M_s is approaching Minkowski space, so there $I^+(p)$ will come to resemble a ordinary Euclidean cone. But going inward, M_s becomes more curved, hence $I^+(p)$ becomes more complicated. (We do know that timelike curves cannot twice cross any achronal hypersurfaces t const in I and, similarly, r constant in II.)

However, it is easy to show that $I^+(p)$ contains all future events that are "far" from p—the decisive fact being that since III is vicious, if $I^+(p)$ meets III, then it contains III. Suppose p is in a Boyer–Lindquist domain $\mathcal{D}(r_+)$. Regardless of which quadrant p is in, the results above show that there is a future-pointing timelike curve from p to some point q in block II of $\mathcal{D}(r_+)$. Then other such curves run from q into each of the adjacent blocks III and III'. This process continues, covering all future blocks.

If p is not in an r_+ domain $\mathcal{D}(r_+)$, then (neglecting two short horizons) it is in either a block III or III', and this case is simple.

Lemma 3.10.5   (1) *If $p\epsilon\mathcal{D}(r_+) \subset M_s$, then $I^+(p)$ contains the future sides C^+ of both the long horizons $\mathbb{H}_-$ that meet the boundary of $\mathcal{D}(r_+)$. (See Figure 3.10.)*

(2) *If p is in a block III or III', then $I^+(p)$ is exactly the future side of the long horizon $\mathbb{H}_-$ that separates III or III' from its adjacent block II.*

It follows that an observer in M_s can reach every event in the Boyer–Lindquist blocks that Figure 3.10 suggests are in her future, but cannot describe her travels to anyone who has crossed fewer horizons. The past sets $I^-(p)$ can be dealt with analogously.

Now we consider violations of chronology, extending to M_s results from Chapter 2.

Definition 3.10.6   *A spacetime M is* vicious *at a point p if there is a closed timelike curve in M through p. The set of all such points is denoted by $\mathcal{V}(M)$.*

If $\mathcal{V}(M)$ is empty, M is said to be *chronological* (roughly speaking, time behaves reasonably in M). M is vicious, in the sense used earlier, if $\mathcal{V}(M) = M$. These definitions apply, of course, to connected open sets in M since these are spacetimes in their own right. Thus, we say that M is chronological at points not in $\mathcal{V}(M)$.

For example, consider the Lorentz cylinders in Figure 3.26. The product manifold $\mathbf{R}_1^1 \times S^1$ can also be expressed as the orbit manifold $\mathbf{R}_1^2/(\phi)$, where $\phi(x, t) = (x + 1, t)$, and analogously, $S_1^1 \times \mathbf{R}^1 = \mathbf{R}_1^2/(\psi)$, where $\psi(x, t) = (x, t + 1)$. The former is chronological, the latter vicious.

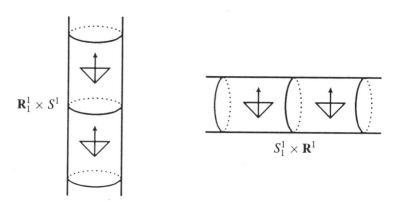

FIGURE 3.26.   The Lorentz cylinders: On the left, all timelike future-pointing curves rise steadily, so $\mathbf{R}_1^1 \times S^1$ is chronological. On the right, cross-sectional circles are timelike, so $S_1^1 \times \mathbf{R}^1$ is vicious.

The case in which some but not all points of M are vicious is clarified by the following description of the vicious set.

Lemma 3.10.7 *In any spacetime M the connected components of the vicious set $\mathcal{V}(M)$ have the form $I^-(p) \cap I^+(p)$.*

Proof. Note that $I^-(p) \cap I^+(p)$ can be described as the set of points lying on closed timelike curves through p. Hence, $I^-(p) \cap I^+(p) \subset \mathcal{V}(M)$. The sets $I^-(p) \cap I^+(p)$ are clearly connected (if nonempty), and they are open since $I^-(p)$ and $I^+(p)$ are each open.

It remains only to show that if there is a point m in both $I^-(p) \cap I^+(p)$ and $I^-(q) \cap I^+(q)$, then these two sets are equal. By symmetry it suffices to show that if $z \epsilon I^-(p) \cap I^+(p)$ then $z \epsilon I^-(q) \cap I^+(q)$. These hypotheses imply that there are timelike future-pointing curves from z to p, from p to m, and from m to q (hence $z \epsilon I^-(q)$), and from q to m, from m to p, and from p to z (hence $z \epsilon I^+(q)$). Thus, $z \epsilon I^-(q) \cap I^+(q)$.   □

Note that the component $C = I^-(p) \cap I^+(p)$ has the property that any two of its points (i.e., events) can be joined by both a future-pointing and a past-pointing curve. In fact, given any sequence of events $p_1, p_2, \ldots, p_n$ in C there is a future-pointing timelike curve running through them in order (so intricate time-travel is possible.)

It is easy to describe the vicious set of M_s. In Chapter 2 we found that, considered as independent spacetimes, blocks I and II are chronological, while block III is vicious. Viciousness is an intrinsic property so III blocks will remain vicious in M_s, and there are no other such points.

Corollary 3.10.8 *The connected components of the vicious set $\mathcal{V}(M_s)$ of M_s are its Boyer–Lindquist III blocks (so M_s is chronological elsewhere).*

Proof. If $p \epsilon$ III, we know there is a closed timelike curve in III through p, so III $\subset \mathcal{V}(M_s)$. Now we show that M_s is chronological at all other points.

First, suppose that p is in a block $B =$ I or II. Since B is chronological as a spacetime in its own right, this single point constitutes an achronal set $\{p\}$ in B. As mentioned earlier, it follows that $\{p\}$ is achronal in M_s. Thus, M_s is chronological at p.

Now suppose p is in a long horizon $\mathbb{H}$. Since $\mathbb{H}$ is achronal in M_s it cannot meet $\mathcal{V}(M_s)$. Consequently, all points outside III blocks are chronological.   □

The chronology of extreme Kerr spacetime is a simpler variant of that of M_s.

The small Kerr spacetimes S_e and S_s defined in the preceding section are both vicious. To show this, it suffices to find a closed timelike curve through each point $p \notin \mathrm{III}$ (see Figures 3.23 and 3.24). Using approximations of the ingoing and outgoing principal null geodesics through p we get timelike future-pointing curves from p to $q \epsilon \mathrm{III}$ and from $q' \epsilon \mathrm{III}$ to p. Since III is vicious, there is a future-pointing timelike curve from q to q'. These three segments then give the required curve.

If a spacetime M is not chronological the situation may be improved by passing to a covering spacetime $\tilde{M}$. For example, the vicious 2-dimensional spacetime $S_1^1 \times \mathbf{R}^1$ in Figure 3.26 is covered by the chronological Minkowski plane $\mathbf{R}_1^2$.

However, the chronology violations in the slow Kerr spacetime M_s cannot be cured in this way: no spacetime that covers M_s is chronological. To prove this it suffices to exhibit a closed timelike curve τ in M_s that is homotopic to a constant, for then every lift of τ into a covering spacetime remains closed and timelike. Such a curve can be gotten by the construction in Section 2.4 that gave a timelike curve joining arbitrary points p and q of III. Taking $p = q$ in the region $r < 0$ we can oblige the resulting closed timelike curve τ from p to p to remain in the region $r < 0$. Topologically, this region is $(-\infty, 0) \times \mathbf{R}^1 \times S^2$. Hence it is simply connected, so τ is homotopic to a constant.

For general treatments of causality see, for example, Beem, Ehrlich (1981), Carter (1971), and Geroch (1970).

KERR GEODESICS

Throughout differential geometry, geodesics are a geometric invariant second in importance only to curvature. For a spacetime, geodesics provide the description of light (future-pointing null geodesics) and freely falling material particles (future-pointing timelike geodesics). To understand the Kerr black holes we must understand, in general terms at least, how geodesics behave with respect to their distinctive features—Boyer–Lindquist blocks and horizons, axis, ring singularity, infinities $r = \pm\infty$, and so on. There must certainly be bound orbits analogous to Keplerian ellipses in the Kerr exterior, Boyer–Lindquist block I, but are there any on "white hole" side $r < 0$? Or between the horizons in block II? Will an astronaut falling through the horizon $r = r_+$ into a slow Kerr black hole inevitably be destroyed? Or is it possible to travel from the exterior through the core of the black hole to $r < 0$?

It is rarely possible to find explicit solutions to the geodesic differential equations $x^{k\prime\prime} + \Sigma\Gamma^k_{ij}x^{i\prime}x^{j\prime} = 0$. The best information is provided by *first-integrals:* expressions that involve only first derivatives of coordinates and are constant on each geodesic γ. There is always one: $q = \langle\gamma', \gamma'\rangle$. Every Killing vector field gives another, so for Kerr spacetime we get two more—energy E and angular momentum L about the axis. The surprising fact is that there is a fourth, called the *Carter constant Q*. Four independent first-integrals on a (four-dimensional) spacetime—*complete integrability*—provide rich information about geodesics that makes a thorough analysis possible.

A crucial preliminary task is show that the maximal Kerr spacetimes are *geodesically complete mod* Σ; explicitly, every geodesic γ can be extended (with affine parametrization) over the entire real line—unless γ hits the ring singularity.

Next comes a classification of geodesics into *orbit types* (e.g., bound, flyby). To relate orbit type and first-integrals the fundamental tool is an r–L *plot*, which displays the relation between angular momentum L and the range of the radial coordinate $r(s)$. (Fortunately, the behavior of the colatitude coordinate $\vartheta(s)$ turns out to be rather simple for all Kerr geodesics.) To avoid a proliferation of special cases we often concentrate on timelike geodesics ($q = -1$) in the slow Kerr black hole. The qualitative character of the r–L relation (in Section 4.8) is then determined by an explicit formula relating energy E and Carter constant Q (Section 4.7). Thereafter, knowledge of r-coordinates is used to find the global trajectories of geodesics through the pattern of Boyer–Lindquist blocks that constitute the black hole (Section 4.10).

Finally, we consider some special classes of geodesics—including those in horizons, axis, or equatorial plane; those with negative Carter constant $Q < 0$; those that approach the center $r = 0$ of the Kerr black hole; and, in particular, those that meet the ring singularity.

4.1   First-Integrals

A *first-integral* for the geodesics of a semi-Riemannian manifold M is a smooth real-valued function k on the tangent bundle TM of M such that for each geodesic γ in M the function $s \to k(\gamma'(s))$ is a constant, $k(\gamma)$. Equivalently, the (smooth) assignment of a constant $k(\gamma)$ to each geodesic γ in M determines such a function $TM \to \mathbf{R}$, namely, $v \to k(\gamma_v)$, where γ_v is the geodesic with initial velocity v. Expressed in terms of coordinates, a first-integral becomes a first-order ordinary differential equation for the coordinates of γ, with additive constant $k(\gamma)$.

The line element ds^2 of M is a first-integral since evaluated on γ' it gives the constant $\langle \gamma', \gamma' \rangle = q$. Because the orbit of a geodesic is unaffected by reparametrization, we usually take q to be $-1, 0$, or $+1$. Recall that when a timelike geodesic γ is interpreted as a freely falling particle, $q = -1$ means parametrization by proper time and γ' is four-velocity. For physical interpretations it is sometimes convenient to use a parametrization incorporating the mass m of the particle, with γ' as energy-momentum vector and $q = \langle \gamma', \gamma' \rangle = -m^2$ (see Remark 1.6.4).

A major source of first integrals is the conservation lemma (Lemma 1.3.7) which asserts that if X is a Killing vector field on M and γ is a geodesic of M, then $<\gamma', X>$ is constant along γ. Thus the function $v \to <v, X_p>$ for all $v\epsilon T_p(M) \subset TM$ is a first-integral, as above. Every Kerr spacetime has, by definition, two independent Killing vector fields, the extensions $\tilde{\partial}_t$ and $\tilde{\partial}_\varphi$ of the Boyer–Lindquist coordinate vector fields ∂_t and ∂_φ. These provide two more first-integrals.

Definition 4.1.1 *For a Kerr geodesicc γ, (1) the constant $E = -<\gamma', \tilde{\partial}_t>$ is called its* energy, *and (2) the constant $L = <\gamma', \tilde{\partial}_\varphi>$ is its* angular momentum *around the axis.*

These physical terms derive from measurement of a freely falling material particle α taken by the Kerr stationary observers in the region $r \gg 1$ of the Kerr exterior (Boyer–Lindquist block I) where the metric is approximately Minkowskian. Recall that if α has mass $m > 0$ and proper time τ, then $\alpha' = d\alpha/ds = m\,d\alpha/d\tau$. Thus for $r \gg 1$,

$$L = <\alpha', \partial_\varphi> \sim mg_{\varphi\varphi}d\varphi/d\tau \sim mr^2 \sin^2 \varphi\, d\varphi/d\tau \sim mR^2 d\varphi/d\tau,$$

where R is orthogonal distance to the axis. This is essentially the classical Newtonian formula for angular momentum. Next,

$$E = -<\alpha', \partial_t> \sim -m\, g_{tt}dt/d\tau \sim m\,dt/d\tau - 2m\text{M}/r.$$

We saw in Section 1.6 that the term $m\,dt/d\tau$ comprises both the *kinetic* energy of α and its *mass* energy. The term $-2m\text{M}/r$ represents (but for the 2) the usual Newtonian *potential* energy of the particle at distance r from a star of mass M.

Thus the first integrals L and E provide relativistic versions of conservation of energy and axial angular momentum for freely falling material particles in Kerr spacetime, and in physical contexts, first integrals are often called *constants of motion*. However, E and L will be used for all geodesics, not just particles. Note: In the literature, L is also written as L_z or Φ.

Remark 4.1.2 Signs of E and L.

1. Let γ be a nonspacelike geodesic in a region where ∂_t is timelike, namely in blocks I and III outside the ergosphere (Section 2.4). Then $E \neq 0$. Furthermore, if both γ and ∂_t are future-pointing timelike, γ has positive energy, $E > 0$.

2. Since $\tilde{\partial}_\varphi = 0$ on the axis A, any geodesic that meets A has $L = 0$. Geodesics are said to have *direct* orbit if $L > 0$, and *retrograde* orbit, if $L < 0$. The sign of the invariant L need not be the same as that of the coordinate object $d\varphi/ds$ (see the equation for L below.)

3. For $c, d \in \mathbf{R}$ the effect of geodesic reparametrization $\tilde{\gamma}(s) = \gamma(cs + d)$ on energy and angular momentum is just $\tilde{E} = cE$, $\tilde{L} = cL$. In particular, reversing the parametrization ($c = -1$), reverses the signs of E and L. For timelike or spacelike geodesics, taking $q = -1$ or $+1$, respectively, normalizes L and E. For null geodesics, only the ratio $E: L$ is important.

Our goal now is to use the first-integrals L, E to find *first-order* differential equations for the Boyer–Lindquist coordinates $\varphi(s)$ and $t(s)$ of an arbitrary geodesic. Substituting $\gamma' = r'\partial_r + \vartheta'\partial_\vartheta + \varphi'\partial_\varphi + t'\partial_t$ into the definitions of E and L gives

$$-E = t'g_{tt} + \varphi'g_{t\varphi},$$
$$L = t'g_{t\varphi} + \varphi'g_{\varphi\varphi}.$$

These equations can be solved for t' and φ' since the matrix of coefficients has determinant $g_{tt}g_{\varphi\varphi} - g_{\varphi t}^2 = -\Delta S^2$, which is nonzero off the axis. However, the resulting formulas depend on the complicated metric components $g_{\varphi\varphi}$, g_{tt}, $g_{\varphi t}$ and are not very informative.

Experience in Chapter 2 suggests that better results can be gotten by involving the metric identities (Lemma 2.1.1), as expressed in terms of the canonical vector fields $V = (r^2 + a^2)\partial_t + a\partial_\varphi$ and $W = \partial_\varphi + aS^2\partial_t$. Since V and W are the "rotational versions" of ∂_t and ∂_φ, respectively, the scalar products of γ' with $-V$ and W bear comparison with E and L. We assign them notation as follows.

Definition 4.1.3 *If γ is a geodesic with energy E and angular momentum L, let*

$$\mathbb{P}(r) = -<\gamma', V> = (r^2 + a^2)\,E - La, \quad and \quad \mathbb{D}(\vartheta) = <\gamma', W> = L - EaS^2.$$

(The formulas follow by substituting the definitions of V and W, and here, as always, we use the abbreviations $S = \sin\vartheta$, $C = \cos\vartheta$.) These analogues of E and L are not constant, but each depends on only a single globally defined function, r or ϑ.

Lemma 4.1.4 *For a geodesic expressed in Boyer–Lindquist coordinates,*

(1) $-at' + (r^2 + a^2)\varphi' = \mathbb{D}/S^2$,

(2) $t' - aS^2\varphi' = \mathbb{P}/\Delta$.

Proof. It suffices to substitute $\gamma' = r'\partial_r + \vartheta'\partial_\vartheta + \varphi'\partial_\varphi + t'\partial_t$ into the definitions of $\mathbb{P}$ and $\mathbb{D}$, and then use the metric identities—in the vector form given after Lemma 2.1.1. For equation (1),

$$\mathbb{D} = <\gamma', W> = \varphi'<\partial_\varphi, W> + t'<\partial_t, W> = \varphi'((r^2 + a^2)S^2) + t'(-aS^2)$$

and for (2),

$$-\mathbb{P} = <\gamma', V> = \varphi'<\partial_\varphi, V> + t'<\partial_t, V> = \varphi'(\Delta a S^2) + t'(-\Delta).$$

$\square$

Equation (1) in this lemma remains valid on the axis, where $S = 0$, because if γ ever meets the axis, then $L = 0$; hence $\mathbb{D}/S^2 = (L - aS^2E)/S^2$ reduces to $-aE$. In equation (2), division by Δ is not yet a problem since only Boyer–Lindquist cooordinates are involved.

Proposition 4.1.5 *The Boyer–Lindquist first-order geodesic equations for φ and t are*

$$\rho^2\varphi' = \mathbb{D}/S^2 + a\mathbb{P}/\Delta,$$
$$\rho^2 t' = a\mathbb{D} + (r^2 + a^2)\mathbb{P}/\Delta$$

where $\mathbb{D} = \mathbb{D}(\vartheta) = L - EaS^2$, and $\mathbb{P} = \mathbb{P}(r) = (r^2 + a^2)E - La$.

Proof. The equations in Lemma 4.1.4 can be solved for φ' and t', since the determinant of coefficients is $\rho^2 > 0$. $\square$

Consequently, at any point $\gamma(s_0)$ there are (linearly isomorphic) correspondences $L, E \leftrightarrow \mathbb{P}_0, \mathbb{D}_0 \leftrightarrow \varphi'_0, t'_0$.

The equations in this proposition express φ' and t' as functions of r and ϑ only. This is valuable in the later construction of geodesics since, given $r(s)$ and $\vartheta(s)$, the equations give explicit integral formulas for $\varphi(s)$ and $t(s)$.

The discovery of the ergosphere suggested to Roger Penrose (1969) an ingenious thought experiment in which energy is extracted from a black hole. Give the Kerr exterior its usual time-orientation, under which the coordinate vector field ∂_t is future-pointing where it is timelike, that is, outside the ergosphere $\mathcal{E}$. Let α be a material particle that falls from $r \gg 1$ into $\mathcal{E}$. As noted above, α necessarily has positive energy $E_\alpha > 0$, but within $\mathcal{E}$ particles can have negative energy since (as we saw in Section 2.2) ∂_t is spacelike there. Now suppose that α splits into two freely falling material particles β and γ. At any such splitting, energy-momentum is conserved, that is, at the splitting event, $\alpha' = \beta' + \gamma'$. Taking scalar products shows that energy and angular momentum are also conserved: $E_\alpha = E_\beta + E_\gamma$, $L_\alpha = L_\beta + L_\gamma$. The Penrose splitting is arranged so that β has negative energy $E_\beta < 0$ and continues on through the horizon $r = r_+$, while γ returns through the stationary limit to ordinary exterior regions. But *the emerging γ has more energy than the infalling α*. In fact, since $E_\gamma = E_\alpha - E_\beta$, the gain in energy is $|E_\beta|$. We envisage here that within the horizon the black hole has a fiery core that devours the particle β and thus has its total mass M reduced by $|E_\beta|$.

Furthermore, since the canonical vector field V is timelike and future-pointing in block I,

$$0 > <\beta', V> = (r^2 + a^2)<\beta', \partial_t> + a<\beta', \partial_\varphi> = -(r^2 + a^2)E_\beta + aL_\beta.$$

Hence $L_\beta < E_\beta < 0$, so γ has larger angular momentum than α, balanced by a decrease in the total angular momentum Ma of the black hole. For more details, see Wald (1984), Chandrasekhar (1983).

4.2   The Carter Constant

Our goal is to find a fourth first-integral for Kerr geodesics. We begin by expressing the simplest one, $q = <\gamma', \gamma'>$, in terms of the coordinate functions $r(s)$ and $\vartheta(s)$.

Lemma 4.2.1   *For a geodesic γ in Kerr spacetime,*

$$q\rho^2 = \frac{\rho^4}{\Delta}r'^2 + \rho^4\vartheta'^2 - \frac{\mathbb{P}^2}{\Delta} + \frac{\mathbb{D}^2}{S^2},$$

where, as usual, $q = <\gamma', \gamma'>$, $\mathbb{P} = (r^2 + a^2)E - aL$, and $\mathbb{D} = L - aES^2$.

Proof.   In Boyer–Lindquist coordinates, $\gamma' = r'\partial_r + \vartheta'\partial_\vartheta + \varphi'\partial_\varphi + t'\partial_t$. Since ∂_r, ∂_ϑ, V, and W are mutually orthogonal, $\varphi'\partial_\varphi + t'\partial_t$ can be replaced by $<\gamma', V>/<V, V>V + <\gamma', W>/<W, W>W$. By definition, $\mathbb{P} = -<\gamma', V>$ and $\mathbb{D} = <\gamma', W>$, and by Lemma 2.1.3, $<V, V> = -\Delta\rho^2$ and $<W, W> = \rho^2 S^2$. Thus,

$$\gamma' = r'\partial_r + \vartheta'\partial_\vartheta - \frac{\mathbb{P}}{\Delta\rho^2}V + \frac{\mathbb{D}}{\rho^2 S^2}W.$$

Hence

$$q = <\gamma', \gamma'> = \frac{\rho^2}{\Delta}r'^2 + \rho^2\vartheta'^2 + \frac{\mathbb{P}^2}{\Delta\rho^2} + \frac{\mathbb{D}^2}{\rho^2 S^2}.$$

The result follows.   □

This lemma remains valid, by continuity, even when γ crosses horizons or axes. In the latter case, since $L = 0$, the threatening term $\mathbb{D}^2/S^2$ reduces to $a^2 E^2 S^2$.

The equation in Lemma 4.6.1 is almost separable, for—with its left side split into $qr^2 + qa^2C^2$—it can be rewritten as

$$-(1/\Delta)\rho^4 r'^2 + qr^2 + \mathbb{P}^2/\Delta = \rho^4 \vartheta'^2 - qa^2C^2 + \mathbb{D}^2/S^2.$$

On the left, Δ and $\mathbb{P}$ depend solely on r, while on the right, S, C and $\mathbb{D}$ depend solely on ϑ. Thus, except for the factor ρ^4, the left side involves only functions of r, the right side only functions of ϑ. If the equation were fully separable, then each side would necessarily be *constant*. A remarkable result, found by Brandon Carter (1968), asserts that, even though separability fails, the two sides are constant anyway.

The following statement of the result introduces two functions $R(r)$ and $\Theta(\vartheta)$ that dominate the study of Kerr geodesics.

Theorem 4.2.2 *Associated with each Kerr geodesic γ is a constant $\mathcal{K} = \mathcal{K}_\gamma$ such that the coordinates r and ϑ of γ satisfy*

(1) $\rho^4 r'^2 = R(r) = \Delta(qr^2 - \mathcal{K}) + \mathbb{P}^2$,

(2) $\rho^4 \vartheta'^2 = \Theta(\vartheta) = \mathcal{K} + qa^2C^2 - \mathbb{D}^2/S^2$.

These first-order geodesic equations are called the r-equation (or radial equation) and ϑ-equation (or colatitude equation).

As an introduction to a proof of Carter's theorem, we consider a special case. Only coordinates r and ϑ are involved in the theorem, so a good place for experiment is a *polar plane P*—these being totally geodesic surfaces coordinatized by r, ϑ (see section 2.3).

Restricted to P, the Boyer–Lindquist form of the Kerr line-element becomes $ds^2 = \rho^2/\Delta dr^2 + \rho^2 d\vartheta^2$. The geodesic differential equations, as found by Lagrangian methods (section 1.3), are

$$((\rho^2/\Delta)r')' = \frac{1}{2}[\partial_r(\rho^2/\Delta)r'^2 + \partial_r(\rho^2)\vartheta'^2],$$

$$(\rho^2\vartheta')' = \frac{1}{2}[\partial_\vartheta(\rho^2/\Delta)r'^2 + \partial_\vartheta(\rho^2)\vartheta'^2]$$

where ∂/∂_r is abbreviated to ∂_r. When these partial derivatives are evaluated the first equation is unpleasant, but $\rho^2 = r^2 + a^2C^2$ and Δ depends only on r, so the second is

$$(\rho^2\vartheta')' = -a^2 SC[r'^2/\Delta + \vartheta'^2].$$

This is an obvious relative of $q = \langle \gamma', \gamma' \rangle = \rho^2[r'^2/\Delta + \vartheta'^2]$, which we use to eliminate r', leaving

$$(\rho^2 \vartheta')' = -q\rho^{-2}a^2 SC.$$

This equation is *integrable*, as is evident when it is written as

$$2(\rho^2 \vartheta')(\rho^2 \vartheta')' = -2qa^2C(-S\vartheta').$$

Its indefinite integral, with arbitrary constant $\mathcal{K}$, is the second equation in the following lemma.

Lemma 4.2.3 *Associated with each geodesic in a polar plane P is a constant $\mathcal{K}$ such that*

$$(1) \quad \rho^4 r'^2 = \Delta(qr^2 - \mathcal{K}),$$
$$(2) \quad \rho^4 \vartheta'^2 = qa^2C^2 + \mathcal{K}.$$

Proof. To get equation (1), multiply $q = \rho^2[r'^2/\Delta + \vartheta'^2]$ by ρ^2/Δ and substitute from equation (2) to obtain

$$q\rho^2\Delta = \rho^4 r'^2 + \Delta(qa^2C^2 + \mathcal{K}).$$

Hence, $\rho^4 r'^2 c = \Delta[q\rho^2 - a^2C^2 - \mathcal{K}] = \Delta(qr^2 - \mathcal{K})$.   □

The differential equations in this lemma are special cases of the r- and ϑ-equations in Theorem 4.2.2. In fact, if γ is a geodesic in a polar plane, then γ' is orthogonal to both ∂_φ and ∂_t, hence $L = E = 0$. It follows that $\mathbb{P} = \mathbb{D} = 0$. Then the functions $R(r)$ and $\Theta(\vartheta)$ in Theorem 4.2.2 reduce to $\Delta(-\mathcal{K} + qr^2)$ and $\Theta(\vartheta) = \mathcal{K} + qa^2C^2$, respectively.

This proof for polar planes extends directly to the general case; the details appear in Appendix C.

To express the first-integral $\mathcal{K}$ as a function $\mathcal{K}: TM \to \mathbf{R}$, recall that ϑ is globally defined on any Kerr spacetime and, though not smooth at the poles, has well-defined directional derivatives $v[\vartheta]$ there. Also, as mentioned earlier, $\mathbb{D}^2/S^2$ is always finite. Thus the ϑ-equation gives the formula $\mathcal{K}(v) = \rho^4 v[\vartheta]^2 + \mathbb{D}^2/S^2 - qa^2C^2$. (See also Lemma 4.2.7.)

Adding the two equations in Theorem 4.2.2 gives the equation in Lemma 4.2.1; thus, *any two of these three equations implies the third.*

Example 4.2.4 We find the first integrals q, L, E, $\mathcal{K}$ for principal null geodesics α. Certainly, $q = 0$. For the Boyer–Lindquist parametrizations, $\alpha' = \pm\partial_r + \Delta^{-1}V$. Since $\langle V, \partial_\varphi \rangle = \Delta aS^2$ and $\langle V, \partial_t \rangle = -\Delta$,

$$L = \langle \alpha', \partial_t \rangle = aS^2 \quad \text{and} \quad E = -\langle \alpha', \partial_\varphi \rangle = +1.$$

(Recall that only the ratio L: E is significant; here $E = +1$ is merely a convenient choice.) There are two special cases:

First, every null geodesic in a horizon H is principal, and all have $L = E = 0$. In fact, they parametrize integral curves of V, which, on H, is orthogonal to both ∂_t and ∂_φ. Second, the principal nulls in the axis have $L = 0$. But they have $E \neq 0$, except for those already in H: the restphotons held at $H \cap A$.

Finally, *every principal null geodesic has* $\mathcal{K} = 0$. In fact, the cases above give $\mathbb{D} = L - aES^2 = 0$, so the ϑ-equation in Theorem 4.2.2 reduces to $\mathcal{K} = 0$. (Lemma 4.2.8 will show that this is the smallest possible value for a null geodesic.)

A geodesic γ is completely determined by its initial position $\gamma(s_0)$ and tangent $\gamma'(s_0)$, and in a spacetime, four coordinate values will determine each of these vectors. In Kerr spacetime, the four first-integrals of γ can usefully replace the coordinates of $\gamma'(s_0)$.

Lemma 4.2.5 *A geodesic γ starting off-axis in a Boyer–Lindquist block, is uniquely determined by the initial data r_0, ϑ_0, φ_0, t_0; sgn r_0', sgn ϑ_0' and the first integrals q, E, L, $\mathcal{K}$.*

Proof. These numbers give us the initial values ρ_0^2, $\mathbb{P}_0$, $\mathbb{D}_0$, Δ_0 and the constant $\mathcal{K}$. Then Theorem 4.2.2 gives $(r_0')^2$ and $(\vartheta_0')^2$, and the signs determine r_0' and ϑ_0'. Finally, Proposition 4.1.5 gives φ_0' and t_0', so the coordinate initial conditions are complete.   □

Now consider the corresponding *existence* result. The first-integrals q, L, E, $\mathcal{K}$ are well-defined functions on the tangent bundle TM; hence so are $\mathbb{P}$ and $\mathbb{D}$, and then R and Θ. For example, if v is a tangent vector v at a point p, then $E(v) = -<v, \tilde{\partial}_t>$, and $\mathbb{D}(v) = L(v) - aE(v)\sin^2(\vartheta(p))$.

Proposition 4.2.6 *Let p_0 be a point off-axis in a Boyer–Lindquist block. Given any numbers q^*, L^*, E^*, $\mathcal{K}^*$ such that the corresponding values of R and Θ are non-negative at p_0, there exists a geodesic γ starting at p_0 for which q^*, L^*, E^*, $\mathcal{K}^*$ are the first-integrals. Furthermore, when $r'(0)^2 \neq 0$ both signs can be realized for $r'(0)$; similarly for $\vartheta'(0)$.*

Proof. It suffices to find a tangent vector v at p for which $q(v) = q^*$, $L(v) = L^*$, $E(v) = E^*$, $\mathcal{K}(v) = \mathcal{K}^*$. Write $v = A\partial_r + B\partial_\vartheta + C\partial_\varphi + D\partial_t$. Then the geodesic γ_v with $\gamma'(0) = v$ will have $A = r'(0)$, $B = \vartheta'(0)$, $C = \varphi'(0)$, $D = t'(0)$.

As in section 4.1, L^* and E^* uniquely determine C and D so that $<v, \partial_t> = -E^*$ and $<v, \partial_\varphi> = L^*$. Thus, they also determine $\mathbb{P}_0$ and $\mathbb{D}_0$. Now use the definitions of R and Θ in Theorem 4.2.2—with q and $\mathcal{K}$ replaced by q^* and $\mathcal{K}^*$—to find A^2 and B^2. (The hypotheses $R \geq 0$, $\Theta \geq 0$ insure that these numbers are nonnegative.) When they are nonzero, either sign can used independently for A and for B.

Adding the r and ϑ equations gives the formula in Lemma 4.2.1 that expresses, in coordinate terms, the equation $<v, v> = q^*$. Finally, the formula above for $\mathcal{K}: TM \to \mathbf{R}$ (with $v[\vartheta]$ replaced by B) gives $\mathcal{K}(v) = \mathcal{K}^*$.   □

Note that this proposition shows that—away from axes, equator, and horizons—the four first-integrals are independent.

Suppose that β is a curve that satisfies the four first-integral equations (Proposition 4.1.5 and Theorem 4.2.2) expressed in terms of some numbers q, E, L, $\mathcal{K}$. Is β a geodesic? The answer is almost always *yes*, but the exceptions occur since the first-integral equations are *first* order differential equations. Thus, they can—and do—have singular solutions that are envelopes of generic solutions. These singular solutions do not correspond to geodesics and will be easily found in Section 4.3. We conclude that in Kerr spacetime nothing is lost by using by the first-integral equations. Indeed, the gain is enormous, since the geodesic equations

$$x^{k''} + \Sigma \Gamma_{ij}^k x^{i'} x^{j'} = 0 \qquad (1 \leq k \leq 4)$$

are a system of second-order nonlinear differential equations that are almost always too complicated to be of much practical use. By contrast, the first-integral equations are surprisingly simple: Those for φ and t in Proposition 4.1.5 are immediately integrable, given $r(s)$ and $\vartheta(s)$. The equations for $r(s)$ and $\vartheta(s)$ in Carter's Theorem (Theorem 4.2.2) are, if not trivial, at least tractable. Their coupling—by the same function $\rho^2 > 0$—is minimal, the function $R(r)$ is a polynomial of degree four, and $\Theta = \Theta(\vartheta)$ depends in an elementary way on $S = \sin \vartheta$ and $C = \cos \vartheta$.

One geometrical interpretation of the first-integral $\mathcal{K}$ of a geodesic γ is as a measure of the relation of γ' to the principal planes Π along γ. In the special case where γ' is always in Π, we say that γ is a *principal* geodesic. In general, let γ'_Π and $\gamma'_\perp$ be the components of γ' in Π and $\Pi^\perp$, respectively. Recall that Π is timelike and $\Pi^\perp$ spacelike; to emphasize that the scalar product on $\Pi^\perp$ is positive definite, we often write $<x, x>$ as $|x|^2$ when $x \in \Pi^\perp$.

Lemma 4.2.7   *If γ is a geodesic with first integrals $\mathcal{K}$ and q, then*

$$\mathcal{K} = \rho^2 |\gamma'_\perp|^2 - qa^2 C^2 = qr^2 - \rho^2 <\gamma'_\Pi, \gamma'_\Pi>.$$

where ρ^2 and $C = \cos \vartheta$ depend, of course, on the coordinates r, ϑ of γ.

Proof. In Boyer–Lindquist terms, the principal null geodesics have tangents $\pm \partial_r + \Delta^{-1} V$, so $\Pi = \operatorname{span}\{\partial_r, V\}$. Thus the equation for γ' in the proof of Lemma 4.2.1 shows that

$$\gamma'_\Pi = r'\partial_r - \frac{\mathbb{P}}{\Delta \rho^2} V, \qquad \gamma'_\perp = \vartheta' \partial_\vartheta + \frac{\mathbb{D}}{\rho^2 S^2} W.$$

Consequently,

$$<\gamma'_\Pi, \gamma'_\Pi> = (\rho^2/\Delta) r'^2 - \mathbb{P}^2/(\rho^2 \Delta^2), \qquad |\gamma'_\perp|^2 = \rho^2 \vartheta'^2 + \mathbb{D}^2/(\rho^2 S^2).$$

Then the ϑ first-integral equation for $\mathcal{K}$ gives

$$\mathcal{K} = \rho^4 \vartheta'^2 + \mathbb{D}^2/S^2 - qa^2 C^2 = \rho^2 |\gamma'_\perp|^2 - qa^2 C^2.$$

The other expression follows from $q = <\gamma'_\Pi, \gamma'_\Pi> + |\gamma'_\perp|^2$.   □

Since r and ϑ are defined globally so are these formulas for $\mathcal{K}$. Both $|\gamma'_\perp|$ and $<\gamma'_\Pi, \gamma'_\Pi>$ can be computed using orthonormal bases, for example, E_0, E_1 for Π and E_2, E_3 for $\Pi^\perp$, where $\{E_i\}$ is the Boyer–Lindquist frame field (Section 2.6). Also if n_1, n_2 is a basis for Π consisting of null vectors (say, $\pm \partial_t + \Delta^{-1} V$), then the expansion

$$\gamma'_\Pi = <\gamma', n_2>/<n_1, n_2> n_1 + <\gamma', n_1>/<n_1, n_2> n_2.$$

leads to the symmetric expression $<\gamma'_\Pi, \gamma'_\Pi> = <\gamma', n_1><\gamma', n_2>/<n_1, n_2>$. (Compare Walker, Penrose 1970.)

As with q, the reparametrization $\gamma(cs + d)$ changes $\mathcal{K}$ to $c^2 \mathcal{K}$. Replacing γ' above by an arbitrary tangent vector v expresses $\mathcal{K}$ as a function on TM. There, like q, it can readily be polarized to give a bilinear form $\mathcal{K}(v, w)$ on each (tangent space) fiber of TM.

Corollary 4.2.8 *Let γ be a Kerr geodesic.*
(1) *If γ is timelike, then $\mathcal{K} \geq 0$, and $\mathcal{K} = 0 \Leftrightarrow \gamma$ is a principal geodesic in the equator.*
(2) *If γ is null, then $\mathcal{K} \geq 0$, and $\mathcal{K} = 0 \Leftrightarrow \gamma$ is principal.*
(3) *If γ is spacelike, then $\mathcal{K} \geq -qa^2$, and $\mathcal{K} = -qa^2 \Leftrightarrow \gamma$ lies entirely in the axis.*

Proof. (1) Since $q < 0$, $\mathcal{K} = \rho^2 |\gamma_\perp|^2 + |q|a^2 C^2 \geq 0$, and equality holds if and only if both $|\gamma_\perp|$ and C are 0. (2) If $q = 0$, then $\mathcal{K} = \rho^2 |\gamma_\perp| \geq 0$. (3) Since $q > 0$, $\mathcal{K} = \rho^2 |\gamma'_\perp|^2 - qa^2 C^2 \geq -qa^2$, with equality if and only if $|\gamma_\perp|^2 = 0$ and $C^2 = 1$.   □

There are surprisingly many Kerr geodesics devoid of both energy and angular momentum. These lazy geodesics γ appear only in the totally geodesic submanifolds: polar plane, horizon, or axis. In fact, if γ is not contained in H or A, Boyer–Lindquist coordinates are applicable, and from Proposition 4.1.5,

$$L = E = 0 \Leftrightarrow \mathbb{P} = \mathbb{D} = 0 \Leftrightarrow t' = \varphi' = 0.$$

Thus γ is tangent to a polar plane, and hence remains in it.

The null case can be settled as follows (recall that a restphoton is a null geodesic that lies in a horizon).

Lemma 4.2.9   *For a Kerr null geodesic γ,*
(1) $\mathcal{K} = 0 \Leftrightarrow \gamma$ *is principal.*
(2) $L = E = 0$*, but* $\mathcal{K} \neq 0$ *(hence* $\mathcal{K} > 0$*)* $\Leftrightarrow \gamma$ *is in a (timelike) polar plane.*
(3) $\mathcal{K} = L = 0$ *but* $E \neq 0 \Leftrightarrow \gamma$ *is in* $A - H$*.*
(4) $L = E = \mathcal{K} = 0 \Leftrightarrow \gamma$ *is a restphoton.*

Proof.   (1) This equivalence follows from Corollary 4.2.8.

(2) If γ is in a polar plane P, then $L = E = 0$, but γ is not principal since the vectors $\pm \partial_r + \Delta^{-1} V$ are transversal to P. Conversely, $L = E = 0$ implies γ is in P, A, or H. But all null geodesics in the latter two are principal.

(3) If γ is in A, then it is principal and $L = 0$. The converse follows from Corollary 4.2.8. But we know that axial nulls have $E \neq 0$ except those also in H.

(4) A restphoton, as we have seen, is principal and has $L = E = 0$ (since V is orthogonal to both ∂_t and ∂_φ on horizons). Conversely, γ is principal, and $L = E = 0$ implies that γ is in H or A. But, as just noted, it is not in A unless it is also in H. Hence γ is a restphoton. $\square$

Thus the restphotons are uniquely characterized as the geodesics whose first-integrals are all zero.

Any function of first-integrals is again a first-integral. A particularly useful case is the first-integral Q defined by the relation $\mathcal{K} = Q + (L - aE)^2$. We call Q, as well as $\mathcal{K}$, *the Carter constant*. Each form has its advantages. Often $\mathcal{K}$ gives simpler formulas, and $\mathcal{K} \geq 0$ *holds for nonspacelike geodesics* (by the ϑ-equation), but the geometric meaning of Q is usually clearer. For example, later in this chapter we see that Q is crucial for the behavior of geodesics near $r = 0$ (e.g., hitting the ring singularity, passage from $r < 0$ to $r > 0$), and, as the following result suggests, Q decisively influences the ϑ motion of geodesics.

Lemma 4.2.10   *In terms of the Carter constant Q, the function Θ of Theorem 4.2.2 can be written as* $\Theta(\vartheta) = Q + C^2[a^2(E^2 + q) - L^2/S^2]$*.*

4.3   Equations and Extensions

One goal of this chapter is to show that if M is a maximal Kerr spacetime, M_f, M_e, or M_s, then all the geodesics of M are complete—except for those that hit the ring singularity Σ. This assertion is described concisely as follows.

Theorem 4.3.1   *The maximal Kerr spacetimes are complete mod Σ.*

It suffices to show that every geodesic $\gamma: [s_0, b) \to M$, with $b < \infty$, can be extended, as a geodesic, over $[s_0, \infty)$—unless some smaller extension hits the ring singularity. This result applies to $(a, s_0]$ via the reparametrization $s \to 2s_0 - s$, and hence proves the theorem. In this section we show that any geodesic can be extended at least until it hits a horizon or the ring singularity.

The following local extension property, valid in any semi-Riemannian manifold, is fundamental: For a geodesic $\gamma: [s_0, s_1) \to M$, *if $\gamma(s)$ approaches a point $p \in M$ as $s \to s_1$, then γ has a geodesic extension to a larger interval $[s_0, s_1 + \delta)$, and $\gamma(s_1) = p$.*

Also, recall from Chapter 1 that if a geodesic γ is tangent to a closed totally geodesic submanifold P at a single point, then γ lies entirely in P. Since the axis A and horizon H of Kerr spacetime are closed totally geodesic submanifolds, Kerr geodesics separate into two quite different types:

• Those that are entirely contained in either A or H,

• All others, which we call *regular*.

Regular geodesics may well meet A or H, but the remarks above show that if they do, they will cut through transversally. Explicitly, if a regular geodesic $\gamma: [s_0, s_1) \to M$ has $\gamma(s)$ approach a point p of A (or H) as $s \to s_1$, then γ can be geodesically extended past s_1, and at $p = \gamma(s_1)$ the vector $\gamma'(s_1)$ is not tangent to A (or H).

Consequently, the geodesic extension problem separates into:

• The case where γ, and hence its extensions, lie entirely in either A or H.

• The regular case, where we can assume that $\gamma: [s_0, b) \to M$ starts at a point $p = \gamma(s_0)$ in a Boyer–Lindquist block but not in the axis. (γ must meet such points, and a reparametrization will start it there.)

The regular case is proved in Section 4.4; the special cases, which require separate treatment, are deferred until Section 4.11 (axis) and Section 4.12 (horizons).

We begin by establishing some basic geodesic equations that will be useful in constructing extensions. In this section Boyer–Lindquist coordinates suffice, but to extend a geodesic through a horizon, Kerr coordinates will be needed.

Corollary 4.3.2 *The Kerr-star first-order geodesic equations for the coordinates*
φ^* *and* t^* *are*

$$\rho^2 \varphi^{*\prime} = S^{-2} \mathbb{D} + a\Delta^{-1}[\mathbb{P} \oplus (\pm\sqrt{R})],$$

$$\rho^2 t^{*\prime} = a\mathbb{D} + (r^2 + a^2)\Delta^{-1}[\mathbb{P} \oplus (\pm\sqrt{R})],$$

where $\pm$ *denotes the sign of* r'. *Here the sign* $\oplus$ *is plus,* $(+)$, *but for star-Kerr*
coordinates $^*\varphi$, $^*t$ *it becomes a minus,* $(-)$.

Proof. In Proposition 4.1.5, substitute $t = t^* - T(r)$ and $\varphi = \varphi^* - A(r)$, where,
as usual, $dT/dr = (r^2 + a^2)/\Delta$ and $dA/dr = a^2/\Delta$. The derivative r' now
appears, but the square root of the geodesic equation $\rho^4 r'^2 = R$ in Theorem 4.2.2
gives $\rho^2 r' = \pm\sqrt{R}$. □

The following integral formulas are valid for both Boyer–Lindquist and Kerr
coordinates. Recall that $\mathbb{D} = L - aES^2$ and $\mathbb{P} = (r^2 + a^2)E - aL$.

Proposition 4.3.3 *Let* γ *be a Kerr geodesic defined on an interval* $[s_0, s_1)$. *Sup-*
pose $R(r) > 0$ *and* $\Theta(\vartheta) > 0$ *on* $[s_0, s_1)$, *so* r' *and* ϑ' *are nonvanishing there. Let*
$r_0 = r(s_0)$, $\vartheta(s_0) = \vartheta(s_0)$, *and* $\epsilon = sgn(r')sgn(\vartheta') = \pm 1$. *Then on* $[s_0, s_1)$,

$$(1) \quad \int_{r_0}^{r(s)} \frac{dr}{\sqrt{R}} = \epsilon \int_{\vartheta_0}^{\vartheta(s)} \frac{d\vartheta}{\sqrt{\Theta}}$$

$$(2) \quad s - s_0 = \left| \int_{r_0}^{r(s)} \frac{r^2 dr}{\sqrt{R}} \right| + \left| \int_{\vartheta_0}^{\vartheta(s)} \frac{a^2 C^2 d\vartheta}{\sqrt{\Theta}} \right|.$$

Proof. (1) Taking square roots of the first-order geodesic equations in Theorem
4.2.2 gives $\rho^2|r'| = \sqrt{R}$, and $\rho^2|\vartheta'| = \sqrt{\Theta}$. (Here both R and Θ are evaluated on
γ, hence are functions of s). These equations imply $|r'|/\sqrt{R} = 1/\rho^2 = |\vartheta'|/\sqrt{\Theta}$.
Since $\epsilon = sgn(r')sgn(\vartheta')$, integrating from s_0 to s yields

$$\int_{s_0}^{s} \frac{r' ds}{\sqrt{R}} = \epsilon \int_{s_{s_0}} \frac{\vartheta' ds}{\sqrt{\Theta}},$$

Then a change of variables in these integrals gives equation (1).
   (2) Since $\rho^2|r'|/\sqrt{R} = 1$ and $\rho^2 = r^2 + a^2 C^2$,

$$s - s_0 = \int_{s_0}^{s} ds = \int_{s_0}^{s} \frac{\rho^2 |r'| ds}{\sqrt{R}} = \int_{s_0}^{s} \frac{r^2 |r'| ds}{\sqrt{R}} + \int_{s_0}^{s} \frac{a^2 C^2 |r'| ds}{\sqrt{R}}.$$

But from equation (1), $|r'|/\sqrt{R} = |\vartheta'|/\sqrt{\Theta}$. Substituting this relation into the preceding integral gives

$$s - s_0 = \int_{s_0}^{s} \frac{r^2 |r'| ds}{\sqrt{R}} + \int_{s_0}^{s} \frac{a^2 C^2 |\vartheta'| ds}{\sqrt{\Theta}}.$$

Again a change of variables gives the required result.   □

Equation (1) in the preceding proposition can be expressed in terms of the functions

$$F(r) = \int_{r_0}^{r} \frac{dr}{\sqrt{R}}, \qquad G(\vartheta) = \int_{\vartheta_0}^{\vartheta} \frac{d\vartheta}{\sqrt{\Theta}},$$

as $F(r(s)) = \epsilon G(\vartheta(s))$ on $[s_0, s_1)$. Under hypotheses as above, both F and G are strictly increasing functions, so they have inverse functions. Hence equation (1) can be solved for

$$r(s) = F^{-1}(\epsilon G(\vartheta(s))) \quad \text{or} \quad \vartheta(s) = G^{-1}(\epsilon F(r(s))).$$

Thus a knowledge of either coordinate function $r(s)$ or $\vartheta(s)$ determines the other. From a more practical viewpoint, on an interval where $R(r) > 0$ we get a separable first-order differential equation

$$d\vartheta/dr = \vartheta'/r' = \pm\sqrt{\Theta}(\vartheta)/\sqrt{R}(r),$$

the sign to be adjusted at zeros of $\Theta(\vartheta)$. This equation can be integrated—numerically if not explicitly—to show the dependence of colatitude ϑ on radius r or the reverse.

Proposition 4.3.4 *Let γ be a Kerr geodesic defined on an interval $[s_0, s_1)$. If $R(r) > 0$, $\Theta(\vartheta) > 0$, and $\Delta(r) \neq 0$ hold for all s in $[s_0, s_1)$, then for Kerr-star coordinates,*

$$(1) \quad \varphi^*(s) - \varphi_0^* = \pm \int_{\vartheta_0}^{\vartheta(s)} \frac{\mathbb{D}S^{-2} d\vartheta}{\sqrt{\Theta}} + \int_{r_0}^{r(s)} \frac{a}{\Delta} \left[\frac{\pm\mathbb{P}}{\sqrt{R}} \oplus 1 \right] dr.$$

$$(2) \quad t^*(s) - t_0^* = \pm \int_{\vartheta_0}^{\vartheta(s)} \frac{a\mathbb{D}}{\sqrt{\Theta}} d\vartheta + \int_{r_0}^{r(s)} \frac{r^2 + a^2}{\Delta} \left[\frac{\pm\mathbb{P}}{\sqrt{R}} \oplus 1 \right] dr.$$

Before each ϑ integral, $\pm$ is $\mathrm{sgn}\,\vartheta'$*; in each r integral $\pm$ is* $\mathrm{sgn}\,r'$*. As usual, $\oplus$ means (+) here, but for star-Kerr coordinates* $*\varphi$*, $*t$ becomes (−).*

Proof. For equation (1), integrating the φ equation in Corollary 4.3.2 gives

$$\varphi^*(s) - \varphi_0^* = \int_{s_0}^{s} \frac{1}{\rho^2}\left[\frac{\mathbb{D}}{S^2} + \frac{a}{\Delta}(\mathbb{P} \oplus (\pm\sqrt{R}))\right] ds.$$

But we saw above that

$$\frac{1}{\rho^2} = \frac{\pm\vartheta'}{\sqrt{\Theta}} = \frac{\pm r'}{\sqrt{R}}.$$

When this is used in the integral we obtain

$$\varphi^*(s) - \varphi_0^* = \pm\int_{s_0}^{s} \frac{\mathbb{D}/S^2}{\sqrt{\Theta}}\vartheta'\,ds + \int_{s_0}^{s} \frac{a}{\Delta}\left[\frac{\pm\mathbb{P}}{\sqrt{R}} \oplus 1\right]t'\,ds.$$

Again, a change of variables gives the result. The case of t^* is similar.   □

Now we show that a regular Kerr geodesic can be indefinitely extended as long as it stays in a single Boyer–Lindquist block and, in block III, does not hit Σ. (For geodesics, "regular," as above, is equivalent to starting off-axis in a Boyer–Lindquist block.)

Lemma 4.3.5 (Existence). *Let s_0; r_0, ϑ_0, φ_0, t_0; q, E, L, Q be regular initial data (as in Lemma 4.2.5). If r_0' and ϑ_0' are both nonzero, there exists a largest interval $[s_0, s_1)$, $s_1 \le \infty$, on which*
1. *there is a unique geodesic γ satisfying these initial conditions, and*
2. *$R(r) > 0$, $\Theta(\vartheta) > 0$, and $\Delta(r)$ is nonvanishing.*

Proof. Let $[s_0, b)$, $b \le \infty$, be the largest interval on which a geodesic γ can be defined that satisfies the initial conditions. Since $\gamma(0)$ is in a Boyer–Lindquist block, $\Delta_0 \neq 0$. And since r_0' and ϑ_0' are nonzero, the first-integral equations for r and ϑ show that R_0 and Θ_0 are positive. Then let $[s_0, s_1)$ be the largest subinterval of $[s_0, b)$ on which these three functions retain these properties.   □

In this lemma, if $s_1 = \infty$, that is, if the geodesic is defined on $[s_0, \infty)$, there is nothing to be proved, so we are free to assume that s_1 is finite. Then, if $\rho^2 = r^2(s) + a^2\cos^2\vartheta(s)$ approaches 0 as $s \to s_1$, any geodesic with these coordinates hits the ring singularity. This is the other of the conclusions we wish to reach, so we can always assume $\rho^2 \not\to 0$.

Note that since $\rho^4 r'^2 = R > 0$ and $\rho^4 \vartheta'^2 = \Theta > 0$, both coordinates $r(s)$ and $\vartheta(s)$ of γ are strictly monotonic on $[s_0, s_1)$. Hence the following limits exist:

$$r_1 = \lim_{s \to s_1} r(s), \qquad \vartheta_1 = \lim_{s \to s_1} \vartheta(s).$$

(Here s_1, r_1, and ϑ_1 may be infinite.) Whether or not the geodesic can be extended to s_1, the continuous functions R, Θ, ρ^2, Δ (evaluated on $r(s)$, $\vartheta(s)$) approach (possibly infinite) limits as $s \to s_1$. Note that if one of these functions, say Δ, does *not* approach 0 as $s \to s_1$, that is, if $\Delta(r_1) \neq 0$, then $\Delta(r(s))$ is bounded away from 0 on $[s_0, s_1)$.

Lemma 4.3.6 *With notation as above, suppose s_1 is finite. Then, (1) both r_1 and ϑ_1 are finite, and (2) if $\Delta(r_1) \neq 0$, the geodesic γ of Lemma 4.3.5 can be extended to a strictly larger interval $[s_0, s_1 + \delta)$.*

Proof. (1) For s_1 finite we assume, as mentioned above, that ρ^2 is bounded away from zero. Write the geodesic equation $\rho^2 r' = \pm \sqrt{R}$ as

$$\frac{dr}{ds} = \frac{\pm R}{\rho^2}$$

where R and ρ^2 are evaluated on $r(s)$, $\vartheta(s)$. As a polynomial in r, ρ^2 has degree 2, and R has degree at most 4. It follows that $|dr/ds|$ is bounded on $[s_0, s_1)$; hence r_1 is finite. The case of ϑ is similar. (If γ does not cross the axis, then $0 < \vartheta < \pi$, so no proof is required.)

(2) In integral form the equations in Proposition 4.1.5 become

$$\varphi - \varphi_0 = \int_{s_0}^{s} \frac{1}{\rho^2} \left[\frac{\mathbb{D}}{S^2} + \frac{a\mathbb{P}}{\Delta} \right] ds, \qquad t - t_0 = \int_{s_0}^{s} \frac{1}{\rho^2} \left[a\mathbb{D} + \frac{r^2 + a^2}{\Delta} \mathbb{P} \right] ds.$$

The functions in these denominators present no problems since both ρ^2 (by assumption) and Δ (by hypothesis) are bounded away from zero. Recall also that if the geodesic has $L \neq 0$, then $S = \sin \vartheta$ is never zero, and if $L = 0$, then $\mathbb{D}/S^2$ reduces to $-a E$. Thus, the geodesic coordinates $\varphi(s)$ and $t(s)$ both approach finite limits as $s \to s_1 < \infty$.

Since its coordinates all approach finite limits, the geodesic γ approaches a point of its Boyer–Lindquist block, and hence, as noted earlier, γ can be extended past s_1. □

Assertion (1) in this lemma is a major step toward the proof of completeness mod Σ, since if the r or ϑ coordinate of a geodesic races to infinity on a finite interval the geodesic cannot be further extended. This assertion is not obvious; it holds because r and ϑ are globally defined functions on M and is not true in general for arbitrary coordinate functions.

Now we consider the significance of zeros of R and Θ.

Proposition 4.3.7 *With notation as above, assume r_1 and ϑ_1 are finite.*

(1) *If r_1 is at most a simple zero of R and ϑ_1 is at most a simple zero of Θ, then s_1 is finite.*

(2) *If r_1 is a repeated zero of R or if ϑ_1 is a repeated zero of Θ, then s_1 is infinite.*

Proof. In equation (2) of Proposition 4.3.3, let $r(s) \to r_1$ and $\vartheta(s) \to \vartheta_1$, obtaining improper integrals as follows:

$$
s - s_0 = \left| \int_{r_0}^{r_1} \frac{r^2 dr}{\sqrt{R}} \right| + \left| \int_{\vartheta_0}^{\vartheta_1} \frac{a^2 C^2 d\vartheta}{\sqrt{\Theta}} \right| .
$$

Write $R(r) = A + B(r_1 - r) + \cdots + E(r_1 - r)^4$.

(1) If r_1 is a simple zero of R, then $A = 0$, but $B \neq 0$. Thus, in the first improper integral above, the integrand behaves like $(r_1 - r)^{-1/2}$; hence the integral is convergent. Evidently the same is true if r_1 is not a zero, that is, if $A \neq 0$.

A similar argument shows that the second integral is convergent (though not a polynomial, Θ is analytic in ϑ). Thus s_1 is finite.

(2) If r_1 is a repeated zero of R, then $A = B = 0$. It follows that the integrand in the first improper integral behaves like $(r_1 - r)^{-1}$; hence the integral is divergent. A similar argument applies to the second integral. $\square$

This result has the following geometric meaning.

Corollary 4.3.8 *Suppose $R(r_0) = 0$. Let γ be a geodesic whose r-coordinate satisfies the initial conditions $r(s_0) = r_0$ and $r'(s_0) = 0$.*

(1) *If r_0 is a simple zero of R, then r_0 is a turning point of r, that is, $r'(s)$ changes sign at s_0.*

(2) *If r_0 is a repeated zero of R, then γ has constant coordinate $r = r_0$. Corresponding results hold with r and R replaced by ϑ and Θ.*

Proof. (1) Since R has a simple zero at r_0, there is a sequence $\{r_n\} \to r_0$ with $R(r_n) > 0$ for all n. Let γ_n be the geodesic with the same initial conditions as γ except that $r_n(0) = r_n$. In particular, then, $r'_n(0) \neq 0$.

Differentiation of $\rho^4 r'^2 = R(r)$ yields

$$(\rho^4)' r'^2 + 2\rho^4 r' r'' = (dR/dr) r'.$$

Taking r to be the r-coordinate r_n of γ_n gives

$$(\rho^4)' r_n'^2 + 2\rho^4 r_n' r_n'' = (dR/dr) r_n'.$$

Evaluate at $s = s_0$. Since $r_n'(s_0) \neq 0$, it can be cancelled , leaving

$$(\rho^4)'|_0 \, r_n'(s_0) + 2\rho^4|_0 \, r_n''(s_0) = (dR/dr)|_0.$$

Because geodesics depend smoothly on their initial conditions, we can take the limit as $n \to \infty$, getting

$$2\rho^4|_0 \, r''(s_0) = (dR/dr)(r_0).$$

By hypothesis R has a simple zero at $r_0 = r(s_0)$; hence $r''(s_0) \neq 0$. Thus, the zero of r' at s_0 is isolated. But r cannot move to the side of r_0 on which $R < 0$, so r' must change sign at s_0.

(2) Assume that r_0 is a repeated zero but that $r(s)$ is not constant. Then there exists an $s_1 \neq s_0$ such that $r(s_1) \neq r_0$. Viewing s_1 as the initial point of the (reparametrized) geodesic, we see it reaching the repeated zero r_0 of R on a finite parameter interval—but this contradicts assertion (2) of Proposition 4.3.7. □

Our progress so far can be summarized as follows:

Proposition 4.3.9 *A regular Kerr geodesic (that is, one that is not entirely contained in the axis or horizon) can be extended over the entire real line—unless it hits the ring singularity or approaches a horizon.*

Furthermore, if the coordinate function r(s) of the geodesic approaches a repeated zero r_0 of R, the approach is asymptotic, that is, as s continues to infinity, r approaches r_0 without reaching it. But when $r(s)$ approaches a simple zero r_0 of R, then r will reach r_0 and bounce back: turning point. Here as mentioned in section 4.2, the constant function $r = r_0$ is a singular solution of the first integral equation $\rho^4 r'^2 = R$ and does not correspond to a geodesic. Analogous results hold for ϑ and Θ.

4.4   Crossing Horizons

To prove that a regular Kerr geodesic γ is complete unless it hits the ring singularity, there remains the following case: $\gamma : [s_0, s_1) \to M$ lies in a single Boyer–Lindquist block, but approaches a horizon as $s \to s_1$. Thus $\Delta(r_1) = 0$ at $r_1 = \lim_{s \to s_1} r(s)$, so $r_1 = r_\pm$ for the slow Kerr black hole and $r_1 = M$ for the extreme one.

As the constructions in Chapter 3 show, γ lies in *two* Kerr patches: one Kerr-star and one star-Kerr.

Following Carter (1968) we now distinguish two cases in terms of the function $\mathbb{P} = (r^2 + a^2)E - La$.

$$\text{Case 1:}\quad \mathbb{P}(r_1) \neq 0, \quad\text{and}\quad \text{Case 2:}\quad \mathbb{P}(r_1) = 0.$$

If γ can actually be extended to cross the (totally geodesic) horizon at s_1, the crossing is transverse. Hence $\mathbb{P} = -{<}\gamma', V{>}$ is nonzero at $r_1 = \gamma(s_1)$ since the canonical vector field V is normal to the horizon. To prove the converse the essential step is the following fact.

Lemma 4.4.1 *In Case 1, that is, if $\mathbb{P}(r_1) \neq 0$, then either the Kerr-star or the star-Kerr coordinates of γ approach a finite limit as $s \to s_1$.*

Proof. Since s_1 is finite, Lemma 4.3.6 shows that the limit values r_1 and ϑ_1 are finite. Now consider the two integral formulas in Proposition 4.3.3. In both, the ϑ integral has a finite limit as $s \to s_1$. In fact, since s_1 is finite, Θ has at most a simple zero at ϑ_1. Consequently both the ϑ integrals are bounded by a constant times $\int (\vartheta - \vartheta_0)^{-1/2} \, d\vartheta$, which is finite.

For the r integrals, the terms a and $r^2 + a^2$ in the numerators are irrelevant, so the question is whether the following improper integral is finite:

$$\int_{r_0}^{r_1} \frac{1}{\Delta} \left[\frac{\pm \mathbb{P}}{\sqrt{R}} \oplus 1 \right] dr.$$

(Here, as usual, the sign $\oplus$ is $(+)$ for Kerr-star coordinates, $(-)$ for star-Kerr.) Write $R = \mathbb{P}^2 + f\Delta$, where $f = f(r) = qr^2 - \mathcal{K}$. As $s \to s_1$, Δ approaches zero and $\mathbb{P}$ does not, hence R does not. Thus, $\mathbb{P}^2/R = 1 - f\Delta/R$, and by the binomial theorem, $\mathbb{P}/\sqrt{R} = \pm 1 \mp \frac{1}{2}f\Delta/R + O(\Delta^2)$. Substituting this into the preceding integral gives

$$\int_{r_0}^{r_1} \frac{1}{\Delta} \left[\pm 1 \mp \frac{1}{2}f\Delta/R + O(\Delta^2) \oplus 1 \right] dr.$$

Naturally we choose the sign $\oplus$ to be *opposite* the sign $\pm$ since this produces the finite integral

$$\int_{r_0}^{r_1} \left[\mp 1 \frac{1}{2} f / R + O(\Delta) \right] dr.$$

Consequently, either the K^* or the $^*K$ coordinate of γ can be extended to s_1—and on past it. $\square$

If it is, say, the Kerr-star coordinates that are extendible, then evidently the geodesic γ meets and crosses the short horizon $r = r_\pm$ in the K^* patch. (See Figure 4.1.) Every short horizon is in exactly one Kerr patch, and the sign choice for $\oplus$ above tells (as specified in Proposition 4.4.6) whether the horizon crossed by γ is in a K^* or a $^*K$ patch.

For Case 2 (and later on) the following unpleasant technicality is needed:

Lemma 4.4.2 *Let* $\gamma: [s_0, s_1) \to M$ *be a geodesic with coordinate functions* $r(s)$ *and* $\vartheta(s)$.

(1) *If* $r(s) \to r_1$ *as* $s \to s_1 < \infty$, *there exists a* $\delta > 0$ *such that* $r'(s)$ *is either identically zero or never zero on* $[s_1 - \delta, s_1)$.
(2) *The corresponding assertion holds for* $\vartheta(s)$.
(3) *If* γ *is regular and* $\Delta(r_1) = 0$, *there exists an* $\epsilon > 0$ *such that* $\Delta(r(s)) \neq 0$ *on* $[s_1 - \epsilon, s_1)$.

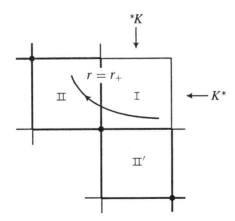

FIGURE 4.1.   A geodesic γ crossing a short horizon in Kerr spacetime. Star-Kerr coordinates suggest that γ has raced off to infinity, but Kerr-star coordinates show that it has merely crossed a horizon in the K^* patch.

Thus nonconstant $r(s)$ and $\vartheta(s)$ approach limits strictly monotonically.

Proof. (1) A priori there is a third possibility, namely: there is an increasing sequence $\{s_n\} \to s_1$ such that $r'(s_{2k}) = 0$, $r'(s_{2k+1}) \neq 0$. To rule this out, consider the r equation $\rho^4 r'^2 = R(r)$. The polynomial $R(r)$ has at most four zeros, hence by passing to a subsequence we can assume $r(s_{2k})$ always has the same value. This value can only be the limit r_1. Since $r'(s_{2k+1}) \neq 0$, the function $r(s)$ is not constant on (s_{2k}, s_{2k+2}) so there must be another point s_{2k+1}^* in this interval at which $r'(s_{2k+1}^*) = 0$ and $r(s_{2k+1}^*) \neq r_1$. Since there are at most three values for $r(s_{2k+1}^*)$, it follows that $r(s)$ does not converge as $s \to s_1$, contrary to hypothesis.

(2) Although $\Theta(\vartheta)$ is not a polynomial the preceding argument applies to $\vartheta(s)$ after the following modification. Even in the exceptional case that $\vartheta(s)$ is not restricted to $(0, \pi)$, since $\vartheta(s)$ approaches a (necessarily finite) limit, the values of ϑ on $[s_0, s_1)$ are contained in a finite interval $[-k, k]$. Because $\Theta(\vartheta)$ is a analytic function of ϑ and is not identically zero, it has only finite number of zeros in $[-k, k]$. Thus the argument for $r(s)$ remains valid for $\vartheta(s)$.

(3) We have $\Delta'(s) = (d/ds)(\Delta(r(s))) = 2(r(s) - \mathrm{M})r'(s)$. If $r(s)$ is constant on $[s_1 - \delta, s_1)$, then $\Delta(s) = 0$ there, so γ is in a horizon, contradicting its regularity. If r' is nonvanishing on $[s_1 - \delta_1, s_1)$, then $\Delta'(s) = 0$ only when $r(s) = \mathrm{M}$. Again since $r(s)$ is strictly monotone there is at most one such point in $[s_1 - \delta, s_1)$. If there is one, call it $s_1 - \epsilon$; otherwise, take $\epsilon = \delta$.   □

Proposition 4.4.3 *Regular geodesics in maximal extreme Kerr spacetime M_e are complete mod Σ.*

Proof. Let γ be a geodesic in M_e with maximal domain $[s_0, s_1)$. We must show that if $s_1 < \infty$ and $\lim_{s \to s_1} \rho^2 \neq 0$, then γ can be extended past s_1. Proposition 4.3.9 provides this extension unless $\Delta(r_1) = 0$, where as usual, $r_1 = \lim_{s \to s_1} r(s)$. By Lemma 4.4.2, $\Delta(r(s)) \neq 0$ for s near s_1, so we can suppose that γ lies in a single Boyer–Lindquist block.

We assert that for M_e, Case 1 always obtains; that is, $\mathbb{P}(r_1) \neq 0$, where now $r_1 = \mathrm{M}$. By definition, $R = \mathbb{P}^2 + \Delta(qr^2 - \mathcal{K})$. Since s_1 is finite, Proposition 4.3.7 implies that R has at most a simple zero at r_1. But $\Delta = (r - \mathrm{M})^2$, so the zero of Δ at $r_1 = \mathrm{M}$ is double. Then, since $\mathbb{P}$ is squared in the formula for R, it cannot be zero at r_1.

Thus Lemma 4.4.1 applies to show that, as $s \to s_1$, γ approaches a point of a horizon in a Kerr patch, and hence has a geodesic extension past s_1.   □

Now suppose that the geodesic γ lies in slow Kerr spacetime M_s. In Case 1, γ is approaching a short horizon. This is the only possibility in M_e, but Case 2 can occur in M_s. Then we will show that γ is approaching a crossing sphere S. $\mathbb{P}(r_1) = 0$ is a necessary condition for this because S consists of critical points of r, so if $\gamma \to S$, then $r' = dr(\gamma') \to 0$. The r equation $\rho^4 r'^2 = \mathbb{P}^2 + \Delta(qr^2 - \mathcal{K})$ then shows that $\mathbb{P}(r) \to 0$.

Proposition 4.4.4 *Let* $\gamma : [s_0, s_1) \to M_s$ *be a regular geodesic. If* $\lim_{s \to s_1} r(s) = r_\pm$ *and* $\mathbb{P}(r_\pm) = 0$, *then* γ *has a geodesic extension that meets, hence cuts through, the crossing sphere* $S(r_\pm)$ *in a KBL domain* $\mathcal{D}(r_\pm)$.

Proof. For definiteness, suppose $\lim_{s \to s_1} r(s) = r_+$. As before, Lemma 4.4.2 shows that for s near s_1, γ lies in a single Boyer–Lindquist block—hence in single domain $\mathcal{D}(r_+)$. For the *KBL* coordinates of γ we must show that, as $s \to s_1$, ϑ and φ^+ approach finite limits, and U^+ and V^+ approach 0.

As before, Lemma 4.3.6 takes care of ϑ, and φ^+ is handled below. In the definition of U^+ and V^+, signs vary with Boyer–Lindquist blocks. Assume first that γ lies in block I of $\mathcal{D}(r_+)$. In the formula for t^* (and $^*t$) in Proposition 4.3.4 the ϑ integral is bounded on $[s_0, s_1)$, as before. For the r integral we again consider $R = \mathbb{P}^2 + f\Delta$, where $f(r) = qr^2 - \mathcal{K}$. Since $\mathbb{P}$ and Δ are zero at r_+, so is R. But since s_1 is finite, R has only a simple zero there. Then we write $R \sim r - r_+$, the tilde ($\sim$) indicating systematic neglect of bounded *positive* factors. Similarly, $\Delta = (r - r_+)(r - r_-) \sim r - r_+$. It follows that f cannot be zero at r_+, hence $f\Delta \sim r - r_+$. $\mathbb{P}$ has the form $Er^2 + $ const, and $\mathbb{P}(r_+) = Er_+^2 + $ const $= 0$; hence

$$\mathbb{P} = Er^2 - Er_+^2 = E(r - r_+)(r + r_+) \sim E(r - r_+).$$

(We cannot neglect E since we do not know if it is nonzero.)

Now substitute these results into the t^* formula in Proposition 4.3.4. We can also neglect summands, such as the ϑ integral, that approach finite limits as $s \to s_1$. Then

$$t^*(s) - t_0^* \sim \int_{r_0}^{r(s)} \frac{1}{r - r_+} \left[\frac{\pm E(r - r_+)}{\sqrt{r - r_+}} \oplus 1 \right] dr$$

$$\sim \pm E \int_{r_0}^{r(s)} \frac{dr}{\sqrt{r - r_+}} \oplus \int_{r_0}^{r(s)} \frac{dr}{r - r_+}.$$

Thus our ignorance of E is not a problem since its integral is negligible. We conclude that $t^*(s) - t_0^* \sim \oplus \ln(r - r_+)$. Explicitly,

$$t^*(s) - t_0^* \sim \ln(r - r_+), \quad \text{and} \quad {}^*t(s) - {}^*t_0 \sim -\ln(r - r_+).$$

Then using Definition 3.4.5, we find $U^+ = \exp(-\kappa_+^* t) \sim (r - r_+)^{\kappa_+}$, and $V^+ = \exp(\kappa_+ t^*) \sim (r - r_+)^{\kappa_+}$. Since $\kappa_+ > 0$, both U^+ and V^+ approach 0 as $s \to s_1$ and hence $r \to r_+$.

The other quadrants are similarly checked. For example, on block II, signs change since $r < r_+$, but it remains true that

$$t^*(s) - t_0^* \sim \int_{r_0}^{r(s)} \frac{dr}{\Delta}.$$

Since both $r(s) < r_0$ and $\Delta \sim r - r_+ < 0$, we get $t^*(s) \sim -\ln(r_+ - r)$. Thus, $U^+ = -\exp(-\kappa_+^* t) = -\exp(\kappa_+ \ln(r_+ - r)) = -(r_+ - r)^{\kappa_+}$, which approaches 0 as before.

Returning to the φ^+ coordinate of γ, note that Proposition 4.3.4 shows that the difference between φ^* and φ_0^* is negligible in the sense above. On block I, for example, $\varphi^*(s) - \varphi_0^* \sim \oplus \ln(r - r_+)$. Since the sign $\oplus$ reverses between K^* and $^*K$, the differences $\varphi^* - {}^*\varphi$ and $t^* - {}^*t$ are both negligible, that is, approach finite limits as $s \to s_1$. By definition, on $\mathcal{D}(r_+)$,

$$\varphi^+ = \frac{1}{2}[\varphi^* + {}^*\varphi - a(r_+^2 a^2)^{-1}(t^* + {}^*t)],$$

so the same is true for φ^+.

These limits for the *KBL* coordinates of γ as $s \to s_1$ show that γ has a geodesic extension that meets, hence crosses, the crossing sphere $S: U^+ = 0$, $V^+ = 0$ of $\mathcal{D}(r_+)$.

$\square$

Corollary 4.4.5 *Regular geodesics in slow Kerr spacetime M_s are complete mod Σ.*

The information found in this section about the crossing of horizons in Kerr spacetime can be summarized as follows:

Proposition 4.4.6 *For a geodesic $\gamma(s)$ in M_s define the function $\oplus(s) = -\mathrm{sgn}(r'(s)\mathbb{P}(s))$. If γ meets $r = r_\pm$ (from outside) at the point $\gamma(s_1)$, then:*
(1) *If $\oplus(s_1) = +1$, then γ cuts through the short horizon $H_\pm$ in a Kerr-star patch K^*.*
(2) *If $\oplus(s_1) = -1$, then γ cuts through the short horizon $H_\pm$ in a star-Kerr patch $^*K$.*
(3) *If $\oplus(s_1) = 0$, then γ cuts through a crossing sphere $S(r_\pm)$.*

4.5   Control of the ϑ Coordinate

A geodesic γ is completely determined by its initial position and velocity; to understand the geometry of any manifold M we would like to know, for given initial conditions, where the geodesic goes in M. In Kerr black holes an immense amount of such information can be derived from a simple observation about the first-integral equations in Theorem 4.2.2:

For a geodesic $\gamma(s)$, the functions $R(r(s))$ and $\Theta(\vartheta(s))$ are equated to squares, hence they can never be negative.

In this section we consider the Θ inequality and lay out the range of possibilities for the qualitative behavior of the ϑ coordinate of a Kerr geodesic. But first, to illustrate the power of these conditions, we use them to prove a notable result.

Corollary 4.5.1 *If a Kerr geodesic γ approaches the ring singularity, then its Carter constant Q is zero.*

Proof. We must show that if $\rho^2(\gamma(s)) \to 0$ as $s \to b$ (finite or infinite), then $Q = 0$. The hypothesis means that $r(s) \to 0$ and $\vartheta(s) \to \pi/2$ as $s \to b$. Since $R(r(s)) \geq 0$ and $\Theta(\vartheta(s)) \geq 0$ for all s, it follows by continuity that $R(0) \geq 0$ and $\Theta(\pi/2) \geq 0$. (We emphasize that the functions R and Θ are defined on the entire real line regardless of the domain of the particular geodesic they derive from.) But $R(0) = (a^2 E - La)^2 + a^2(-\mathcal{K}) = -a^2 Q$ and $\Theta(\pi/2) = Q$. Hence Q can only be 0. $\square$

The proof actually shows (1) if γ approaches the throat T: $r = 0$, then $Q \leq 0$, and (2) if γ approaches the equator $\vartheta = \pi/2$, then $Q \geq 0$.

Thus, as first observed for principal null geodesics, it is rare for a Kerr geodesic to hit the ring singularity. By contrast, in Schwarzschild spacetime every particle falling through the horizon inexorably meets the central singularity (unless it perishes earlier). The converse of Corollary 4.5.1 fails. In fact, it is immediate from the ϑ-equation—with Θ in terms of Q—that *every geodesic in the equatorial plane has $Q = 0$*. But Section 4.14 will show that most equatorial geodesics avoid the ring singularity.

In studying the ϑ motion of a geodesic γ its energy E and causal character are significant only as they contribute to the first-integral $a^2(E^2 + q)$, which we call the *rotational energy* of γ and (in this section only) denote by $\tilde{E}$. In Schwarzschild spacetime, with its nonrotating star, $a = 0$, hence all its geodesics have $\tilde{E} = 0$. In view of Lemma 4.2.10 the ϑ-equation can now be written as

$$\rho^4 \vartheta'^2 = \Theta(\vartheta) = Q + C^2(\tilde{E} - L^2/S^2).$$

Then the definition

$$V(\vartheta) = C^2(-\tilde{E} + L^2/S^2)$$

lets the ϑ-equation be written as an *energy equation*

$$\rho^4 \vartheta'^2 + V(\vartheta) = Q.$$

This is interpreted as conservation of mechanical energy for a point moving on the ϑ-axis with $V(\vartheta)$ as potential energy and with $\rho^4 \vartheta'^2$ in the role of kinetic energy. The latter has the essential property of positive definiteness: $\rho^4 \vartheta'^2 \geq 0$, and if $\rho^4 \vartheta'^2 = 0$ then $\vartheta' = 0$.

Caution: For a Kerr geodesic, $\rho^2 = r^2(s) + a^2 \cos^2 \vartheta(s)$ is variable, by contrast with its (constant) Newtonian analogue. Thus, information about $\rho^4 \vartheta'^2$ does not automatically give information about ϑ'. However, if $\rho^4 \vartheta'^2 \to 0$ and ρ^2 is bounded away from 0, then $\vartheta' \to 0$.

The same graphical methods as used in the Newtonian conservation of energy readily show the influence of the first-integrals of a geodesic γ on the global behavior of its colatitude function $\vartheta(s)$. Sketch the graph of $V(\vartheta)$, and draw a horizontal line at any height Q. Since $\rho^4 \vartheta'^2 \geq 0$, the ϑ coordinate is restricted to a largest ϑ-interval I_γ on which the graph of V is below (or at) the Q line. (The interval may be infinite or shrink to a point.) By the completeness theorem, the values of $\vartheta(s)$ fill the interval—always assuming γ does not depart for $r = \infty$. Suppose $\vartheta(s)$ is moving toward an endpoint $\vartheta_1 \neq \pi/2$ of I_γ, so $\Theta(\vartheta_1) = 0$. We saw in Section 4.4 that (assuming $\rho^2 \not\to 0$) if $d\Theta/d\vartheta|_{\vartheta_1} \neq 0$, then $\vartheta(s)$ reaches ϑ_1, turns, and goes back in the opposite direction, but if $d\Theta/d\vartheta|_{\vartheta_1} = 0$, then $\vartheta(s)$ only approaches ϑ_1 asymptotically.

Definition 4.5.2 *The rotational energy $\tilde{E} = a^2(E^2 + q)$ of a geodesic (as compared to its angular momentum L around the axis) is said to be*

high *if $\tilde{E} > L^2$,*   critical *if $\tilde{E} = L^2$,*   *and*   low *if $\tilde{E} < L^2$.*

Note that these cases depend only on the *squares* of E and L. Although $\tilde{E}$ increases with E^2, it ranges from $a^2 q$ to $+\infty$. Thus when γ is timelike, $\tilde{E}$ can be negative.

Consider first the generic condition $L \neq 0$.

Remark 4.5.3 Properties of the function $V(\vartheta) = C^2(-\tilde{E} + L^2/S^2)$. Here ϑ is merely a variable on the open interval $(0, \pi)$, and $\tilde{E}$ and $L = 0$ are arbitrary constants.

1. *Endpoints.* $V(\vartheta) \to +\infty$ as $\vartheta \to 0$ or π.

2. *Zeros.* If $\tilde{E} \leq L^2$, then $\pi/2$ is the unique zero; if $\tilde{E} > L^2$ there are two more zeros, $\vartheta_{\pm}$, symmetric about $\pi/2$: $0 < \vartheta_- < \pi/2 < \vartheta_+ < \pi$.

3. *Slope.* $dV/d\vartheta = 2SC(\tilde{E} - L^2 S^{-4})$. Hence, $dV/d\vartheta \to -\infty$ as $\vartheta \to 0$, and $dV/d\vartheta \to +\infty$ as $\vartheta \to \pi$.

4. *Critical Points.* $dV/d\vartheta = 0$ if $\vartheta = \pi/2$, and if $\tilde{E} \leq L^2$ this is the only critical point. If $\tilde{E} > L^2$ there are two more critical points $\vartheta_{c\pm}$, such that

$$0 < \vartheta_{0-} < \vartheta_{c-} < \pi/2 < \vartheta_{c+} < \vartheta_{0+} < \pi.$$

5. *Minimum Points.* If $\tilde{E} \leq L^2$, the only critical point is at $\vartheta_c = \pi/2$; hence by (1) it is a minimum. If $\tilde{E} > L^2$, then at a critical point $\vartheta_c \neq \pi/2$, we find $V(\vartheta_c) = -\tilde{E}\cos^4 \vartheta_c = -(\tilde{E}^{1/2} - |L|)^2$, an unusual expression first noted by Carter (1968). Since $V(\pi/2) = 0$, it follows that V has a minimum at $\vartheta_{c\pm}$.

Proposition 4.5.4 *Let γ be a Kerr geodesic with low or critical rotational energy $\tilde{E} \leq L^2 \neq 0$, hence Carter constant $Q \geq 0$. If $Q > 0$, the ϑ coordinate of γ oscillates symmetrically about $\pi/2$. If $Q = 0$, then γ remains stably in the equator $\vartheta = \pi/2$.*

Proof. When $\tilde{E} \leq L^2$, the energy equation expresses Q as a sum of nonnegative terms. By the previous discussion, Figure 4.2 is virtually a proof of the case $Q > 0$. If $Q = 0$, the other two terms in the energy equation must be zero. But $\rho^4 \vartheta'^2 = 0$ implies $\vartheta' = 0$, hence ϑ is constant.

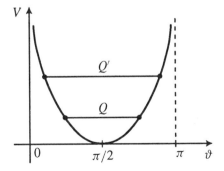

FIGURE 4.2.   Graph of $V(\vartheta)$ for $\tilde{E} \leq L^2 \neq 0$. The horizontal lines at height $Q \geq 0$ determine the ϑ-interval I for geodesics with Carter constant Q.

Now consider $C^2(-\tilde{E} + L^2/S^2) = 0$. If $C^2 = 0$, then $\vartheta = \pi/2$. And if $-\tilde{E} + L^2/S^2 = 0$, that is, $\tilde{E} = L^2/S^2$, then $\tilde{E} \leq L^2$ implies $S^2 \geq 1$ so $\vartheta = \pi/2$ again. Thus, γ lies in the equator $\vartheta = \pi/2$. Furthermore, it is there *stably* since (as Figure 4.2 indicates) a small change in E or L can result in no more than a small oscillation of ϑ about $\pi/2$, and not a major departure.   □

The behavior of the ϑ coordinate in the low energy case is qualitatively the same as in Newtonian theory. For a satellite in Keplerian elliptical orbit in a (nonpolar) plane through the center of earth, its colatitude oscillates this way as it proceeds around the earth. But novelties appear in the high energy case, as can be seen in Figure 4.3, which derives immediately from Remark 4.5.3 .

Proposition 4.5.5   *For a Kerr geodesic γ with high rotational energy $\tilde{E} > L^2 \neq 0$, the smallest possible value of the Carter constant is $Q_{\min} = -(\tilde{E}^{1/2} - |L|)^2$. Further, there are four cases:*
(1) *If $Q > 0$, then ϑ oscillates symmetrically about $\pi/2$.*
(2) *If $Q = 0$, then either (a) γ lies (unstably) in the equator, or (b) γ does not meet the equator, has one ϑ turning point, and either asymptotically approaches equator or approaches the ring singularity.*
(3) *If $Q_{\min} < Q < 0$, then ϑ oscillates between a and b, where*

> *either $\vartheta_- < a < \vartheta_{c-} < b < \pi/2$*   *(northern hemisphere),*
>
> *or $\pi/2 < a < \vartheta_{c+} < b < \vartheta_+$*   *(southern hemisphere).*

(4) *If $Q = Q_{\min}$, then ϑ is constant at $\vartheta_{c\pm} \neq 0$ and $L^2 = \tilde{E}S^4$.*

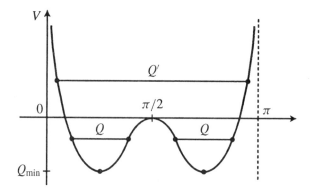

FIGURE 4.3.   Graph of $V(\vartheta)$ in the high energy case, $\tilde{E} > L^2 \neq 0$. For $Q' > 0$ there is a single ϑ-interval; for $Q < 0$, there are two, one in each hemisphere. The case $Q = 0$ is exceptional.

Proof. Evidently, the geodesics with minimum Carter constant Q occur at minimum points of V other than $\pi/2$. Then assertion (5) in Remark 4.5.3 gives Q_{min}.

Cases (1) and (2) of the proposition can be read from the Figure 4.3. In particular, a high energy geodesic in the equator is there *unstably* since arbitrarily small changes in L or E can turn it into one of the radically different cases (1) or (3).) The case $Q = Q_{min}$ derives again from Remark 4.5.3.   □

If a geodesic has negative Carter constant $Q < 0$ we call it *vortical* (de Felice 1968). The ϑ-equation shows at once that vortical geodesics have high rotational energy and cannot cross the equator, hence remain entirely over the northern or southern hemisphere, as for case (3) of the preceding proposition. In Newtonian gravitation and even in the Schwarzschild black hole, this restriction is possible only for particles that fall directly toward the center of the star, and hence have $L = 0$. So vortical geodesics must be regarded as a distinctive feature of relativistic rotation.

The previous cases establish a strong relation between the Carter constant Q of a geodesic and its latitudinal motion.

Corollary 4.5.6 *Let γ be a Kerr geodesic with $L \neq 0$.*

(1) *If $Q > 0$, then $\vartheta(s)$ oscillates around the equator $\pi/2$.*

(2) *If $Q = 0$, then γ either lies in the equatorial plane, asymptotically approaches it, or approaches the ring singularity.*

(3) *If $Q < 0$, then $Q_{min} \leq Q < 0$ where $Q_{min} = -(\tilde{E}^{1/2} - |L|)^2$, and γ does not meet the equator.*

Now we consider geodesics with $L = 0$. The potential function reduces to $V(\vartheta) = -\tilde{E}C^2$, so the energy equation is

$$\rho^4 \vartheta'^2 - \tilde{E}C^2 = Q.$$

As before, there are three basic cases: low, critical, and high rotational energy. For each we sketch the graph of V and read off a number of subcases, some quite Newtonian and others not.

Case 1. $\tilde{E} < 0$. (See Figure 4.4.) The next section shows that such geodesics have bounded r-coordinate.

(a) $Q > -\tilde{E}$. Then $\rho^4 \vartheta'^2 \geq b > 0$; and since $r(s)$ is bounded, so is ρ^2, hence $|\vartheta'| \geq c > 0$. Thus, γ is in permanent polar orbit, passing repeatedly through the north and south axes.

(b) $Q = -\tilde{E}$. Either γ remains in the axis (unstably), or asymptotically departs from the axis, and, after passing across the equator, asymptotically approaches the axis in the other hemisphere.

(c) $0 < Q < -\tilde{E}$. Here ϑ oscillates in familiar fashion around $\pi/2$, bounded away from the axis.

(d) $Q = 0$. γ remains stably in the equator.

Case 2. $\tilde{E} = 0$. Then $V = 0$, so the energy equation is just $\rho^4 \vartheta'^2 = Q$.

(a) $Q > 0$. See Case 1(a).

(b) $Q = 0$. ϑ is constant at arbitrary ϑ_0.

Case 3. $\tilde{E} > 0$. (See Figure 4.4.)

(a) $Q > 0$. Colatitude $\vartheta(s)$ circles through all values (and $r(s)$ need not be bounded).

(b) $Q = 0$. Either γ remains in the equator or, as in assertion (2) of Proposition 4.5.5, γ crosses the axis once and at each endpoint either approaches Σ or asymptotically approaches Eq. (Contrast with case 1(b).)

(c) $-\tilde{E} < Q < 0$. γ oscillates symmetrically around the axis, bounded away from the equator. (Contrast with case 1(c).)

(d) $Q = -\tilde{E}$. γ remains stably in the axis.

Corollary 3.5.6 and the various $L = 0$ cases have the following geometric consequence.

Corollary 4.5.7 *A Kerr geodesic γ has Carter constant $Q = 0$ if and only if at least one of the following holds: (1) γ lies entirely in the equator, (2) γ asymptotically approaches the equator, (3) γ approaches the ring singularity.*

It is easy to find the principal null geodesics among the preceeding cases, since these geodesics have ϑ constant. For those in the axis, $-Q = \tilde{E} = a^2 > L^2 = 0$, the minimum points in Figure 4.4. Those in the equator have $\tilde{E} = a^2 = L^2$ and $Q = 0$, the minimum point in Figure 4.2. All others have $\tilde{E} > L^2 > 0$ and $Q = Q_{\min}$, the minimum points in Figure 4.3.

Nonnull principal geodesics are scarce. Every geodesic in the axis A is principal since the tangent planes to A are all principal planes Π. But *the only other nonnull principal geodesics are equatorial geodesics with* $L = aE$. To see this, note first that outside the axis, $\Pi^{\perp}$ is spanned by ∂_ϑ and the canonical vector field $W = \partial_\varphi + aS^2 \partial_t$. Thus a geodesic γ is principal if and only if $\langle \gamma', \partial_\vartheta \rangle = 0$ and $\langle \gamma', W \rangle = 0$. The former gives $\vartheta' = 0$; that is, $\vartheta(s)$ constant, and the latter gives $L = aES^2$. Since $\vartheta = \pi/2$ in the equator, this criterion becomes $L = aE$.

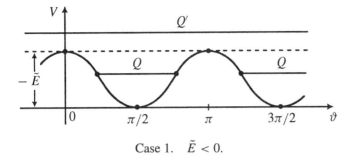

Case 1. $\tilde{E} < 0$.

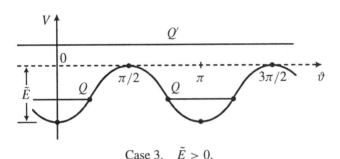

Case 3. $\tilde{E} > 0$.

FIGURE 4.4.   Graph of $V(\vartheta)$ for $\tilde{E} < 0$ and $\tilde{E} > 0$. The ϑ-intervals of asymptotic orbits are indicated by dashed lines.

The only other case where $\vartheta(s)$ is constant is at the minimum points of $V(\vartheta)$ in the high rotational energy case. But there only the null geodesics are principal. In fact, by Proposition 4.5.5, $L^2 = \tilde{E}S^4$, while the formula $L = aES^2$ gives $L^2 = a^2E^2S^4$; hence $\tilde{E} = a^2(E^2 + q) = a^2E^2$, so $q = 0$.

For a study of $\vartheta(s)$ (and also $r(s)$, in the exterior) in terms of dynamical systems see Krivenko et al. (1976).

4.6   Control of the *r* Coordinate

We now consider the effect of the inequality $R(r) \geq 0$ on the *r* coordinate of a Kerr geodesic γ. Though the effect is not as immediate as for $\Theta(\vartheta) \geq 0$, the radial coordinate *r* is more important than the colatitude coordinate ϑ, determining as it does such critical issues as escape to infinity and passage through horizons.

Substituting $\Delta = \rho^2 - 2Mr + a^2$ and $\mathbb{P} = (r^2 + a^2)E - aL$ into the definition $R(r) = \mathbb{P}^2 + \Delta(qr^2 - \mathcal{K})$ gives this fundamental result.

Proposition 4.6.1 *For a geodesic γ with $q = \langle\gamma',\gamma'\rangle$, energy E, angular momentum L, and Carter constant Q,*

$$R(r) = (E^2 + q)r^4 - 2Mqr^3 + \mathcal{X}r^2 + 2M\mathcal{K}r - a^2Q,$$

where $\mathcal{X} = a^2(E^2 + q) - L^2 - Q$ and, as usual, $\mathcal{K} = Q + (L - aE)^2$.

In view of the r-equation $\rho^4 r'^2 = R$, it is clear that the dependence of the radial motion of Kerr geodesics on the basic first-integrals q, L, E, Q is expressed by the way they appear in the coefficients of the polynomial $R(r)$.

Evidently $R(r)$ is unchanged if the pair E, L is replaced by $-E$, $-L$, as happens, for example, when parametrizations are reversed: $\tilde{\gamma}(s) = \gamma(-s)$. However, $R(r)$ *is* changed (provided $EL \neq 0$) if only *one* of the two sign changes $E \to -E$, $L \to -L$ is made, since $\mathcal{K} = Q + L^2 - 2aLE + a^2E^2$. Accordingly, in Kerr spacetime direct and retrograde orbits with the same q, E, $\mathcal{K}$ and $|L|$ usually differ in their radial motion (except in the special cases $E = 0$ or $L = aE$). This distinction stems from Kerr rotation and is absent in the Schwarzschild black hole (see Remark 4.14.11.).

The first-integrals q, E, L, Q of a Kerr geodesic γ determine the polynomial $R(r)$, and the radial coordinate $r(s)$ of γ is restricted to a maximal interval J_γ on which $R(r) \geq 0$. This interval may be infinite or degenerate to a single point. At an ordinary endpoint r_1 of J_γ, $R(r_1)$ is zero; and by Corollary 4.3.8 if this is a simple zero of $R(r)$, then γ has an r-turning point at r_1, while a repeated zero implies asymptotic approach.

As Figures 4.5–4.7 make clear, a single function $R(r)$ may have more than one such interval of positivity: these represent (in general) geodesics with the same first-integrals but different initial values r_0.

Remark 4.6.2 For the polynomial $R(r)$ there is always a (possibly degenerate) interval J on which $R(r) \geq 0$ since the formula $R(r) = \mathbb{P}^2 + \Delta(qr^2 - \mathcal{K})$ shows that $R(r_\pm) \geq 0$. Then if $Q \geq 0$, there are geodesics γ with $J_\gamma = J$. (To prove this, use Proposition 4.2.6 and Section 4.5). But if $Q < 0$, results from Section 4.5 impose the further conditions: $E^2 + q > L^2$ and $Q \geq Q_{min} = -\left[a(E^2 + q)^{1/2} - |L|\right]^2$ (see Section 4.9).

Since J_γ has at most two endpoints, *a Kerr geodesic γ has at most two r-turning points*; that is, there are at most two values of r at which $dr/ds = 0$ (though these may be taken on at many parameter values).

Definition 4.6.3 *For a Kerr geodesic γ: $\mathbf{R} \to M$ the following* orbit types *have 0, 1, and 2 r-turning points, respectively:*

(0) Transit. *As s traverses $\mathbf{R}$, $r(s)$ goes from limit value $+\infty$ to $-\infty$ (or vice versa).*

(1) Flyby. *As s traverses $\mathbf{R}$, $r(s)$ goes from $+\infty$ to $+\infty$ (or $-\infty$ to $-\infty$).*

(2) Interval-bound. $r_1 \le r(s) \le r_2$ *with both values $r_1 < r_2$ taken on.*

Geodesics of these types are said to be ordinary. *All other geodesics are* exceptional.

To say that a geodesic has *bound orbit* can only mean that its radial function $r(s)$ is bounded. As we shall see there are a variety of such orbits whose behaviors differ radically—thus the terminology in the two-turn case above.

Flyby orbits are familiar in Newtonian gravitation as Keplerian hyperbolas and so are bound orbits, as Keplerian ellipses. However, *transit orbits* are quite relativistic, and even in the Schwarzschild case there are none. But there are many in Kerr black holes, for example, nonequatorial principal null geodesics.

The simplest exceptional orbits are the *spherical orbits*, those for which the r-coordinate is constant, $r(s) = r_0$. Then, as for ϑ in Section 4.5, adjacent to r_0 there will be at most two r-intervals representing (exceptional) *asymptotic orbits* for which $r(s)$ asymptotically approaches r_0. If R is a maximum at a repeated zero r_0, no asymptotic approach is possible. If r_0 is an inflection point of R, then γ is approached asymptotically from one side or the other, and if r_0 is a minimum, from both sides. (See Figures 4.5–4.7, in which such asymptotic orbits are rendered as dashed lines.)

All other exceptional orbits derive from the ring singularity Σ, and thus depend also on the ϑ coordinate: We say that a geodesic γ has *crash-escape orbit* if in one parameter direction, γ hits Σ; in the other it approaches $r = \pm\infty$. There are no turns. Depending on the sign of $r' \ne 0$, an astronaut with this orbit type either crashes into Σ or escapes entirely from the black hole. Principal null geodesics in the equator all have such orbits.

In the worst case, a geodesic defined on a finite interval has *crash-crash orbit* if it meets Σ at both (finite) endpoints.

These Σ-related orbits, though exceptional, are not particularly strange. From a common sense viewpoint, a pebble tossed straight upward from the earth is of crash-crash type if it lacks escape energy; otherwise it has crash-escape type.

We consider some stability properties of the orbit types. If v is a tangent vector to M, recall that γ_v denotes the maximal geodesic with initial tangent $\gamma_v'(0) = v$. In this context we ignore parameter changes $\gamma(s) \to \gamma(cs + d)$; thus every geodesic in M has many expressions as γ_v.

A property of geodesics is *open* if the set of all $v \in TM$ such that γ_v has the property is an open set of TM. For example, being timelike (or spacelike) is a open property of geodesics, but being null is not. Roughly speaking, two geodesics that are "nearby" with this notion of open set are somewhere close together and have almost the same first-integrals.

Two stability properties of polynomials are needed. Fix n, and for each point $A = (a_0, a_1, \ldots, a_n)$ in $\mathbf{R}^{n+1}$ let $p_A(r)$ be the polynomial $a_n r^n + \cdots + a_1 r + a_0$. Let $N_\delta(A)$ be the δ neighborhood of A, that is, set of all $B \in \mathbf{R}^{k+1}$ such that $|A - B| < \delta$. Then the following properties are easily proved:

P1. If $p_A(r)$ has no roots in the (possibly infinite) closed interval I, then there exists a $\delta > 0$ such that if $B \in N_\delta(A)$ then $p_B(r)$ has no roots in I. (All the roots of $p_A(r)$ are in some finite $[-k, k]$.)

P2. If $p_A(r)$ has a simple root at r_1, then there exists a $\delta > 0$ such that if $B \in N_\delta(A)$ then $p_B(r)$ also has exactly one root in $(r_1 - \delta, r_1 + \delta)$ and that root is simple. (Apply P1 to dR/dr.)

Lemma 4.6.4 *In a maximal Kerr spacetime, each of the ordinary orbit types (flyby, interval-bound, transit), is open.*

Proof. The transit case is clear from property P1, and the flyby case is a simpler variant of the bound orbit case.

So suppose γ_v has interval-bound orbit. Then the polynomial $R_v(r)$ for γ_v is positive on the open interval (r_1, r_2), and has simple roots at r_1 and at r_2.

The first integrals q, L, E, and $\mathcal{K}$ are continuous functions on TM. Accordingly, for vectors w in a sufficiently small neighborhood of v in TM, the values of q, L, E, and $\mathcal{K}$ at w will be arbitrarily close to their values at v. The formula in Proposition 4.6.1 shows that the polynomials $R_w(r)$ will be arbitrarily close to $R_v(r)$—that is, their coefficient 5-tuples will be arbitrarily close in $\mathbf{R}^5$.

Then using property P2 we can pick $\delta_1 > 0$ so that if $w \in N_{\delta_1}(v)$, then $R_w(r)$ has a unique and simple root in $(r_1 - \delta_1, r_1 + \delta_1)$; similarly, for δ_2 and $(r_2 - \delta_2, r_2 + \delta_2)$. Pick δ_3 using P1 so that if $w \in N_{\delta_3}(v)$, then $R_w(r) > 0$ on $[r_1, r_2]$. Let $\delta = \min\{\delta_1, \delta_2, \delta_3\}$. Then $w \in N_\delta(v)$ implies that γ_w has bound orbit. $\qquad\square$

Thus if a geodesic γ has, say, flyby orbit type and β is a geodesic such that some $\beta'(s_1)$ is sufficiently near some $\alpha'(s_0)$, then β also has flyby orbit.

The exceptional types are not open. For example, a geodesic with either kind of crash orbit has $Q = 0$, and hence can be saved by a small change to $Q \neq 0$ (so such orbits are nowhere dense).

A geodesic γ with bound orbit has *stably bound* orbit if every nearby geodesic has bound orbit. By Lemma 4.6.4 interval-bound orbits are stably bound, and so are spherical orbits at a maximum point r_0 of $R(r)$. The other spherical orbits—those with adjacent asymptotic orbits—are unstably bound. The significance of this distinction among spherical orbits is illustrated later, by Proposition 4.8.1.

Except perhaps for its causal character the most important single property of a Kerr geodesic is the sign of the leading coefficient in its polynomial $R(r)$.

Lemma 4.6.5 *A Kerr geodesic such that $|r(s)| \to \infty$ as $S^2 \to \pm\infty$ has $E^2 + q \geq 0$. (Here, as always, $S = \sin\vartheta$.)*

Proof. If $E^2 + q < 0$, then $R(r) < 0$ for large $|r|$. $\square$

For a freely falling material particle with $q = -m^2$, this means that $E^2 \geq m^2$ is a necessary condition for escape from the Kerr gravitational field. Thus we say in general that if $E^2 + q \geq 0$ then γ has *escape energy: strict* if $E^2 + q > 0$, *marginal* if $E^2 + q = 0$. Evidently, geodesics of transit or flyby orbit type have escape energy.

The class of geodesics with strict escape energy is quite large. It includes all spacelike geodesics, all null geodesics with $E \neq 0$, and many timelike geodesics, in particular, those with high rotational energy $a^2(E^2 - 1) > L^2$. Evidently, having strict escape energy is an open property.

To visualize the r-intervals in this case we draw the graph of $R(r)$ and look for maximal r-intervals J on which $R(r) \geq 0$. (So these intervals are *beneath* the graph rather than above, as in the "effective potential" game.) By not specifying the position of the (horizontal) r-axis a single graph can describe many polynomials. In Figure 4.5 for example, any horizontal line would represent a possible position of the r-axis; we draw only the maximal $R \geq 0$ intervals in a few such lines.

As Figure 4.5 correctly suggests, though escape energy is necessary for escape to infinity, it is not sufficient. By contrast, having less than escape energy is severely restrictive.

Corollary 4.6.6 *A geodesic γ with less than escape energy, $E^2 + q < 0$, is timelike, has bound orbit and Carter constant $Q \geq 0$, and does not meet the region $r < 0$. Furthermore, $Q = 0$ if and only if γ is in the equator.*

Proof. Obviously, $q < 0$, and in $\Theta(\vartheta) = Q + C^2[a^2(E^2 + q) - L^2/S^2]$ the second summand is ≤ 0, so $Q \geq 0$. Then $Q = 0$ implies $C = \cos\vartheta = 0$, which means that γ is in the equator. (The converse is always true.)

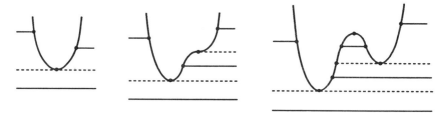

FIGURE 4.5.   Graphs of $R(r)$ for strict escape energy, $E^2 + q > 0$. Any horizontal line is a possible position of the r-axis. Here solid line segments denote ordinary orbits, dashed lines exceptional orbits. Reflection in a vertical axis gives two more cases.

Now consider the formula

$$R(r) = (E^2 + q)r^4 - 2\text{M}qr^3 + \mathfrak{X}r^2 + 2\text{M}\mathcal{K}r - a^2 Q$$

from Proposition 4.6.1. Evidently, $R(r) < 0$ for r large. We assert that $R(r) < 0$ for all $r < 0$. Since $Q \geq 0$ and $E^2 + q \leq 0$ we have $\mathfrak{X} = a^2(E^2 + q) - L^2 - Q \leq 0$. Also $\mathcal{K} = Q + (L - aE)^2 \geq 0$. So for $r < 0$, every term in the formula above is nonpositive, and $2\text{M}r^3$ is negative; thus $R(r) < 0$.   □

The graphs of R in Figure 4.6 merely invert those above, but the pattern of r-intervals is quite different, displaying the impossibility of escape to infinity.

Since $E^2 + q$ is the coefficient of r^4 in $R(r)$, the cases with $E^2 + q \neq 0$ are those for which $R(r)$ has degree 4. The lower degree possibilities are as follows.

- *Degree $R(r) = 3$.* Here $E^2 + q = 0$ but $q \neq 0$, hence $q < 0$. Thus γ is a timelike geodesic with marginal escape energy. In general, if a geodesic has $|r| \to \infty$, then at infinity

$$r'^2|_\infty = \lim_{|r| \to \infty} R(r)/\rho^4 = E^2 + q.$$

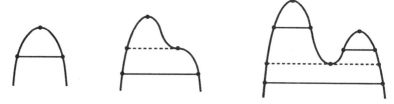

FIGURE 4.6.   Some typical graphs of $R(r)$ for less than escape energy, $E^2 + q < 0$.

Thus if γ does manage to escape it reaches infinity with radial speed zero. We can see at once that $\Theta(\vartheta) \geq 0$ implies $Q \geq 0$. Furthermore, $R(r) \geq 0$ implies $R(r) < 0$ for $r < 0$, so these geodesics do not enter the region $r < 0$. (See Figure 4.7)

- *Degree $R(r) = 2$*. Such geodesics are very special. Degree $R(r) \leq 2$ already implies that γ is null and has $E = 0$. Then $\Theta(\vartheta) \geq 0$ implies $Q \geq 0$, so $\mathcal{K} = -(L^2 + Q) < 0$. Then

$$R(r) = -(L^2 + Q)r^2 + 2\mathrm{M}(L^2 + Q)r - a^2 Q$$
$$= -L^2 r(r - 2\mathrm{M}) - Q\Delta.$$

If the degree of R is exactly 2, then $L^2 + Q \neq 0$, and the graph of R is a downward-opening parabola with vertex at $r = \mathrm{M}$. These orbits are bound, and none enters $r < 0$. (For an example in the equator, see Section 4.14.)

- *Degree $R(r) = 1$ or 0*. There are no such geodesics. $R(r)$ cannot have degree 1, because in the computations above, $\mathcal{K} = L^2 + Q = 0$ implies both $R(r) = -a^2 Q$ and $Q \leq 0$. This would give degree 0 if $Q < 0$, but again $\Theta(\vartheta) \geq 0$ implies $Q \geq 0$.

There remains only the degreeless case, $R(r)$ identically zero. Since the co-efficients of $R(r)$ all vanish, it follows that all four first-integrals are zero. Such geodesics do exist; as shown in Section 4.2 they are the restphotons. (Note that here it is the *polynomial $R(r)$* that is zero not the *function $s \mapsto R(r(s))$*. Evidently, $R(r(s)) = 0$ can occur for spherical geodesics.)

Working from the graphs above, a list could be made of all possible orbits involving asymptotic approach.

The previous results demonstrate a striking fact about the Kerr black hole.

Corollary 4.6.7 *Every geodesic γ that meets the region $r < 0$ has strict escape energy $E^2 + q > 0$.*

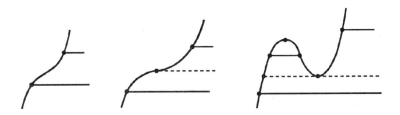

FIGURE 4.7. Graphs of $R(r)$ for degree 3. Most orbits are flyby, but only from $r = +\infty$.

Proof.   If $E^2 + q < 0$, then Corollary 4.6.6 asserts that γ does not meet $r < 0$. If $E^2 + q = 0$, then $R(r)$ has degree 2 or 3 and we have seen that the same conclusion holds.   □

Thus, crossing the throat $r = 0$ is not just a matter of dropping a pebble into the black hole. High energy is necessary, but as we shall see, by no means always sufficient.

4.7 r–L Plots

The goal of the next few sections is to show how the first-integrals of a Kerr geodesic determine its global trajectory in a Kerr black hole. To avoid a proliferation of cases we concentrate on timelike geodesics (though some results are independent of causal character). Null geodesics are analogous and are usually easier to deal with, but causality is weaker for spacelike geodesics, so their trajectories are more varied, as will be evident in the case of equatorial geodesics in Section 4.14.

There is a graphical way to see how first-integrals control orbit types. At a general level, suppose the Kerr parameters M, a and the causal character parameter q are fixed. Then the function $R(r)$ of Proposition 4.6.1 depends also on E, L, Q, giving a function $R: \mathbf{R}^4 \to \mathbf{R}$. For given values of L, E, Q the resulting r-line cannot enter the region $R < 0$. Thus the set $Z: R = 0$ in $\mathbf{R}^4$ is the key to the situation since it supplies the boundary points of $R < 0$. Fortunately, it is possible to determine the qualitative character of orbits by considering R only as a function of the variables r and L—for a few selected constant values of E and Q. (Of course, we must prove what the pictures can only suggest.)

The complement of Z in the r–L plane consists of open, connected regions, said to be *allowed* if $R > 0$, *forbidden* if $R < 0$. A diagram of this situation is called an r–L *plot* (see figures in Section 4.8). For any L_0, the intersection of the horizontal line at height $L = L_0$ with the allowed regions gives the maximal intervals J that describe orbit types (Section 4.6).

The set $Z: R = 0$ is usually a curve, that is, a smooth one-dimensional submanifold of the r–L plane. Corners can appear only when a point (r, L) of Z is a critical point of R.

Conventions:

1. The abbreviation $e = E^2 + q$ is often convenient. This first-integral is called *effective energy*, and in view of Lemma 4.6.5 could also be called the *escape parameter*. (The notation Γ has also been used in the literature.)

2. When a sign choice is required for the energy E, we assume $E > 0$. The r–L plots for $E < 0$ would differ by reflection in the r-axis since $R(r)$ is unchanged under the double switch $E, L \to -E, -L$.

3. As usual, $q = -1$ is used for timelike geodesics, so $e = E^2 - 1$. For example, the set of all timelike r–L plots is parametrized by part of the e–Q plane (see Figure 4.8).

We consider first some properties of r–L plots that are independent of causal character.

It is natural to regard the function in Proposition 4.6.1 as a polynomial $R(r)$ in r, but in dealing with r–L plots we of course understand R to be a function $R(r, L)$ of r and L. In some contexts it is valuable to view this function as a polynomial $R(L)$ in L, with coefficients determined by the remaining first-integrals and r. Explicitly,

$$R(L) = r(2\mathrm{M} - r)L^2 - 4\mathrm{M}aErL + R(r, 0),$$

where $R(r, 0)$ is the restriction of the function R to the r-axis, $L = 0$.

Since the polynomial $R(L)$ has degree at most 2 and $R(r)$ has degree at most 4, it follows that in any r–L plot, *every vertical line $r = $ const meets the set $R = 0$ in at most two points, and every horizontal line $L = $ const meets $R = 0$ in at most four points.* For future reference we call this the *two–four rule.*

Using $R(L)$ we derive explicit formulas for the implicitly defined set $R = 0$.

Lemma 4.7.1   *In any r–L plot the set $R = 0$ is the union of the graphs of (the real values of) the two functions*

$$L_\pm(r) = (2\mathrm{M}aEr \pm D(r)^{1/2})/(r(2\mathrm{M} - r)),$$

for $r \neq 0, \neq 2\mathrm{M}$, where $D(r) = r\Delta(r)\phi(r)$ is one-fourth of the discriminant of $R(L)$, and the discriminant polynomial $\phi(r)$ *is*

$$\phi(r) = (E^2 + q)r^3 - 2\mathrm{M}qr^2 - Qr + 2\mathrm{M}Q.$$

Proof.   Solve the equation $R(L) = 0$ for L by the quadratic formula. If $\mathcal{D}$ is the discriminant of $R(L)$, this gives the formula above for $L_\pm$ with $D(r) = \mathcal{D}(r)/4$. Thus,

$$D(r) = 4a^2 E^2 \mathrm{M}^2 r^2 + r(r - 2\mathrm{M})R_0(r)$$
$$= r[4a^2 E^2 \mathrm{M}^2 r + (r - 2\mathrm{M})R_0(r)].$$

We could show that $D(r_\pm) = 0$, hence that $D(r)$ is divisible by $\Delta(r)$. However, a direct computation gives $D(r) = r\Delta(r)\phi(r)$, with $\phi(r)$ as stated.   □

Except for $r = 0$ and $r = 2M$ the polynomial $R(L)$ is quadratic in L, hence its graph is parabolic with *vertex* at $L_v(r) = 2MaE/(2M - r)$. The number $\mu(r)$ of points on each vertical line with r-coordinate r is 0, 1, or 2 as the modified discriminant $D(r)$ is negative, zero, or positive, respectively. Evidently, when $\mu(r) = 2$ the two points are placed symmetrically on either side of $L_v(r)$. Accordingly, we call the curve $r \to (r, L_v(r))$ in an r–L plot the *midline*. This curve is a rectangular hyperbola with asymptotes the r-axis and the line $r = 2M$, and (for $E > 0$) is above the r-axis for $r < 2M$, below it for $r > 2M$.

When $\mu(r) = 1$, that is, when there is just one point of $R = 0$ on the line $L \to (r, L)$, that point is the vertex $(r, L_v(r))$. The vertex is also the unique point on the line at which $\partial R/\partial L = 0$. Such a point is either a critical point of $R(r, L)$ or a point at which $R = 0$ is smooth with vertical tangent. Thus the intersection of the midline with $R = 0$ consists of the vertical tangent points of $R = 0$ and critical points that lie on $R = 0$. Armed with these properties one can vizualize the midline in the r–L plots in figures of the next section.

Another simple but useful property of allowed/forbidden regions follows immediately from the formula for $R(L)$: If $0 < r < 2M$, then $r(2M - r) > 0$, so $R(r, L) > 0$ for $|L|$ large. Thus, whatever happens for $|L|$ small, for $|L|$ large the r-line is in an allowed region. Correspondingly, if $r < 0$ or $r > 2M$, then $r(2M - r) < 0$, so $R(r, L) < 0$ for $|L|$ large, and these points are in a forbidden region. (See Figures 4.9 and 4.12.)

The formula for $R(L)$ shows that $r = 0$ and $r = 2M$ are special. When $r = 0$, R has constant value $-a^2 Q$, and the behavior of R on and near the line $r = 2M$ is as follows.

Lemma 4.7.2 *In an r–L plot the function $L \to R(2M, L)$ is linear with negative slope (if $E > 0$). Thus there is a critical momentum L_0 such that $R(2M, L) < 0$ if $L > L_0$ and $R(2M, L) < 0$ if $L < L_0$. Furthermore, if $0 < \delta \ll M$, then $R(2M + \delta) < 0$ and $R(2M - \delta) > 0$ for all sufficiently large $|L|$.*

Proof. Substituting $r = 2M$ in $R(L)$ gives a linear function $L \to R(2M, L)$, with slope $-8M^2 aE < 0$ if $E > 0$. This function is zero at $L_0 = R(2M, 0)/(8M^2 aE)$. The final assertion follows by remarks above or by direct substitution of $r = 2M \pm \delta$ into $R(L)$. ☐

In terms of r–L plots this means that the vertical line $r = 2M$ is an asymptotic line for the curve $R(r, L) = 0$ as $L \to +\infty$ and as $L \to -\infty$. Furthermore (always for $E > 0$), the approach is from the left ($r < 2M$) as $L \to +\infty$ and from the right ($r > 2M$) as $L \to -\infty$. This behavior also is evident in Figures 4.9 and 4.12

Investigation of r–L plots tends to come in vertical strips: the Boyer–Lindquist block II strip $r_- < r < r_+$ is particularly simple for nonspacelike geodesics; they cannot turn in II, hence R is never zero there.

Lemma 4.7.3 *In the r–L plot of a timelike (or null) geodesic γ in slow Kerr spacetime, we have $R > 0$ on the closed vertical strip $r_- \leq r \leq r_+$ except that $R = 0$ at the two points (r_-, L_-) and (r_+, L_+), where*

$$L_{\pm} = L_v(r_{\pm}) = 2MaE/(2M - r_{\pm}) = 2MaE/r_{\mp} = 2Mr_{\pm}E/a.$$

At these points the curve $R = 0$ is smooth and tangent (from outside the strip) to the lines $r = r_-$ and $r = r_+$.

Proof. For any geodesic the formula $R = \mathbb{P}^2 + \Delta(qr^2 - \mathcal{K})$ gives $R(r_{\pm})^2 = \mathbb{P}(r_{\pm})^2$, where $\mathbb{P}(r_{\pm}) = (r_{\pm}^2 + a^2)E - La$. Thus, $R(r_{\pm}) > 0$ except when $\mathbb{P}(r_{\pm}) = 0$, which is at $L_{\pm}$. The other formulas follow since $r_- + r_+ = 2M$ and $r_-r_+ = a^2$. In particular, these are the points at which the midline meets the lines $r = r_{\pm}$.

As previously noted, $R > 0$ eventually holds on, say, the line $r = M$. Thus, $R > 0$ on the entire strip $r_- < r < r_+$ since R is never zero there.

Evidently $\partial R/\partial L = 0$ at the points $(r_{\pm}, L_{\pm})$ since $L_{\pm}$ is a minimum point on $r = r_{\pm}$. It remains only to show that $\partial R/\partial r \neq 0$ at these points. But

$$\partial R/\partial r = 2\mathbb{P}\,\partial\mathbb{P}/\partial r - 2(r - M)(r^2 + \mathcal{K}) - 2r\Delta,$$

which at $(r_{\pm}, L_{\pm})$ reduces to $-2(r_{\pm} - M)(r_{\pm}^2 + \mathcal{K}) \neq 0$. $\square$

Remark 4.7.4 A crucial effect of the Carter constant Q on orbits derives from the fact that the function R, which depends quadratically on L, depends *linearly* on Q. In fact, the formulas $R = \mathbb{P}^2 - \Delta(r^2 + \mathcal{K})$ and $\mathcal{K} = Q + (L - aE)^2$ give $R(r) = -\Delta(r)Q + R_0(r)$, where $R_0(r)$ is obtained by setting $Q = 0$. (Thus $R_0(r)$ governs the r-equation for geodesics in the equator.) It follows that when the other first-integrals are held constant, *as Q increases, forbidden regions expand*. In fact, the strip $r_- \leq r \leq r_+$, where $\Delta \leq 0$, is (but for two boundary points) always in a forbidden region; so elsewhere, as Q increases, $R(r)$ decreases.

We now examine the discriminant polynomial $\phi(r)$ of Lemma 4.7.1 in more detail. This polynomial is important because its roots determine those the $D(r)$ and thereby the $\mu = 1$ points of r–L plots; these in turn are crucial to the pattern of forbidden regions in r–L plots and hence to an understanding of orbits.

Lemma 4.7.5 *For timelike geodesics, $\phi(r) = er^3 + 2Mr^2 - Qr + 2MQ$, where $e = E^2 + q$.*

(1) *If $Q > 0$, then ϕ has no roots in $0 \le r \le 2M$. If also $e > 0$, then ϕ has exactly one negative root, hence either zero or two positive roots. If $e < 0$, ϕ has no negative roots, hence either one or three positive roots. For $e = 0$, ϕ has no real roots.*

(2) *If $Q < 0$, so $e > 0$, then ϕ has a unique root in $0 < r < 2M$, and either zero or two roots in $r < 0$.*

Proof. (1) If $e > 0$, then for $r < 0$ the signs for Descartes' rule are $- + + +$, so there is a unique negative root. Since $Q \ne 0$, $r = 0$ is not a root. To show that $\phi(r) > 0$ for $0 < r \le 2M$, note that $er^3 + 2Mr^2 > 0$ for $r > 0$ while $-Qr + 2MQ > 0$ for $r < 2M$.

If $e < 0$, then all terms in $\phi(r)$ are positive for $r < 0$. Since $0 > e \ge -1$, we have $er^3 + 2Mr^2 > 0$ for $0 < r < -2M/e \ge 2M$, and as before, $-Qr + 2MQ > 0$ for $r < 2M$. If $e = 0$, then since M and Q are both positive there can be no real roots.

(2) The Cartesian signs show there is a unique positive root; furthermore, $\phi(0) < 0$ and $\phi(2M) > 0$.   □

The special cases $Q = 0$ and $Q < 0$ will be considered later in this chapter. Note that for $Q > 0$ *the roots of ϕ are distinct from the other roots $0, r_-, r_+$ of $D = r^2 \Delta \phi$.*

The ambiguities 0/2 and 1/3 in the preceding lemma can be settled precisely. Recall that the discriminant d of a real cubic polynomial signals the number of its real roots: one if $d < 0$, three if $d > 0$, and repeated roots if $d = 0$. Applying the general formula

$$\text{discrim}(ar^3 + br^2 + cr + d) = -4ac^3 - 4b^3d + b^2c^2 - 27a^2d^2 + 18abcd,$$

to the polynomial ϕ we find $d = d(e, Q) = \text{discrim}(\phi, r) = -4Qf(e, Q)$, where

$$f(e, Q) = -eQ^2 + QM^2(27e^2 + 18e - 1) + 16M^4.$$

Our goal now is to find the regions in the e–Q plane where $d > 0, d = 0, d < 0$, respectively. Let $\beta = \beta(e) = 27e^2 + 18e - 1$, and solve $f = 0$ for

$$Q_\pm(e) = M^2[\beta \pm (\beta^2 + 64e)^{1/2}]/(2e).$$

Of course these functions are real-valued only when $\beta^2 + 64e \ge 0$. This inequality always holds if $e > 0$. The remarkable factorization

$$\beta^2 + 64e = (27e^2 + 18e - 1)^2 + 64e = (e + 1)(9e + 1)^3$$

shows that $\beta^2 + 64e > 0 \Leftrightarrow e > -1/9$. (The zero at $e = -1$ can be neglected since $Q < 0$ there.) Thus we get the explicit formula:

$$Q_{\pm}(e) = (M^2/2e)[27e^2 + 18e - 1 \pm (e+1)^{1/2}(9e+1)^{3/2}]$$

Consider the functions $Q_-(e)$, $Q_+(e)$ on $e \geq -1/9$. At $e = -1/9$ they have the same value, $12M^2$, but thereafter are quite different. For Q_+, taking the limit value, we define $Q_+(0) = 16M^2$. For $e \gg 1$, $Q_+(e) \sim \beta(e)/e \sim M^2(27e + 18)$, so as Figure 4.8 shows, Q_+ becomes almost linear.

By contrast, Q_- approaches $+\infty$ as $e \to 0$ from negative values. (Note that $Q_-(e) > Q_+(e)$ for $e < 0$.) Since Q_+ is bounded near $e = 0$ this is clear from

$$\frac{1}{2}(Q_-(e) + Q_+(e)) = M^2(27e^2 + 18e - 1)/(-2e).$$

For example, with $M = 1$, at $e = -.01$ we find $Q_+(-.01) \approx 15.6$ but $Q_-(-.01) > 100$.

For $e > 0$, $Q_-(e)$ is always negative, approaching $-\infty$ as $e \to 0$, and approaching 0 as $e \to +\infty$.

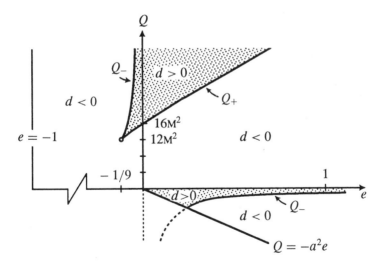

FIGURE 4.8.   This e–Q *chart* shows the sign of $d(e, Q) = \mathrm{discrim}(\phi, r)$ on the relevant portion of the e–Q plane. If $d(e, Q) < 0$, then in the r–L plot determined by (e, Q) the curve $R = 0$ has only one vertical tangent other than those at $(r_{\pm}, L_{\pm})$. For $d > 0$ there are three such points. Repeated roots occur for $d = 0$. (The scale on the negative Q axis is exaggerated.)

Not every point of the e–Q plane corresponds to an r–L plot. There are three restrictions: (1) $e \geq -1$, since $e = E^2 - 1$, (2) $Q > 0$ if $e < 0$, since vortical geodesics have escape energy; and (3) $Q > -a^2 e$ if $e > 0$, by Carter's inequality. Thus in Figure 4.8 the third quadrant is irrelevant, as is the region in the fourth quadrant below the line $Q = -a^2 e$.

Remark 4.7.6   To summarize: The qualitative character of r–L plot given by effective energy $e = E^2 + q$ and Carter constant Q is determined by the discriminant polynomial $\phi(r) = er^3 - 2\mathrm{M}qr^2 - Qr + 2\mathrm{M}Q$, in particular, by the number of zeros ϕ has. For given e and Q, this number is determined by the sign of a second discriminant, namely $d(e, Q) = \mathrm{discrim}(\phi, r)$. Hence d becomes a function on the e–Q chart in Figure 4.8. The regions $d > 0$ and $d < 0$ in the chart are described explicitly since the curves separating them are graphs of the functions

$$Q_{\pm} = (\mathrm{M}^2/(2e))[27e^2 + 18e - 1 \pm (e + 1)^{1/2}(9e + 1)^{3/2}].$$

Consequently, if $e > 0$, we can read from the chart that when $0 < Q < Q_{\pm}$ the negative root of ϕ is its only root, but if $Q > Q_{\pm}$, then ϕ has three roots. These roots are all simple hence, as we see next, represent the vertical tangents in the r–L plot determined by e and Q. Analogously, if $e < 0$, the negative root of ϕ is its only root unless both $-1/9 < e < 0$ and $Q_{+} < Q < Q_{-}$ hold, giving three roots.

Now we show how the discriminant polynomial ϕ distinguishes vertical tangent points of $R = 0$ from critical points of R on $R = 0$. The rest of this section describes the configurations of r–L plots near such critical points.

Lemma 4.7.7   *For $e \neq 0$ and $Q > 0$, let $L_0 = L_v(r_0)$, where r_0 is a root of $\phi(r)$. Then:*

(1) $R = 0$ has a vertical tangent at $(r_0, L_0) \Leftrightarrow r_0$ is a simple root of $\phi(r)$; and

(2) (r_0, L_0) is a critical point of the function $R(r, L) \Leftrightarrow r_0$ is a repeated root of $\phi(r)$. If $\phi''(r_0) > 0$, then (r_0, L_0) is a saddle point, with adjacent allowed regions along the midline, and if $\phi''(r_0) < 0$, then (r_0, L_0) is an isolated maximum point, surrounded by a forbidden region.

Proof.   Consider $L_{\pm}(r) = [2\mathrm{M}aEr \pm (r\Delta(r)\phi(r))^{1/2}]/(r(2\mathrm{M} - r))$ near r_0. Since $\phi(r)$ has no roots in $0 \leq r \leq 2\mathrm{M}$, none of the terms in this expression is zero near r_0 except $\phi(r)$.

If r_0 is a simple root, that is, if $\phi(r_0) = 0$ and $\phi'(r_0) \neq 0$, then ϕ is positive on one side of r_0, negative on the other. For definiteness suppose $\phi'(r_0) > 0$. (Thus, points near (r_0, L_0) on the midline are allowed for $r > r_0$, forbidden for $r < r_0$.) Then for $r > r_0$ near r_0 we can write $\phi(r) \approx (r - r_0) f(r)$, where $f(r_0) > 0$. Thus $L_\pm(r) \approx L_0 \pm (r - r_0)^{1/2} B$, with B real and nonzero. It follows that $R = 0$ has a vertical tangent at (r_0, L_0).

Now suppose r_0 is a double root. If $\phi''(r_0) > 0$, then $\phi(r) > 0$ for $r \neq r_0$ near r_0. Thus $L_\pm(r) \approx L_0 \pm |r - r_0| B$, with $B \neq 0$. This shows that $R(r, L)$ has a saddle point at (r_0, L_0) with allowed regions along the midline on both sides of (r_0, L_0).

If r_0 is a double root with $\phi''(r_0) < 0$, then $\phi(r) < 0$ for $r \neq r_0$ near r_0. Hence there is a strip $r_0 - \epsilon < r < r_0 + \epsilon$ around (r_0, L_0) on which $R < 0$ except for $R(r_0, L_0) = 0$. So (r_0, L_0) is an isolated maximum point of the function $R(r, L)$.   □

A notable timelike *r–L* plot is that determined by the point $(e, Q) = (-1/9, 12\text{M}^2)$ at the tip of the region $d \geq 0$ in Figure 4.8. This point is distinguished algebraically by the fact that its polynomial $\phi(r)$ has a triple root (thus covering the case $\phi''(r_0) = 0$ neglected in Lemma 4.7.7). In fact, equating coefficients in the equation $\phi(r) = e(r - r_0)^3$ gives

$$-3er_0 = 2\text{M}, \qquad 3er_0^2 = -Q, \qquad -er_0^3 = 2\text{M}Q,$$

from which it follows that $r_0 = 6\text{M}$ and $(e, Q) = (-1/9, 12\text{M}^2)$. The resulting *r–L* plot must have a *degenerate* critical point at (r_0, L_0) where, since this point is on the midline, $L_0 = L_v(6\text{M}) = -a\sqrt{2}/9$. This *r–L* plot resembles that in Figure 4.12 but the allowed region has a horizontal spike stretching out to (r_0, L_0).

Critical points on $R(r, L) = 0$ occur in *r–L* plots that have $d(e, Q) = 0$, and for $Q > 0$ the next result describes the location of critical points and vertical tangents.

Proposition 4.7.8 *For $Q > 0$, consider the r–L plot of (e, Q) where $d(e, Q) = 0$ (so $e \geq -1/9$ and $Q = Q_\pm(e) > 0$). Excluding $e = -1/9$, $R = 0$ has a unique vertical tangent point (r_1, L_1) with $r_1 \neq r_\pm$ and a unique critical point (r_2, L_2), both necessarily on the midline, so $L_i = L_v(r_i)$. There are three cases:*

(1) If $e \geq 0$ and $Q = Q_+(e)$, then $r_1 < 0 < 2\text{M} < r_2$. Further, (r_2, L_2) is a saddle point of the function $R(r, L)$, with adjacent allowed regions along the midline.

(2) If $-1/9 < e < 0$ and $Q = Q_+(e)$, then $2\text{M} < r_2 < r_1$, in fact,

$$2\text{M} < r_2 < 2\text{M}/(3|e|) < r_1 < 2\text{M}/|e|,$$

and (r_2, L_2) is a saddle point, as above.

(3) *If* $-1/9 < e < 0$ *and* $Q = Q_-(e)$, *then* $2M < r_1 < r_2$, *in fact,*

$$2M < r_1 < 2M/(3|e|) < r_2 < 2M/|e|,$$

and (r_2, L_2) *is an isolated* $R = 0$ *point in a forbidden region.*

Proof. (1) For $e > 0$ and $d = 0$ it follows, using Lemma 4.7.5, that $\phi(r)$ has a simple root in $r < 0$ and a double root in $r > 0$. Now $\phi(0) = 2MQ > 0$ and $\phi'(0) = -Q < 0$, while $\phi(r) \to +\infty$ as $r \to +\infty$. Thus the double root in $r > 0$ must have $\phi''(r_2) > 0$ (since $\phi'' = 0$ has previously been excluded.) By Lemma 4.7.7, r_2 is a saddle point as specified.

(2) For $e < 0$, Lemma 4.7.5 asserts that there are no negative roots, but there are positive roots of total multiplicity one or three. The former can be excluded since $d = 0$. Since $e = 1/9$ is excluded, for multiplicity three there must be two roots, say, r_1 simple and r_2 double.

We omit the more tedious analysis of case (3).   □

Although the figures in the next section illustrate generic r–L plots, the special case (1) in this proposition can be inferred from Figure 4.11 and the other two special cases from Figure 4.15.

The critical points (r_2, L_2) in Proposition 4.7.8 can be located as follows: r_2 is a solution of both the quadratic equations

$$Mr^2 - Qr + 3MQ = 0, \quad \text{and} \quad er^2 - M(3e - 1)r - 4M^2 = 0,$$

which result from eliminating e and Q, respectively, from the equations $\phi(r) = 0$, $\phi'(r) = 0$. Then, by the general midline formula, $L_2 = L_v(r_2) = 2MaE/(2M - r_2)$.

4.8   First-Integrals and Orbits

Using r–L plots we show how the first-integrals of a timelike Kerr geodesic determine not only its orbit type but also the relation of its radius function $r(s)$ to the horizon radii $0 < r_- < r_+ < 2M$ of the slow Kerr black hole. Section 4.10 will show how this information determines the trajectory of the geodesic through the global pattern of Boyer–Lindquist blocks.

The most important question about a timelike geodesic is whether or not it has escape energy (determined by the sign of $e = E^2 - 1$). Next is the sign of the Carter constant Q. This section deals with the "standard" case $Q > 0$. Such geodesics cannot reach $r = 0$ and have ϑ motion oscillating about $\pi/2$. Both these properties fail in the special cases $Q < 0$ and $Q = 0$. We discuss the vortical case $Q < 0$ in Section 4.9 and $Q = 0$ in Sections 4.14 and 4.15.

The r–L plots for $Q > 0$ fall into four basic configurations determined by the sign of e and the relative size of Q. We consider these in turn.

CONTINENTS CONFIGURATION

Among all r–L plots perhaps the most common is that of escape energy $e > 0$ and moderate Carter constant $0 < Q < Q_+(e)$, where $Q_+(e) \sim 16 + 27e$ is given precisely by the formula in Remark 4.7.6. Figure 4.9 shows such a plot, which is distinguished by two "continents" on the $r > 0$ side. There the situation is more or less Newtonian: Particles infalling from $r = +\infty$ have flyby orbit if their

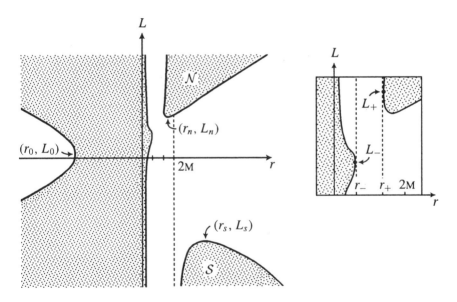

FIGURE 4.9.   Continents configuration: a typical r–L plot for a timelike geodesic with escape energy $e > 0$ (in fact, $E > 1$) and Carter constant $0^+ < Q < Q_+(e)$. Forbidden regions are shaded. On each horizontal line $L = $ const, the maximal segments on which $R \geq 0$ give ranges of $r(s)$ for a geodesics with that angular momentum. The inset shows the region around the block II strip. Most particles infalling from $r = +\infty$ have flyby orbit; on the $r < 0$ side, all do. (Parameters: M $= 1$, $a^2 = .84$, $e = E^2 - 1 = 0.4$, $Q = 1$. *Note:* Our standard choice $a^2/\text{M}^2 = .84$ is large in order to emphasize rotational effects, and it gives convenient values $r_- = .6\text{M}$ and $r_+ = 1.4\text{M}$ for the horizon radii.)

angular momentum $|L|$ is large, while those with smaller $|L|$ continue on through the horizons—with consequences detailed in Section 4.10.

In the strip $0 < r < 2$M there are many bound orbits, both direct and retrograde. No particle—no matter how energetic—can reach $r = 0$. As predicted by Remark 4.7.6 and Lemma 4.7.7 there are no critical points and only one vertical tangent other than those at $r = r_\pm$, namely the one at (r_0, L_0), which marks the closest approach to $r = 0$ in $r < 0$. Here r_0 is the unique real root of the discriminant polynomial $\phi(r)$, and the point is on the midline, which runs also through the two vertical tangent points in the inset of Figure 4.9.

Reasonable questions about orbit types can be answered in terms of polynomials—but the polynomials themselves are not always reasonable. For example, location of spherical orbits is a fundamental problem, since they signal changes in orbit type. Their radii are the simultaneous solutions of the polynomial equations $R = 0$, $dR/dr = 0$, but these horizontal tangents are harder to locate than vertical tangents. In the simpler case of equatorial timelike geodesics, Lemma 4.14.9 gives a single polynomial of (smallest possible) degree six whose real roots are the circular orbit radii. The same approach here yields a polynomial of degree ten, with limited conceptual value. However, for illustrative purposes we sometimes solve the equations $R = 0$, $dR/dr = 0$ numerically. In Figure 4.9, for example, we find $(r_n, L_n) \approx (1.63, 3.06)$ and $(r_s, L_s) \approx (4.56, -6.52)$.

Various properties of the r–L plot in Figure 4.9 hold more generally. In it there are there are no interval-bound or stable spherical orbits in the Kerr exterior (although unstable spherical orbits appear). The following result due to D.C. Wilkins (1972) shows that this holds regardless of the size of Q.

Proposition 4.8.1 *Let γ be a timelike geodesic in the Kerr exterior, block I, of slow or extreme Kerr spacetime. If γ has escape energy $e = E^2 - 1 \geq 0$, then γ does not have interval-bound or stable spherical orbit.*

Proof. Assume that γ has such an orbit and that $e \geq 0$; we deduce a contradiction. Suppose first that the orbit is interval-bound; thus, it has turning points r_1, r_2 such that $r_+ < r_1 < r_2$, and dR/dr is positive at r_1 and negative at r_2. It follows that $R(r)$ is negative just to the left of r_1 and the right of r_2. Whether e is strictly positive or zero, the first two terms in the formula

$$R(r) = er^2 + 2\text{M}r^3 + \mathfrak{X}r^2 + 2\text{M}\mathcal{K}r - a^2 Q$$

show that R has another zero at some $r_3 > r_2$. On the left, Lemma 4.7.3 shows that $R > 0$ on $r_- < r < r_+$, hence there is a fourth zero r_0 with $0 < r_0 < r_1$.

Thus $R(r)$ has four positive zeros. But this is impossible since $R(r)$ has degree four, and examination of $R(r)$ shows that their sum is $-2M/e \leq 0$.

The argument applies equally well when $r_1 = r_2$ is a stable spherical orbit, for then it is also true that $R(r)$ is negative just to the left and just to the right of $r_1 = r_2$. $\square$

Wilkins's proof (1972) is indirect; see also Krivenko, et. al (1976). The r–L plots in this section indicate that, as mentioned above, *unstable* spherical orbits are common in the Kerr exterior.

A second property observable in Figure 4.9 is valid for all $Q \geq 0$.

Proposition 4.8.2 *For a timelike Kerr geodesic with $Q \geq 0$, every orbit in $r < 0$ has flyby type, with turning point at $r_\tau \leq -2M/e$, where $e = E^2 - 1$.*

Proof (Robert Steinberg). For such a geodesic γ the polynomial $R(r)$ has at least one negative root. In fact, if $Q > 0$ then $R(0) = -a^2 Q < 0$, and if $Q = 0$ then $(dR/dr)(0) = 2M\mathcal{K} > 0$. Thus $R(r) < 0$ for small $r < 0$. Since γ has escape energy, $\lim_{r \to -\infty} R(r) = +\infty$, so a negative root exists. We must show that this root is unique and is less than $-2M/e$. (Recall that $e > 0$ for geodesics in $r < 0$.)

Suppose first that $\mathfrak{X} = a^2 e - L^2 - Q$ is nonpositive. Then for $r < 0$ the signs in

$$R(r) = er^4 + 2Mr^3 + \mathfrak{X}r^2 + 2M\mathcal{K}r - a^2 Q$$

are $+- \sim --$ (where $\sim$ represents $\mathfrak{X} \leq 0$). So Descartes' rule of signs guarantees a unique negative root. Furthermore, $R(r) < 0$ on the interval $I: -2M/e < r < 0$ since $er^4 + 2Mr^3 = r^3(er + 2M) \leq 0$ on I and $\mathfrak{X}r^2 + 2\mathcal{K}r - a^2 Q < 0$ for all $r < 0$ (since $\mathfrak{X} \leq 0$, $\mathcal{K} > 0$, and $Q \geq 0$).

Now suppose $\mathfrak{X} > 0$. Write $R(r)$ as the sum of the polynomials

$$R_1(r) = (r^2 + 2Mr/e)(er^2 + \mathfrak{X}), \quad \text{and} \quad R_2(r) = 2M(\mathcal{K} - \mathfrak{X}/e)r - a^2 Q$$

The fact that $R(r) < 0$ on the interval $I: -2M/e < r < 0$ follows from
1. $R_1(r) < 0$ on I. *Proof:* $r^2 + 2Mr/e < 0$ on I, and $er^2 + \mathfrak{X} > 0$ for all r.
2. $R_2(r) \leq 0$ for $r < 0$. *Proof:* Since $Q \geq 0$, it suffices to show that the slope of the line $r \to R_2(r)$ is non-negative, or equivalently that $e\mathcal{K} - \mathfrak{X} \geq 0$.
But

$$e\mathcal{K} - \mathfrak{X} = (E^2 - 1)(Q + (L - aE)^2) - a^2(E^2 - 1) + L^2 + Q$$
$$= E^2 Q + E^2(L - aE)^2 + 2aEL - 2a^2 E^2 + a^2$$
$$= E^2 Q + (E(L - aE) + a)^2 \geq 0.$$

Since the graph of $R(r)$ is below the r-axis from 0 out to $-2\text{M}/e$, it suffices to show that $d^2R/dr^2 \geq 0$ thereafter; that is, that the graph is convex upward for at least $r < -2\text{M}/e$. Since

$$d^2R/dr^2 = 12er^2 + 12\text{M}r + 2\mathfrak{X} = 12r(er + \text{M}) + 2\mathfrak{X},$$

the required convexity in fact holds for $r < -\text{M}/e$.   □

The sloping shores of the continents $\mathcal{N}$ and $\mathcal{S}$ in Figure 4.9 suggest a property that turns out to hold for all Kerr geodesics with escape energy, regardless of causal character: The curve $R = 0$ in the r–L plane is asymptotic (in all four quadrants) to a pair of intersecting straight lines, these lines depending solely on $e = E^2 + q > 0$ and M. Thus, at great distances from the center, rotational effects are negligible.

The explicit formulas $L_\pm(r) = (2\text{M}aEr \pm D(r)^{1/2})/(r(2\text{M}-r))$ for these shores were found in Lemma 4.7.1.

Lemma 4.8.3 *For an arbitrary Kerr geodesic γ with escape energy, let $A_\pm(r) = \mp(E^2 + q)^{1/2}(r + ME^2)/(E^2 + q)$. Then $L_\pm(r) - A_\pm(r) \to 0$ as $|r| \to \infty$.*

Proof. These asymptotic lines are derived as follows. For $r \gg 1$, the term $2\text{M}aEr \ll r^2$ in the numerator of $L_\pm$ (above) can be dropped. Thus, $L_\pm(r) \approx \pm D(r)^{1/2}/(r(r - 2\text{M}))$, where

$$D(r) = r\Delta(r)\phi(r) \quad \text{and} \quad \phi(r) = (E^2 + q)r^3 - 2\text{M}qr^2 - Qr + 2\text{M}Q.$$

Moving the denominator $r(r - 2\text{M})$ under the square root, we find $L_\pm(r) \approx F(r)^{1/2}$, where

$$F(r) = \Delta(r)\phi(r)/(r(r - 2\text{M})^2) = (E^2 + q)r^2 + 2ME^2r + \text{const}.$$

The binomial theorem shows that that $r \to [ar^2 + br + c]^{1/2}$ is asymptotic to the line $\sqrt{a}(r + b/(2a))$, and this gives the formula for $A_\pm$.

Verification of the limit assertion is similar.   □

We can now characterize the orbit types with escape energy and moderate Carter constant.

Proposition 4.8.4 *In an r–L plot for timelike geodesics with escape energy $e > 0$ and Carter constant $0 < Q < Q_+(e) \sim 16 + 27e$, there are three forbidden regions. In $r > \text{M}$, the (connected) forbidden regions $\mathcal{N}$ and $\mathcal{S}$ have extreme* *boundary points: For $\mathcal{N}$, (r_n, L_n) has smallest L; for $\mathcal{S}$, (r_s, L_s) has largest L. Furthermore, $L_s < L_n$.*

All geodesics in $r < 0$ have flyby orbits. In $r > 0$, a geodesic with angular momentum $L > L_n$ or $L < L_s$ has bound orbit meeting both horizons if r is small, and flyby orbit not meeting a horizon if r is large. If $L_s < L < L_n$, then—except for small bound orbits near $(0, aE)$ when Q is small—γ has flyby orbit meeting both horizons.

Proof. By Remark 4.7.6 the discriminant polynomial ϕ has only one root, and it is negative. For $r \gg 1$, Lemma 4.8.3 shows that $\mu(r) = 2$, that is, there are two point of $R = 0$ on the line $r = \mathrm{const}$. Since ϕ has no positive roots, $\mu = 2$ for all $r > r_+$ except $r = 2\mathrm{M}$ (the latter described in Lemma 4.7.2). Thus, in $r > 2\mathrm{M}$ the midline does not meet $R = 0$; in fact, it lies in $R > 0$, since $R < 0$ for $|L|$ large.

Let $\mathcal{N}$ be the set of points (r, L) such that $R(r, L) < 0$ and either (1) $r > \mathrm{M}$ and $L > L_v(r)$ or (2) $r_- < r \le 2\mathrm{M}$. This set is open; to prove it is connected we show that any two of its points, say (r_1, L_1) and (r_2, L_2), can be joined by a curve in $R < 0$.

Case 1. $L_1 = L_2 \gg 1$. Consider the horizontal line $L = L_1 = L_2$. Since the leading coefficient e of $R(r)$ is positive this line meets $R = 0$ once in $r < 0$ and again in $r > 2\mathrm{M}$. Since $R = 0$ is asymptotic to the L-axis and to $r = 2\mathrm{M}$, the line also meets $R = 0$ in the strip $0 < r < r_-$ and again in $r_+ < r < 2\mathrm{M}$. Since the line can meet $R = 0$ only four times it follows that the segment of the line between r_1 and r_2 remains in $R < 0$.

Case 2. $r_1 > 2\mathrm{M}$, $r_2 > 2\mathrm{M}$. Then the points (r_1, L_2) and (r_2, L_2) are, by definition, above the midline. Thus, on each line $r = r_i$ ($i = 1, 2$) and for any $L > \max\{L_1, L_2\}$, the segment $[L_i, L]$ is in $R < 0$. For $L \gg 1$, Case 1 shows that the points (r_1, L) and (r_2, L) can be connected in $R < 0$. Hence (r_1, L_1) and (r_2, L_2) can be joined in $R < 0$, by $[L_1, L] \cup [r_1, r_2] \cup [L_2, L]$.

Case 3. One or both of the points (r_1, L_1) or (r_2, L_2) has r-coordinate $r_+ < r_i \le 2\mathrm{M}$ (see Figure 4.9). As before there are already four zeros of R on the line $L = L_i$ so a segment of this line will join (r_i, L_i) to the connected region of Case 2 without leaving $R < 0$.

Thus $\mathcal{N}$ is connected. The definition of $\mathcal{S}$ is simpler: It consists of the points (r, L) below the midline in $r > 2\mathrm{M}$. The proof that $\mathcal{S}$ is connected is essentially as above except that Case 2 is already the general case.

By construction $\mathcal{N}$ and $\mathcal{S}$ contain all points of $R < 0$ in $r > r_+$. Since $\mu = 2$ for $r > 2\mathrm{M}$ the set $\mathcal{N}$ extends to $L = +\infty$, but by Lemma 4.8.3 the part of $\mathcal{N}$ below any horizontal line has compact closure and hence contains a limit point (r_n, L_n) of minimum L. Analogously, there is a maximum point (r_s, L_s) for $\mathcal{S}$.

By the two–four rule these points are unique and are isolated zeros of R, hence they correspond to unstable spherical orbits.

The inequality $L_n > L_s$ holds (that is, $\mathcal{N}$ and $\mathcal{S}$ do not overlap) because since $R(r)$ always has a negative root, $L_n \leq L_s$ would produce five roots. The orbit assertion follow.   □

Before proceding to the next major case,we consider the exceptional bound orbits near $(0, aE)$. These occur for all energy levels when $|Q| > 0$ is small, and all are located in a neighborhood of the point $(0, aE)$. Figure 4.9 shows a narrow forbidden region ("shore") along the L-axis that swells out gradually to the line $r = r_-$. To see how this picture changes when $|Q|$ is sufficiently small we start with $Q = 0$, then vary Q slightly.

For $Q = 0$, the point $(0, aE)$ is a critical point of $R = R(r, L)$. Indeed, $\partial R/\partial L = 0$ on the L-axis since $R = 0$ there, and when $r = 0$, $\partial R/\partial r = 2\mathcal{K} = 2(L - aE)^2$, which is zero at $L = aE$ only. Thus, we anticipate that near $(0, aE)$ the function $R(r, L)$—and hence the shape of forbidden regions—is sensitive to small changes in first-integrals.

Lemma 4.8.5 *For a timelike r–L plot, if $Q = 0$, then for all E there is a unique forbidden region $\mathcal{U}$ in the strip $0 < r < r_-$. $\mathcal{U}$ is bounded and stretches from $r = 0$ to $r = r_-$. In fact, its boundary (part of $R = 0$) is a smooth simply closed curve that is tangent to the L-axis at $L = aE$ and tangent to the line $r = r_-$ at $L_- = 2MEr_-/a$. (See $Q = 0$ in Figure 4.10.)*

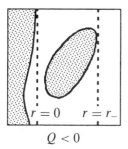
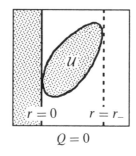
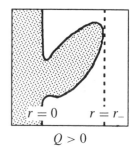

FIGURE 4.10.   Forbidden regions for small $|Q|$ near the critical point $(0, aE)$. When $Q = 0$, the oval $\mathcal{U}$ is tangent to the L-axis. As Q decreases slightly, $\mathcal{U}$ becomes an off-shore island. As Q increases slightly $\mathcal{U}$ merges with the shore forming a knob. Further increases in Q lead to a smooth shore line like those in Figures 4.9 and 4.12. For all Q the curve $R = 0$ remains pinned to the point (r_-, L_-) by Lemma 4.7.3.

Proof. Consider first the behavior of R near the critical point $(0, aE)$. Write $R(r) = rF(r)$, where $F(r) = er^3 + 2\mathrm{M}r^2 + \mathfrak{X}r + 2\mathfrak{K}$. At $(0, aE)$,

$$\partial F/\partial r = \mathfrak{X} = a(E^2 - 1) - a^2 E^2 = -a^2 < 0.$$

Thus, by the implicit function theorem, near $(0, aE)$ the curve $F = 0$ and hence $R = 0$ can be expressed as the graph of a smooth function $r = f(L) \leq 0$. Thus $R = rF$ is tangent to the L-axis at $(0, aE)$.

Lemma 4.7.3 shows that $R = 0$ is similarly tangent to the line $r = r_-$. On $0 \leq r \leq r_-$, the midline runs through the critical point $(0, aE)$ and the vertical tangent point (r_-, L_-). Consequently, to prove that $\mathcal{U}$ stretches from the L-axis to the line $r = r_-$ it suffices to show that the discriminant D of Lemma 4.7.1 is positive on $0 < r < r_-$. Now $D(r) = r\Delta(r)\phi(r)$, and $Q = 0$ simplifies $\phi(r)$ to $r^2(er + 2\mathrm{M})$; so even if $e < 0$ we have $\phi(r) > 0$ for $0 \leq r < 2\mathrm{M}$. Thus, in $0 < r < r_-$ the set $R = 0$ is the union of the graphs of L_- and L_+. But near each of the endpoints 0 and r_- we saw above that $R = 0$ is a smooth curve tangent to vertical lines $r = 0$ and $r = r_-$. Hence the boundary of $\mathcal{U}$ is a smooth simply closed curve. In particular, L is bounded on the set $R < 0$ in $0 < r < r_-$.   □

For $Q = 0$ the region $\mathcal{U}$ described in this lemma is the oval in Figure 4.10. Remark 4.7.4 shows that when Q increases slightly, $\mathcal{U}$ expands slightly forming a "knob," as in Figure 4.10.

Furthermore, on the L-axis, $R = -a^2 Q < 0$ so the knob is smoothly joined to a thin forbidden region along the L-axis. Evidently this shape gives rise to small bound orbits near the throat $r = 0$ of the black hole.

As Q continues to increases, first the bound orbits with $L < aE$ are smoothed away, then those with $L > aE$, so the forbidden region resembles those in Figure 4.9. It is the lower orbits that disappear first, for $E > 0$, because $aE < L_- = 2\mathrm{M}r_+ E/a$, which causes the upward tilt of the initial oval $\mathcal{U}$. For example, numerical computations as mentioned earlier show that for $e = .4$ (the value in Figure 4.9) one must take $Q > 0$ less than about 0.005 to find all four of the spherical orbits suggested by Figure 4.10.

The fact that $\mu(r) = 2$ for $0 < r < r_-$ is enough to show that there can be only one such knob and that the "shoreline" near the L-axis must asymptotically approach this axis as $|L| \to \infty$.

BARRIER CONFIGURATION

As Q increases from small values as above, we know that (like all forbidden regions) the continents $\mathcal{N}$ and $\mathcal{S}$ enlarge, so the southernmost point (r_n, L_n) of $\mathcal{N}$

and the northernmost point (r_s, L_s) of S move toward each other. When Q reaches $Q_+(e) \sim 16 + 27e$ the two extremities meet—at a critical point of the function $R(r, L)$ (see assertion (1) of Proposition 4.7.8.) Thereafter they merge into a single *barrier*, separating the block $\rm{II}$ strip $r_- \le r \le r_+$ at the core of the black hole from the exterior, as shown in Figure 4.11.

The existence of the barrier when Q is large can be seen directly from the formula $R(r) = -\Delta(r)Q_+R_0(r)$ in Remark 4.7.4 for if $r_0 > 2\rm{M}$ then $R < 0$ on the entire line $r = r_0$ when $Q > R_0(r_0)/\Delta(r_0)$.

The structure of forbidden regions is simple in this case.

Proposition 4.8.6 *In an r–L plot for timelike geodesics with escape energy $E > 1$ and Carter constant $Q > Q_+$, the set $R = 0$ consists of the graphs of four smooth functions $r_i = f_i(L)$, where $f_0 \le c_0 < 0 < f_1 \le r_- < r_+ \le f_2 < f_3$. Hence there are just two forbidden regions: between f_0 and f_1, and between f_2 and f_3. (The latter is the barrier.)*

All orbits in $r < 0$ or meeting $r \gg 1$ are flyby. The latter type do not reach a horizon. All other orbits are bound, meet both r_+ and r_-, and remain in $r > 0$. There are no spherical orbits.

Proof. Section 4.7 shows that since $Q > Q_+(e)$, ϕ has three distinct (simple) roots, with two positive, say $2\rm{M} < r_1 < r_2$. By Lemma 4.7.7, these produce

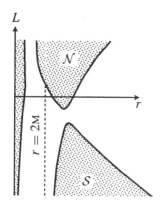
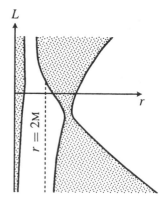

FIGURE 4.11.   As Q increases, the continents $\mathcal{N}$ and S in Figure 4.9 draw closer and merge, forming a barrier between $r = +\infty$ and the horizons. (Parameters: same as in Figure 4.9 except that on the left, $Q = 27.5$, and on the right, $Q = 28$). The critical point occurs when $Q = Q_+(0.4) \approx 27.74$, and is located at about $(3.42, -1.52)$.

vertical tangent points, and since $\phi > 0$ for $2M < r < r_1$ we must have $\phi < 0$ between r_1 and r_2. Thus $\mu = 0$ on $r_1 < r < r_2$, hence $R < 0$ there.

Now consider any horizontal line $L = $ const. We can count four meetings of this line with $R = 0$: one for $r < 0$, one in $0 \le r \le r_-$ (these as before), but now, because of the barrier, one in $r_+ \le r \le r_1$ and one in $r \ge r_2$. Thus, for every L the degree 4 polynomial $R(r, L_0)$ has four distinct (real) roots. Consequently, when these are consistently ordered, each depends smoothly on the coefficients of $R(r)$—in the case at hand, on L. The resulting four smooth functions $f_0 < f_1 < f_2 < f_3$ then have the required properties. (Note that Proposition 4.8.2 gives $c_0 \le -2M/e$.)

The orbit assertions follow at once. In particular, there can be no spherical orbits, since $\partial R/\partial r = 0$ on the set $R = 0$ would imply an infinite derivative for one of the smooth functions f_i.   □

Knobs cannot occur here since for every L there are already four zeros of $R(r)$. Exactly at $Q = Q_+(e)$ the orbit structure is essentially the same as the barrier case, except of course for an unstable spherical orbit at the critical point—and its flanking asymptotic orbits. There are no other spherical orbits.

The continental configuration in Figure 4.9 is in a sense the "ordinary" one for timelike Kerr geodesics with escape energy, comparable to usual Newtonian behavior of free fall toward a gravitational source. Nevertheless, in the rotating case, general relativity shows that complete protection against a crash into a black hole is provided—no matter how small angular momentum L may be—by large values of the Carter constant Q.

BAY CONFIGURATION

The most striking effect of less than escape energy is that all orbits are bound and none meet $r < 0$. Figure 4.12 shows an $r-L$ plot for this case. The allowed region is contained in a strip $0 < r \le r_1$, and most of the resulting bound orbits pass through the horizons.

Comparison with Figure 4.9 shows that for the same value of Q the energy decrease from $e > 0$ to $e < 0$ (that is, from $E > 1$ to $E < 1$) has in effect enclosed the large allowed region on the $r > 0$ side, and pushed the allowed region in $r < 0$ out to $r = -\infty$. But there are no major changes at the core of the black hole, where as before two forbidden regions clamp the block II strip $r_- \le r \le r_+$ at vertical tangent points.

The scarcity of bound orbits in the Kerr exterior is striking—there were none for $e > 0$, and in Figure 4.12 they occur only for a narrow range of positive momenta. (Figure 4.13 shows what the polynomials $R(r)$ are like for these.)

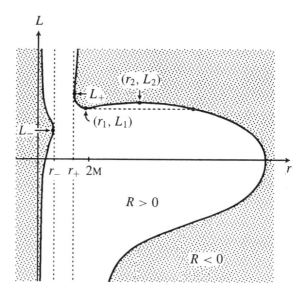

FIGURE 4.12.   Bay configuration. An r–L plot for timelike geodesics with less than escape energy, $e < 0$ and $d(e, Q) < 0$. All orbits are bound, but only those with $L_1 \leq L \leq L_2$ are entirely contained in the Kerr exterior. For $r < r_-$ the plot still resembles the $e > 0$ case in Figure 4.9. (Parameters: $q = -1, e = E^2 - 1 = -0.2, Q = 1, \text{M} = 1, a^2 = .84$.)

For fixed Q, there are retrograde as well as direct orbits in the exterior when $e < 0$ is near zero, but for e nearer -1 all bound orbits disappear. (See Figure 4.14.)

In one aspect, quantitative results are more difficult for $e < 0$ than for $e > 0$ since the latter case depends strongly on vertical tangents ($\partial R/\partial L = 0$) while for $e < 0$ a principal problem is the determination of bound orbits, which (as noted earlier) are usually harder to locate.

The following corollary summarizes previous results.

Corollary 4.8.7 *In a timelike r–L plot for geodesics with $e < 0$ (hence $Q > 0$ and $d(e, Q) < 0$) there is a unique allowed region. This lies in the strip $0 < r < r_1$, where r_1 is the unique root of the discriminant polynomial $\phi(r)$. All orbits are bound. Spherical orbits occur only as follows: For Q sufficiently small there are*

bound orbits (including spherical orbits) near $(0, aE)$ *in* $0 < r < r_-$. *For* $e < 0$
small there are bound orbits (including spherical orbits) in the Kerr exterior.

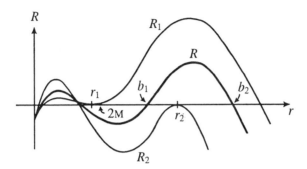

FIGURE 4.13.   Graphs of $R(r)$ for $L_2 < L < L_2$ where L_1 and L_2 are critical values
of angular momentum as in Figure 4.12. For L the function $R(r)$ (heavy line) yields two
bound orbits: $[b_1, b_2]$ in the Kerr exterior and another in the core of the black hole. L_1 has
an unstable spherical orbit at r_1, flanked by asymptotic orbits. The spherical orbit of L_2 is
stable.

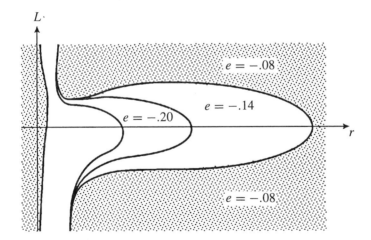

FIGURE 4.14.   Effect of varying $e < 0$ on a timelike r–L plot for constant $Q < 12$. The
darkly shaded forbidden region is for $e = -.08$, $Q = 5$. For $e = -.04$ (not shown) the
forbidden region is scarcely distinguishable from that of the marginal escape energy case
$e = 0$, so it produces retrograde as well as direct bound orbits. As e decreases, the allowed
region shrinks so that for $e = -.08$ there are only direct bound orbits and then none.

LAKE CONFIGURATION

As Figure 4.8 suggests, radical changes from the bay configuration result if $e < 0$ is small and Q is moderately large, to be precise, if $-1/9 < e < 0$ and $Q_+(e) \leq Q \leq Q_-(e)$. To picture these changes, let us start in Figure 4.15 from $e = -.08$, $Q = 5$, so $-1/9 < e < 0$ but $Q < Q_+(e)$. (This plot is qualitatively the same as the one in Figure 4.12.) Now let Q increase, so the forbidden regions enlarge. Much as for the $e > 0$ case the jaws of the bulge draw together. When $Q = Q_+(e)$ they meet—at a saddle point of $R(r, L)$ as predicted by assertion (2) of Proposition 4.7.8— and thereafter merge, forming a "lake" as shown in Figure 4.15.

Every r–L plot with $-1/9 < e < 0$ and $Q_+(e) < Q < Q_-(e)$ has this lake. In fact, since $d > 0$, ϕ has two simple roots $2M < r_1 < r_2$ and $\mu = 2$ between r_1 and r_2. The boundary of the lake is the simple closed curve formed by the two curves $L_-(r)$ and $L_+(r)$, which meet with vertical tangents at r_1 and r_2. The midline runs along the middle of the lake.

Thus there are many bound orbits in the Kerr exterior, given by horizontal line segments within the lake, and clearly all are stably bound.

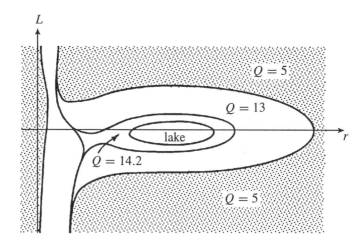

FIGURE 4.15.   The effect of increasing Q on a timelike r–L plot for $-1/9 < e < 0$. Here, as in Figure 4.14, the initial plot has $e = -.08$, $Q = 5$ (forbidden region shaded). As Q increases through $Q = 13$ the bulge in the forbidden region is closed off at $Q_+(.08) = 13.28^+$, and a lake is formed (shown for $Q = 14.2$). As Q increases, the lake shrinks, disapppearing at $Q_-(.08) = 15.06^-$.

Suppose Q continues to increase. By contrast with the $e > 0$ case where there was only one qualitative change, there is here a second: The enlarging forbidden regions shrink the lake until at $Q = Q_-(e)$ it vanishes. However, as specified by assertion (3) of Proposition 4.7.8, there remains an isolated maximum point (r_m, L_m) of $R(r, L)$, at which $R = 0$. This point represents an extremely stable spherical orbit.

Thereafter, for $Q > Q_-(e)$, the bay configuration of Figure 4.12 is restored. As Figure 4.8 indicates, for fixed $12 < Q < 16$ roughly the same changes can be gotten, for example, by holding Q constant and varying energy.

In the r–L plot for the origin $(e, Q) = (0, 0)$ in Figure 4.8 all of $r < 0$ is forbidden, and the only forbidden region in the strip $0 < r < r_-$ is the oval described in Lemma 4.8.5; otherwise the (0,0) plot has general conformation of Figure 4.9. Except that $r < 0$ remains forbidden, the plots for marginal escape energy $e = 0$ are essentially as for $e > 0$. As Q increases, the northern and southern continents draw together and, when $Q = 16M^2$, meet at the critical point $(4M, -M)$ to form a barrier.

The configurations found in this section are summarized in Figure 4.16.

For a survey to its date of the literature on Kerr geodesics, see Sharp (1979). For geodesics in the Kerr exterior, Stewart, Walker (1973) is recommended.

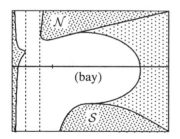

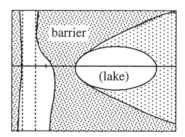

FIGURE 4.16.   The four basic configurations of r–L plots for timelike Kerr geodesics with $Q > 0$. Dark shading indicates forbidden regions for escape energy $e > 0$: continents, $0^+ < Q < Q_+(e)$; barrier, $Q_+(e) < Q$. (For Q small a knob forms near $(0, aE)$.) Lightly shaded regions are added for $e < 0$: lake if $-1/9 < e < 0$ and $Q_+(e) < Q < Q_-(e)$; bay otherwise.

4.9   Vortical Timelike Geodesics

For vortical geodesics—those with $Q < 0$—the barrier between $r > 0$ and $r < 0$ is lifted, and in fact every vortical geodesic infalling from $r = +\infty$ passes through the throat $r = 0$. Some but not all of these continue on to $r = -\infty$, thus achieving transit orbit. (As mentioned earlier, any geodesic sufficiently near a nonequatorial principal null geodesic will share its transit orbit type.).

Corollary 4.9.1   *A timelike* $(\tilde{q} = -1)$ *geodesic* γ *with* $Q < 0$ *has* $e = E^2 - 1 > 0$, $\mathfrak{X} > -Q > 0$, *and* $|L| + |Q|^{1/2} < a\sqrt{e}$. *In particular,* $R(r)$ *has no roots in* $r \geq 0$.

Proof.   We saw in Section 4.5 that $Q < 0$ implies high rotational energy $a^2 e = a^2(E^2 - 1) > L^2$; hence $e > 0$ and $\mathfrak{X} = a^2 e - L^2 - Q > -Q > 0$. Also Q satisfies the Carter inequality $Q \geq Q_{\min} = -[(a\sqrt{e}) - |L|]^2$. This can be written as $|Q|^{1/2} \leq |(a\sqrt{e}) - |L|| = a\sqrt{e} - |L|$, since $a^2 e > L^2$.

Consequently, all coefficients of the polynomial $R(r) = er^4 + 2Mr^3 + \mathfrak{X}r^2 + 2M\mathfrak{K}r - a^2 Q$ are positive, so there can be only negative roots.   □

In particular, every material particle with $Q < 0$ falling into the black hole from anywhere in $r > 0$ must continue on through the horizons and the throat into $r < 0$. Thus in a vortical r–L plot we need only examine the region $r < 0$, and even there the only horizontal lines that actually represent geodesics are those in the *Carter horizontal band* $|L| < L_{\max}$, where $L_{\max} = a\sqrt{e} - |Q|^{1/2}$. (See Remark 4.6.2.) For example, in Figure 4.17, considering $R = 0$ alone would suggest that there is a bound orbit with r-interval between the continent $\mathcal{N}'$ and the small island, but restriction to the strip eliminates this possibility.

Descartes' rule of signs places no restriction of the negative roots of $R(r)$; *a priori*, there could be as many as four; however this limit is never attained.

Proposition 4.9.2   *For a timelike vortical geodesic, the polynomial* $R(r)$ *has at most two roots.*

Before giving the proof of this result we consider its orbital consequences.

Corollary 4.9.3   *The orbit types for timelike vortical geodesic are* (1) *transit* , (2) *flyby orbit in* $r < 0$, (3) *flyby orbit from* $r = +\infty$ *with turn in* $r < 0$, *and* (4) *unstable spherical orbit at* $r_0 < 0$, *with adjacent asymptotic orbits.*

Proof. Since $R(r)$ has no roots in $r \geq 0$, the proposition implies that $R(r)$ has either (1) no real roots, (2,3) one double root $r_0 < 0$, or (4) two real roots, both negative. □

As Figure 4.17 indicates, these possibilities all actually occur. But there are no stable bound orbits, so except for (4) all have $|r| \to \infty$ as τ approaches either $+\infty$ and $-\infty$. Also, there are no flyby orbits confined to the Kerr exterior. Except for (2), none of the listed orbits are possible for $Q > 0$.

To prove Proposition 4.9.2 *it suffices to show that* $R(r)$ *has a unique critical point*. To see this, note that $R(0) = -a^2 Q > 0$, $(dR/dr)(0) = 2M\mathcal{K} > 0$, and $\lim_{r \to -\infty} R(r) = +\infty$. Consequently, as r decreases from 0, $R(r)$ decreases until reaching the unique critical point and thereafter increases toward $+\infty$. Evidently this produces zero, one, or two negative roots depending on the sign of R at the critical point.

Lemma 4.9.4 *For a vortical Kerr geodesic the polynomial*

$$F(r) = \frac{1}{2}dR/dr = 2er^3 + 3Mr^2 + \mathfrak{X}r + M\mathcal{K}$$

has a unique root if $2e\mathcal{K} \geq \mathfrak{X}$.

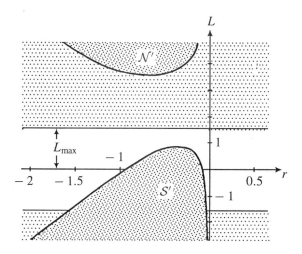

FIGURE 4.17.   In this vortical r–L plot the forbidden regions $R < 0$ are darkly shaded. The Carter inequality further excludes the lightly shaded region outside the horizontal band $|L| < L_{\max}$. Such bands never meet forbidden regions in $r > 0$. (Parameters: $M = 1$, $a^2 = .84$, $E = 3$, $Q = -1$, hence $L_{\max} \approx 1.59$.)

Proof.   We have $F' = dF/dr = 6er^2 + 6\text{M}r + \mathfrak{X}$, with discriminant $12(3\text{M}^2 - 2e\mathfrak{X})$.

Case 1. If $2e\mathfrak{X} > 3\text{M}^2$, then F' has no real roots. Thus $F' > 0$ since $F'(0) = \mathfrak{X} > 0$. Since it is strictly increasing, F has a unique root (necessarily negative).

Case 2. If $2e\mathfrak{X} = 3\text{M}^2$ then F' has a unique root, but F remains strictly increasing.

Case 3. If $2e\mathfrak{X} < 3\text{M}^2$, then F' has two real roots, the critical points of F, namely

$$\rho_\pm = -\text{M}/(2e) \pm (3\text{M}^2 - 2e\mathfrak{X})^{1/2}/(\sqrt{12}e),$$

with $-\text{M}/e < \rho_- < -\text{M}/2e < \rho_+ < 0$. Now $F(0) = \mathcal{K} > 0$ and $F'(0) = \mathfrak{X} > 0$, so as r decreases from 0, F descends to a local minimum at ρ_+, then rises to a local maximum at ρ_-, then decreases toward $-\infty$ as $r \to -\infty$ (since $e > 0$). Thus $F(\rho_+) > 0$ implies that F has a unique root in $r < 0$ (its only root since $F > 0$ for $r \geq 0$).

Since ρ_+ is in the interval $-\text{M}/(2e) < r < 0$, to prove $F(\rho_+) > 0$ it suffices to show that $F(r) > 0$ on $-\text{M}/(2e) < r < 0$. Write $F = F_1 + F_2$, where $F_1(r) = r^2(2er + 3\text{M})$ and $F_2(r) = \mathfrak{X}r + \text{M}\mathcal{K}$. Evidently, $F_1(r) > 0$ on the larger interval $-3\text{M}/(2e) < r < 0$. Similarly, $F_2(r) > 0$ from 0 out $r = -\text{M}\mathcal{K}/\mathfrak{X}$. But $-\text{M}\mathcal{K}/\mathfrak{X} \leq -\text{M}/(2e)$ since we assume $2e\mathcal{K} \geq \mathfrak{X}$. $\square$

Proof of Proposition 4.9.2.   It suffices to show that $2e\mathcal{K} \geq \mathfrak{X}$ always holds. Then the preceding lemma shows that F has a unique root; that is, $R(r)$ has a unique critical point.

In view of Corollary 4.9.1, let $|Q|^{1/2} = ca\sqrt{e}$, where $0 < c < 1$. Then the maximum value of $|L|$ is $(1 - c)a\sqrt{e}$, so $L = \lambda(1 - c)a\sqrt{e}$, where $|\lambda| < 1$. Now

$$\mathfrak{X} = a^2e - Q - \lambda^2L^2 = a^2e(1 + c^2 - \lambda^2(1 - c)^2), \quad \text{and}$$

$$\mathcal{K} = Q + (L - aE)^2 = -c^2a^2e + (\lambda(1 - c)a\sqrt{e} - aE)^2$$

$$= -c^2a^2e + \lambda^2(1 - c)^2a^2e + a^2(e + 1) - 2\lambda a^2(1 - c)\sqrt{e}E.$$

If $\lambda \geq 0$, then using the inequality $\sqrt{e}E = (e^2 + e)^{1/2} < e + \frac{1}{2}$ yields

$$(2e\mathcal{K} - \mathfrak{X})/a^2 > 2(1 - c^2)e^2 + e(1 - c^2)$$

$$+ 2\lambda^2(1 - c)^2e^2 + \lambda^2(1 - c^2)e - 4\lambda(1 - c)e^2 - 2\lambda(1 - c)e$$

$$= e^2[2(1 - c^2) + 2\lambda^2(1 - c)^2 - 4\lambda(1 - c)]$$

$$+ e[1 - c^2 + \lambda^2(1 - c^2) - 2\lambda(1 - c)]$$

$$= (1 - c)(1 + 2e)e[(1 - \lambda)^2 + c(1 - \lambda^2)].$$

This last expression is positive, hence $2eK - \mathfrak{X} > 0$. If $\lambda < 0$, then the inequality $\sqrt{e}E > e$ gives similar result. $\square$

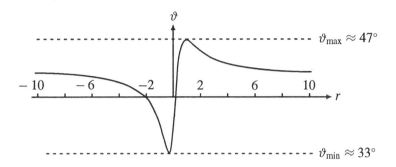

FIGURE 4.18.   A plot of $(r(\tau), \vartheta(\tau))$ for a timelike transit orbit. (Parameters: M $= 1$, $a^2 = .84$, $E = 3$, $L = 1.5$, $Q = -1$, as for Figure 4.17, and $L = 1.5 < L_{\max} \approx 1.59$.) In terms of North latitude, ϑ varies only about 14°. As Figure 4.17 indicates, if the value of L is reduced below about 0.9 the orbit type changes to flyby.

Since only vortical geodesics can have transit orbit, a fundamental problem is to find which vortical geodesics have this orbit type. As background we first show that restricting to $r < 0$ (as Corollary 4.9.1 permits), there are only two configurations for timelike vortical r–L plots. Imagine Q decreasing from 0; so forbidden regions are shrinking. For $Q < 0$ small, the forbidden region in $r < 0$ (like the one in Figure 4.9) pulls away from the L-axis as shown in Figure 4.10. The result—in addition to a small island in $r > 0$—is a barrier in $r < 0$ comparable to those in $r > 0$ for $Q > 0$. As Q decreases, the barrier thins, and at $Q = Q_-(e)$ separates into two continents $\mathcal{N}'$ and $\mathcal{S}'$ as in Figure 4.17. (Compare the $r > 0$ ones in Figure 4.9). In terms of the results of Section 4.7, the vortical r–L plots are parametrized by the points (e, Q) in the fourth quadrant of Figure 4.8 with $-a^2e < Q < 0$, as required by the Carter inequality. For $d(e, Q) > 0$ there are two vertical tangents in $r < 0$: barrier configuration; for $d(e, Q) < 0$ there are no vertical tangents in $r < 0$: continents configuration.

Obviously, the barrier prevents transit orbits, so it is clear from Figure 4.8 that a vortical geodesics whose effective energy e is too small cannot have transit orbit. In fact, if $e_\#$ is the value of e at which the line $Q = -a^2e$ and the curve $Q_-(e)$ intersect, then $e_\#$ is the largest value of e whose r–L plots have barrier configuration for all admissible Q, and hence $e_\#$ is a lower bound for effective energy of a timelike transit geodesic. Using the formula for Q_- in Remark 4.7.6 we can express $e_\#$ as the root of a polynomial in e depending only the Kerr parameters M, a. For example, if M $= 1, a^2 = .84$, then $e_\# \approx .72$. But Proposition 4.9.5

improves this to about 1.09, showing that even after the barrier is separated into continents, one of them can continue to block the Carter band $|L| < L_{max}$.

The situation is complicated by the fact that as $Q < 0$ decreases, forbidden regions shrink (tending to make transit easier), but the band width $L_{max} = a\sqrt{e} - |Q|^{1/2}$ also decreases (tending to make transit harder).

Proposition 4.9.5 *For a Kerr black hole with parameters M, a there exists a timelike transit geodesic with effective energy e if and only if $e > M/a$.*

Proof. There will be four steps.

Step 1. *For any $e > M/a$ there exist transit geodesics of effective energy e.* Proof. First consider, for arbitary $e > 0$, the smallest admissible value of Q, that is, $Q = -a^2e$. Then the Carter inequality is satisfied by $L = 0$, so such geodesics actually exist. (They lie in the Axis since $0 \leq \Theta = -a^2e + C^2a^2e$ implies $C = 1$, and will reappear in Section 4.11.) Consider the function $R(r)$ for these first-integrals. In particular, $\mathcal{K} = Q + (L - aE)^2$ becomes $-a^2e + a^2E^2 = a^2$, so

$$R(r) = er^4 + 2Mr^3 + 2a^2er^2 + 2Ma^2r + a^4e = (r^2 + a^2)(er^2 + 2Mr + ea^2).$$

Thus such a geodesic has transit orbit, that is, $R(r) > 0$, if and only if $er^2 + 2Mr + ea^2 > 0$ for all r. Since $ea^2 > 0$ this is equivalent to negative discriminant: $4M^2 - 4e^2a^2 < 0$, that is, to $e > M/a$.

Thus for $e > M/a$, these geodesics have transit orbits.

Step 2. *For any $e > 0$, if γ is a transit geodesic with first-integrals e, Q_0, L_0, then there is another transit geodesic with the same e, but $Q_1 \leq Q_0$ and $L_1 = a\sqrt{e} - |Q_1|^{1/2}$ (at the top of the Carter band).* Proof. As noted above, if $Q_0 = -a^2e$, then $L = 0$, so there is nothing to prove. And if $e \geq M/a$ we always take $Q_1 = -a^2e$. Thus we can suppose $e < M/a$ and $Q_0 > -a^2e$. Now imagine Q increasing from $-a^2e$ to Q_0. In the r–L plot for $Q = -a^2e$ (where $L = 0$) the r-axis meets a continent since by Step 1 this $L = 0$ geodesic does not have transit type. In fact, it meets the southern continent $\mathcal{S}'$ since the midline is above the r-axis. As Q increases the obstructing continent shrinks but so does the width of the Carter band. By hypothesis, when Q reaches Q_0, some line $L = $ const misses the continents. Thus at some earlier stage there is an algebraically smallest Q such that the top of the band is tangent to $\mathcal{S}'$. But for Q_1 just after that, the top of the band, that is, $L_1 = a\sqrt{e} - |Q|^{1/2}$, is unobstructed, giving transit orbit.

For any $L = L_{max}$ we will soon need the first-integrals $\mathfrak{X}$ and $\mathcal{K}$:

$$\mathfrak{X} = a^2 e - Q - (a\sqrt{e} - |Q|^{1/2})^2 = a^2 e - Q - a^2 e - |Q|$$
$$+ 2a\sqrt{e}|Q|^{1/2} = 2a\sqrt{e}|Q|^{1/2},$$
$$\mathcal{K} = Q + (a\sqrt{e} - |Q|^{1/2} - aE)^2 = Q + a^2(E - \sqrt{e})^2 + |Q|$$
$$+ 2a(E - \sqrt{e})|Q|^{1/2}$$
$$= a^2(E - \sqrt{e})^2 + 2a(E - \sqrt{e})|Q|^{1/2}.$$

Step 3 (the critical step). *No geodesic with $e = \text{M}/a$ has transit orbit.* Proof. For the geodesic in Step 1 with $e = \text{M}/a$ the formula for $R(r)$ becomes

$$R(r) = (\text{M}/a)(r^2 + a^2)(r + a)^2,$$

Thus, the minimum value, 0, is taken on at $r = -a$. We will show that every geodesic with $e = \text{M}/a$ and $Q > -a^2 e = -\text{M}a$ has $R(-a) < 0$, hence does not have transit type. By Step 2 it suffices to assume that these geodesics have $L = a\sqrt{e} - |Q|^{1/2}$. The first two terms of $R(r)$ are $(\text{M}/a)a^4 - 2\text{M}a^3 = -\text{M}a^3$, so

$$R(-a) = -\text{M}a^3 + \mathfrak{X}a^2 - 2\text{M}\mathcal{K}a - a^2 Q,$$

where $\mathfrak{X}$ and $\mathcal{K}$ are as in Step 2. Since $|Q| = -Q < a^2 e = \text{M}a$, the sum of the two outer terms here is $-\text{M}a^3 - a^2 Q < -\text{M}a^3 + \text{M}a = 0$. Thus, it suffices to show that the sum of the other two terms in $R(-a)$ is also negative, or equivalently, that $a\mathfrak{X} - 2\text{M}\mathcal{K} < 0$. Substituting values from Step 2 gives

$$a\mathfrak{X} - 2\text{M}K = 2a^2\sqrt{e}|Q|^{1/2} - 2\text{M}[a^2(E - \sqrt{e})^2 + 2a(E - \sqrt{e})|Q|^{1/2}]$$
$$= -2\text{M}[a^2(E - \sqrt{e})^2] + 2a|Q|^{1/2}[a\sqrt{e} - 2\text{M}(E - \sqrt{e})]$$

When $|Q| = -Q$ has its maximum value, namely $\text{M}a$, we know from Step 1 that $R(-a) = 0$. Also the sum of the two outer terms $-\text{M}a^3 - a^2 Q$ is zero. Hence $a\mathfrak{X} - 2\text{M}\mathcal{K} = 0$ then. (This can be checked by direct computation.) Since $a\mathfrak{X} - 2\text{M}\mathcal{K}$ is linear in $|Q|^{1/2}$ with constant term negative, it suffices to show that $a\sqrt{e} - 2\text{M}(E - \sqrt{e})$ is *positive* when $e = \text{M}/a$. Since $E = (e + 1)^{1/2} = (\text{M} + a)^{1/2}/a^{1/2}$, we find

$$a\sqrt{e} - 2\text{M}(E - \sqrt{e}) = (\text{M}a)^{1/2} - 2\text{M}[(\text{M} + a)^{1/2} - \text{M}^{1/2}]/a^{1/2}.$$

We can multiply by $(\text{M}a)^{1/2}$ getting

$$f(a) = \text{M}a - 2\text{M}\sqrt{\text{M}}(\text{M} + a)^{1/2} + 2\text{M}^2.$$

It remains to show that $f(a) > 0$ for $0 < a < M$. Now $f(0) = 0$, and $df/da = M - M\sqrt{M}/(M+a)^{1/2}$. But $M(M+a)^{1/2} - M\sqrt{M} > 0$ for all $a > 0$, hence $f(a) > 0$. Thus $a\sqrt{e} - 2M(E - \sqrt{e})$ is positive, as asserted.

Step 4. *No geodesic with $e < M/a$ has transit orbit.* The proof is a modification of that for Step 3. From Step 1, the geodesic with $Q = Q_{\min} = -a^2 e$, hence $L = 0$, has $R(r) = (r^2 + a^2)(er^2 + 2Mr + ea^2)$. This has a minimum at $r = -M/e$, where $R(-M/e) < 0$ since $e < M/a$.

We show that every geodesic with $e < M/a$ and $Q > -a^2 e$ also has $R(-M/e) < 0$. By Step 2 we can assume that $L = a\sqrt{e} - |Q|^{1/2}$. The first two terms of $R(-M/e)$ are $e(-M/e)^4 + 2M(-M/e)^3 = -M^4 a^3$, so

$$R(-a) = -M^4/e^3 + M^2 \mathfrak{X}/e^2 - 2M^2 \mathcal{K}/e - a^2 Q.$$

Here $-M^4/e^3 - a^2 Q < -M^4/e^3 + a^4 e = (-M^4 + a^4 e^4)/e^3 < 0$, since $e < M/a$. Hence it suffices to show that $\mathfrak{X} - 2e\mathcal{K} < 0$, for then $R(-1/e) < 0$. The formulas for $\mathfrak{X}$ and $\mathcal{K}$ in Step 1 yield

$$\mathfrak{X} - 2e\mathcal{K} = -2ea^2(E - \sqrt{e})^2 + 2a|Q|^{1/2}[\sqrt{e} - 2e(E - \sqrt{e})].$$

Thus the result follows if $\sqrt{e} - 2e(E - \sqrt{e}) > 0$ for all $e > 0$. When $e = 1$, this expression becomes $3 - 2\sqrt{2} > 0$. Assume $\sqrt{e} - 2e(E - \sqrt{e}) = 0$. Then

$$\sqrt{e}(1 + 2e) = 2eE = 2e(1 + e)^{1/2} \Rightarrow e(1 + 4e + 4e^2) = 4e^2(1 + e) \Rightarrow e = 0.$$

$\square$

As Figure 4.17 suggests, for $E > 0$, transit orbit is easier to achieve for direct orbits then for retrograde:

Remark 4.9.6 If a geodesic γ with $E > 0$, $L < 0$ has transit orbit, then so does any geodesic γ_1 with the same first-integrals except for angular momentum L_1 with $|L_1| \le |L|$. In fact, for such a γ_1 to exist and have transit orbit, we require the Carter inequality $|L_1| < a\sqrt{e} - |Q|^{1/2}$ and $R_1(r) > 0$ for all r. The condition on L_1 is clear since the existence of γ implies $|L| < a\sqrt{e} - |Q|^{1/2}$.

The change from $R(r)$ for γ to $R_1(r)$ for γ_1 affects only the terms $\mathfrak{X}r^2 + 2M\mathcal{K}r$. But $|L_1| \le -L$ and $E > 0$ imply both $\mathfrak{X}_1 - \mathfrak{X} \ge 0$ and

$$\mathcal{K}_1 - \mathcal{K} = L_1^2 - 2aEL_1 - L^2 + 2aEL = L_1^2 - L^2 + 2aE(L - L_1) \le 0$$

Thus $R_1(r) \ge R(r) > 0$ for $r < 0$. Then as noted above, $R_1(r) > 0$ holds for all r. As Figure 4.17 indicates, this comparison fails for $L > 0$.

Nevertheless, there are transit geodesics with retrograde orbits since having transit orbit is an open property, and, as we have seen, those with $e > M/a$ and $L = 0$ have transit type. Then, choosing Q near $-a^2 e$ yields retrograde transit geodesics.

4.10 Timelike Global Trajectories

Observers studying orbits in the exterior of a black hole can afford to ignore what happens to a particle once it falls through the horizon since they will never see it again. But for complete orbits in a slow Kerr black hole, crossing a horizon is only a beginning. In this section we chart global trajectories for the ordinary orbit types of Definition 4.6.3. Although we neglect unstable spherical orbits and their associated asymptotic orbits, the other exceptional case—approach to the ring singularity—is dealt with in Section 4.15. We concentrate on timelike geodesics not contained in the axis (Section 4.11) or a horizon (Section 4.12). (For a study of global null geodesic orbits, see Helliwell, Mollinckrodt 1975).

Three *horizon restrictions*, established earlier, are fundamental:

1. A timelike geodesic γ can cross any horizon, say with $r = r_+$, at most once. Thus, if its coordinate function $r(s)$ reaches r_+ a second time—which, as we have seen, frequently happens—then γ is crossing a different horizon.

2. If a geodesic γ crosses $r = r_\pm$, the values of the function $\oplus$ in Proposition 4.4.6 tell whether the crossing is through $S(r_\pm)$ or through a short horizon in a K^* or a $^*K$ patch.

3. A geodesic γ from outside a short horizon H can never have an r-turning point as it meets H since then γ' would be tangent to H (impossible, since H is totally geodesic). However, when γ meets a crossing sphere $S(r_\pm)$, it always has an r-turning point at the meeting. But as we saw in Chapter 3, the geodesic itself does not turn there, it cuts on transversally through $S(r_\pm)$—only the function $r(s)$ "turns."

Note also that a timelike geodesic with $Q = 0$ cannot cross the throat $r = 0$. (This follows since $R(0) = 0$ and $(dR/dr)(0) = 2M\mathcal{K} > 0$, hence $R(r) < 0$ for small $r < 0$.)

Throughout this section, Boyer–Lindquist blocks I, II, III inherit time-orientations from K^* patches (with their usual time-orientation). Recall that since $<V, V> = -\Delta\rho^2$ the canonical vector field V is timelike future-pointing on blocks I and III, spacelike on II, and null on horizons. Outside ergospheres, in particular for $r \geq 2M$, the same holds for ∂_t. Then on I' and III', where time-orientations have been reversed, *future-pointing* above is replaced by *past-pointing*.

Remark 4.10.1   With the conventions above, a timelike or null geodesic γ obeys the following sign restrictions.

1. *Sign of E.* If γ (future-pointing) starts in the region of block I or III where ∂_t is timelike, it has $E = -<\gamma', \partial_t> > 0$; hence γ can never reach the corresponding region in block I' or III' (where $E < 0$).

2. *Sign of* $\mathbb{P}(r)$. While γ (future-pointing) is in I or III, it has $\mathbb{P}(r) = -<\gamma', V> > 0$. In I' or III' it has $\mathbb{P}(r) < 0$.

With rare exceptions, specified below, any timelike geodesic that crosses a horizon must reach the Kerr exterior, block I. No timelike geodesic can remain entirely in block II or entirely in a horizon. Geodesics that remain entirely in block III are easily described, most having flyby orbits. Thus the class of geodesics that meet the exterior is quite large. For these it is convenient to specify some choices such as future-pointing versus past-pointing, or ingoing versus outgoing, that have no effect on global trajectories.

Definition 4.10.2   A standard particle *in slow Kerr spacetime* M_s *is a timelike* $(q = -1)$ *future-pointing geodesic* γ, *starting, with* $r'(s_0) < 0$, *at a point* $\gamma(s_0)$ *in block I of a* K^* *patch.*

Lemma 4.10.3   *If a standard particle* γ *(as defined above) has* $r(s)$ *reach* r_+ *while* $r' < 0$, *then* γ *crosses the short horizon* H_+ *in the patch* K^*, *thus leaving block I and entering block II of* K^*.

Proof.   Since $r' < 0$, γ can leave block I only through $r = r_+$. While γ is in block I, $\mathbb{P} = -<\gamma', V> > 0$. Thus, when $r(s)$ reaches r_+, the crossing sign, $\oplus = -\text{sgn}(r'\mathbb{P})$ from Proposition 4.4.6 will be sgn $\mathbb{P}(r_+) \geq 0$. Hence γ must leave block I through either the short horizon H_+ in K^* with sgn $\mathbb{P}(r_+) = +1$, or through the crossing sphere $S(r_+)$, with sgn $\mathbb{P}(r_+) = 0$. The latter can be ruled out as follows:

Assume γ leaves through $S(r_+)$ at parameter value s_1. Thus γ has an r-turning point at s_1, so immediately afterwards, $r' > 0$. Since $E > 0$, it follows that the function $\mathbb{P}(r) = (r^2 + a^2)E - La$ has a minimum at s_1; its value is $\mathbb{P}(r_+) = 0$, so $\mathbb{P}(r) > 0$ holds just past s_1. However, at the crossing point, γ is tangent to neither of the (totally geodesic) long horizons that intersect at $S(r_+)$, hence it cuts transversally through both and enters block I' of a patch $^*K$ (see Figures 4.19 and 4.20). But as previously noted, V is past-pointing in block I'; hence $\mathbb{P}(r) = -<\gamma', V> \; < 0$ holds just past s_1, contradicting the inequality above.

$\square$

TIMELIKE FLYBY ORBITS

Since Newtonian theory approximates general relativity, among the timelike Kerr orbits we must find analogues of Keplerian hyperbolas and ellipses. The first visitor to an astronomical black hole will certainly plan to fly past it along some quasi-hyperbolic orbit, but there are other possibilities.

- *Exterior flyby orbit.* This is the Keplerian analogue. Falling in from $r = +\infty$, γ turns before meeting the horizon $r = r_+$ and thus returns to $r = \infty$ in the same block I. All geodesics from $r \gg 1$ have this orbit type if $|L|$ is large enough—or for every L when the barrier exists. Generally, the larger $|L|$, the sooner the turn, and for given $|L|$, the turn is sooner (if $E > 0$) for direct orbits than for retrograde orbits, since $R_{-L}(r) > R_L(r)$ for $L > 0$.

- *Flyby orbit entirely in block* III. The white hole character of block III makes these short flyby orbits more common in block III than in block I. In fact, every $Q > 0$ timelike geodesic that meets $r < 0$ has this orbit type.

- *Long flyby orbits*, that is, flyby orbits that cross horizons. Such timelike orbits are possible *only for geodesics infalling from* $r = +\infty$ (rather than $-\infty$). In fact, geodesics in $r < 0$ with $Q \geq 0$ cannot cross the throat to reach a horizon, and we saw in Section 4.9 that vortical geodesics that meet $r < 0$ have either transit or short flyby orbits.

We can assume that a timelike geodesic γ ingoing from $r = +\infty$ is a standard particle, and Remark 4.10.1 shows that $\mathbb{P} = -<\gamma', V> \; > 0$ while γ is in block I. By Lemma 4.10.3, γ crosses the short horizon H_+ in K^*, so $\mathbb{P}(r_+) > 0$, that is, $(r_+^2 + a^2)E > aL$. Thus, γ enters II, where no r-turns are possible; hence $r' < 0$ continues to hold. Having flyby orbit, γ must leave II, necessarily at $r = r_-$, but through which horizon? This is determined by the crossing sign, $\oplus(r_-) = \mathrm{sgn}\,\mathbb{P}(r_-)$, so there are three cases: (A) $\oplus = +1$, (S) $\oplus = 0$, and (B) $\oplus = -1$.

Case A. $\oplus = +1 \Leftrightarrow \mathbb{P}(r_-) > 0 \Leftrightarrow (r_-^2 + a^2)E > aL \Leftrightarrow L < 2(\mathrm{M}/a)Er_-$. In terms of r–L plots such as Figure 4.9 this means that the L constant line representing γ meets the vertical line $r = r_-$ *below* the unique $R = 0$ point (r_-, L_-) on $r = r_-$. In particular, all flyby $L < 0$ orbits are included (for $E > 0$).

But in terms of trajectories, $\oplus = +1$ means that γ leaves II through the short horizon H_- in this same K^* patch, hence entering block III. Since $r(s)$ is decreasing at this entrance and eventually $r(s) \to +\infty$, γ must turn in III. After this (unique) turn, $r' > 0$ always holds, henceforth $\oplus = -\text{sgn}\,\mathbb{P}$. Thus, as γ leaves III at $r = r_-$, $\oplus = -\text{sgn}\,\mathbb{P}(r_-) = -1$. This means that γ leaves III through H_- in the K^* patch containing III, entering block II' of that patch. Again, γ leaves II' with $\oplus = -\text{sgn}\,\mathbb{P}(r_+) = -1$ (recall from above that $\mathbb{P}(r_+) > 0$. This implies that γ crosses H_+ into block I of the same $^*K$ patch. Now $r' > 0$ so there are no more crossing and γ goes on out to $r = +\infty$ in I as shown in Figure 4.19.

Case S. $\oplus = 0 \Leftrightarrow \mathbb{P}(r_-) = 0 \Leftrightarrow (r_-^2 + a^2)E = aL \Leftrightarrow L = L_- = 2(\text{M}/a)Er_-$. Thus L and E are precisely proportioned, and in an r–L plot such as Figure 4.9, the line $L = L_-$ hits the unique $R = 0$ point (r_-, L_-) on $r = r_-$. Now $\oplus = 0$ at $r = r_-$, which means that γ leaves II through the crossing sphere $S(r_-)$, and since it has an r-turn there, enters the adjoining block II'. (The inset of Figure 4.9 shows graphically that entry into III is impossible.) Flyby orbit means no more turns, so $r' > 0$ holds thereafter, giving $\oplus = -\text{sgn}\,\mathbb{P}$. Evidently γ must leave II' and do so at $r = r_+$. There $\oplus = -\text{sgn}\,\mathbb{P}(r_+) = -1$. Thus γ crosses H_- in the $^*K$

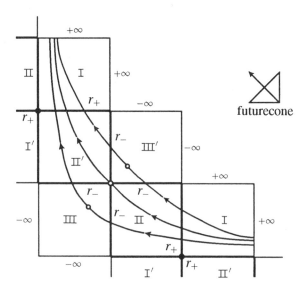

FIGURE 4.19.   Long flyby orbits for timelike geodesics in a slow Kerr black hole. The three possible cases A, S, and B are illustrated: Case A with conventional turning point (hollow dot) in III, Case S passing through the crossing sphere $S(r_-)$, and Case B with turning point in III'.

patch containing II', entering block I. As before, γ proceeds to $r = +\infty$ in this block, as shown in Figure 4.19.

Case B. $\oplus = -1 \Leftrightarrow \mathbb{P}(r_-) < 0 \Leftrightarrow (r_-^2 + a^2)E < aL \Leftrightarrow L > 2(\mathrm{M}/a)Er_-$. Larger angular momentum is required for this case, and in Figure 4.9 the L constant line passes below the northern continent but hits $r = r_-$ *above* the point r_-, L_-). That $\oplus = -1$ means that γ departs through the short horizon H_- in the $*K$ patch containing II, entering block III' of $*K$. (See Figure 4.19.) Since $r(s) > +\infty$, γ must leave III'; since r is decreasing as it enters III', γ must turn in III'. After the turn, $r' > 0$ always holds, so as γ reaches $r = r_-$ again, $\oplus = -\mathrm{sgn}\,\mathbb{P}(r_-) = +1$. Thus γ crosses H_- in the $K^{*\prime}$ patch containing III', entering II'. As γ leaves II', necessarily at $r = r_+$, $\oplus = -\mathrm{sgn}\,\mathbb{P}(r_+) = -1$. Hence the crossing is into block I, and as before, γ proceeds to $r = +\infty$.

All these long flyby orbits turn in block III(or III'), and the location of the turning point $r_\tau < r_-$ is influenced by the sign of the Carter constant Q, with $Q > 0 \Rightarrow r_\tau > 0$, $Q < 0 \Rightarrow r_\tau < 0$, and $Q = 0 \Rightarrow r_\tau \geq 0$.

In these more daring flights through the black hole, explorers return to a I block—but not the one they left. They can never go home again or even send messages back, although they can receive messages and new emigrants from home.

TIMELIKE BOUND ORBITS

Here the Newtonian analogue is a Keplerian ellipse, its radius oscillating between $r_1 = r_{\min}$ *(pericenter)* and $r_2 = r_{\max}$ *(apcenter)*. All timelike geodesics with less than escape energy have bound orbits, and Section 4.8 showed that in the strip $0 < r < 2\mathrm{M}$, bound orbits are common at all energy levels. By contrast there are none in $r < 0$ (unstable spherical orbits excepted) and none oscillating between $r < 0$ and $r > 0$.

Consider first the *small* bound orbits, those that remain in a single Boyer–Lindquist block.

- *Bound orbit entirely in block I.* γ remains in block I, so $r_+ < r_1 < r_2$. This is the Keplerian orbit of our earth around the sun, and Section 4.8 exhibited many such orbits, both direct and retrograde.
- *Bound orbit entirely in block III.* These are rare. As we have seen, there are none in $r < 0$ (unstable spherical orbits excepted), and none cross the throat. However, Section 4.8 showed that the transitory appearance of "knobs" for $Q > 0$ small produces such orbits in the strip $0 < r < r_-$ of block III. For $E > 0$ these are all direct, with L near aE.

- *Large bound orbits.* Let γ be a timelike geodesic with large bound orbit, and assume that γ meets block I. Such geodesics are common in the barrier and bay configurations (see Figure 4.16). Choosing a starting point $r_0 = \gamma(s_0)$ just after $r_2 = r_{\max}$ makes γ a standard particle. As before, γ crosses the short horizon H_+, tranverses II, and leaves through $r = r_-$ with the same three cases, A, S, and B.

Case A. γ enters block III in the same patch K^*. Then γ has its second r-turn, at $r_1 = r_{\min} < r_-$, and hence enters II$'$ of the $^*K$ containing III. This is the case shown in Figure 4.20.

Case S. γ passes through $S(r_-)$ into the same block II$'$ as in Case A (see Figure 4.20). Since the meeting point is an r-turn, $r_1 = r_{\min} = r_-$.

Case C. γ enters block III$'$ of the $^*K$ patch containing II, with $r' < 0$. It is saved from $r = -\infty$ by its second turn, at $r_1 = r_{\min} < r_-$. It then enters the same II$'$ as did γ in cases A and S.

Thus the geodesics of all three cases are reunited in II$'$. Thereafter, γ cannot enter I$'$, and since $\mathbb{P}(r_+) > 0$ cannot hit $S(r_+)$. Hence it must cross H_+ into block I of this patch $^*K$, with $r' > 0$. When $r(s)$ reaches the radius of the initial turn, γ turns again. These trips from block I to the next block I are duplicated in the future and in the past. So these bound orbits pass through every block I in the spacetime, every block II and II$'$, and in case A every block III, in case C every

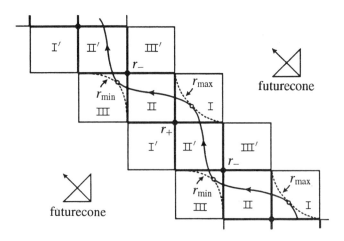

FIGURE 4.20.   A large bound orbit in a slow Kerr black hole. This typical Case A geodesic has $0 < r_{\min} < r_- < r_+ < r_{\max}$, and passes through every block I, II, II$'$, III, but does not cross the throat $r = 0$. Hollow dots indicate its r-turning points, at which γ is tangent to the hypersurfaces $r = r_{\min}$ and $r = r_{\max}$.

block Ⅲ′. All these orbits enter block Ⅲ, but (as shown above) none crosses the throat $r = 0$. Such journeys, of course, are very relativistic. A Newtonian satellite β, after each circuit of its bound orbit, does not return to its starting event, but it may well return to the same point in space, and it can be welcomed back by those who saw its departure. But γ from the case above will, after two circuits, be permanently separated from everyone in the departure crowd who has remained in the original Boyer–Lindquist block.

- *Large bound orbits that do not meet block* Ⅰ. Such orbits occur in all four $Q > 0$ configurations. They all have $L = L_+ = 2(\text{M}/a)Er_+$, and in r–L plots the $L = L_+$ line runs from (r_1, L_+) to (r_+, L_+), with $0 < r_1 < r_-$. Thus, their trajectories travel repeatedly from block Ⅲ through Ⅱ′ then via $S(r_+)$ to Ⅱ and back to Ⅲ (compare Figure 4.20).

Although no timelike orbits can be contained entirely in block Ⅱ, there are some that do not meet either Ⅰ or Ⅲ. These are too relativistic for any reasonable Newtonian analogy.

- *Time traveler.* $r_- = r_1 < r_2 = r_+$. Such a geodesic γ has its r-turns as it cuts alternately through crossing spheres $S(r_-)$ and $S(r_+)$. Since $\mathbb{P}(r_\pm) = (r_\pm^2 + a^2)E - aL = 0$, γ can meet no short horizons, so it remains permanently in blocks Ⅱ and Ⅱ′, traversing all of these and meeting every crossing sphere. The equations $\mathbb{P}(r_+) = \mathbb{P}(r_-) = 0$ imply $(r_+^2 - r_-^2)E = 0$; hence $E = 0$, and then $L = 0$. Thus γ is orthogonal to both ∂_t and ∂_ϑ, so it lies in a polar plane—necessarily the timelike kind, $P_{\overline{Ⅲ}}$ (see Section 3.6).

We can imagine these orbits as providing a delivery service where packages deposited at a convenient $S(r_\pm)$ can be picked up at any future crossing sphere.

TIMELIKE ORBITS OF TRANSIT TYPE

A rough Newtonian analogy is a particle falling radially inward from $r = +\infty$ and just missing a small massive star, hence continuing back out to infinity. The Kerr geodesics run from $r = \pm\infty$ to $r = \mp\infty$, passing safely through the throat of the black hole and hence are vortical. As noted previously, such orbits are not uncommon since any geodesic close to a principle null geodesic has transit type.

Lemma 4.10.4 *A timelike geodesic γ of transit type in a slow Kerr black hole remains in a single Kerr patch.*

Proof. We can suppose that γ comes in from $r = +\infty$ in block Ⅰ and hence is a standard particle. Thus γ crosses H_+ into block Ⅱ of K^*. Since $r(s)$ ranges from

$+\infty$ to $-\infty$, the function $\mathbb{P}(r) = (r^2 + a^2)E - La$ takes on a minimum at $r = 0$, where $\mathbb{P}(0) = a(aE - L)$. Since γ is vortical (and $E > 0$) the Carter inequality yields $|L| < a\sqrt{e} - |Q|^{1/2} \le aE$. Thus $\mathbb{P}(r) \ge \mathbb{P}(0) > 0$ for all r. But then $\oplus(r_-) = -\mathrm{sgn}\,(-\mathbb{P}(r_-)) = \mathbb{P}(r_-) > 0$. Thus as $r(s)$ passes r_-, γ crosses the short horizon H_- in K^*. Since γ has no r-turns, the sign $r' < 0$ is maintained, so γ continues to $r = -\infty$ in K^*. $\square$

4.11   Axial Geodesics

The axis, with its two isometric components, is well-defined in any Kerr spacetime, where it is a closed, totally geodesic timelike surface (Section 2.5). The axis crosses horizons and avoids the ring singularity. Despite its symmetrical location, rotational effects persist.

The first goal of this section is to prove that the axis is geodesically complete; then we consider what its geodesic orbits are like. Chapter 3 showed that the axis in a maximal Kerr spacetime is assembled from the axes of K^* and $^*K$ in the same way the full spacetime is assembled from K^* and $^*K$. It suffices to let A be either component of the full axis.

On A the first-integrals specialize as follows: $L = 0$, since $\partial_\varphi = 0$ on A, and $\mathcal{K} = -qa^2$, by the ϑ-equation. Thus $Q = -a^2(E^2 + q)$, so all geodesics with escape energy are vortical. Since $C = \pm 1$, ρ^2 becomes $r^2 + a^2$; thus $\mathbb{D} = 0$. Consequently, on the axis the r-equation can be written as

$$r'^2 = q\Delta/(r^2 + a^2) + E^2.$$

In this section and the next we need this general fact.

Lemma 4.11.1 *Let $F: \mathbf{R} \to \mathbf{R}$ a smooth function, bounded above. If $F(y_0) \ge 0$, then the initial value problem $y'^2 = F(y)$, $y(s_0) = y_0$ has exactly two smooth solutions, each defined on the entire real line—except in the case $F(y_0) = F'(y_0) = 0$, where the only solution is the constant one, $y = y_0$.*

Proof. First consider locally defined solutions. If $F(y_0) \ne 0$, the only solutions are those of $y' = \pm F(y)^{1/2}$. If $F(y_0) = 0$, then an argument as for Corollary 4.3.8 gives two cases: If $F'(y_0) = 0$, then $y(s) = y_0$ is the only solution, and other solutions can approach y_0 only asymptotically. If $F'(y_0) \ne 0$, there are two smooth solutions, one with $y''(s_0) = F'(y_0)/2 \ne 0$, the other, constant $y = y_0$. (The latter "envelope solution" will have no geodesic significance.)

Evidently one global solution to the differential equation is the constant one, $y = y_0$. Let $y(s)$ be any other smooth solution, defined on an open interval I. We must show that if $I \neq \mathbf{R}$ the solution can be extended. Now I has a finite endpoint b, and since F is bounded, $y(s)$ approaches a finite limit y_1 as $s \to b$. There is no problem unless $y(s)$ is nonconstant and $F(y_1) = 0$; thus $y' \to 0$. Near b, y is a solution of $y' = +F(y)^{1/2}$ or $y' = -F(y)^{1/2}$. Hence y can be extended through s_0 by the unique nonconstant local solution near s_0. $\square$

To prove completeness, let γ be an axial geodesic defined only locally. We can suppose that γ starts outside horizons. (This follows, as in the general case since a geodesic of A that remains in H can only be a restphoton, and these are complete.) The right side of the axial r-equation (preceding Lemma 4.11.1) is a bounded function of r, so by Lemma 4.11.1 there is a solution to this equation that extends the r-coordinate of γ over the entire real line. Thus, the problem is to extend its t coordinate. We deal with increasing parameter s; a change of variables then gives decreasing s.

For definiteness, suppose γ starts in a K^* patch. In terms of the Kerr-star coordinates t^*, r on A, the energy of γ has coordinate expression

$$E = -<\gamma', \partial_t> = -<t^{*\prime}\partial_t + r'\partial_r, \partial_t> = -g_{tt}t^{*\prime} - g^*_{rt}r' = t^{*\prime}\Delta/(r^2+a^2) - r'.$$

Solving for $t^{*\prime}$ yields the t^*-equation,

$$t^{*\prime} = (r' + E)(r^2 + a^2)/\Delta.$$

Unlike the axial r-equation, which holds wherever γ goes, this equation is valid only in K^* patches. But this is enough to show that the axis in the fast Kerr spacetime M_f is geodesically complete. We already have the r coordinate on $\mathbf{R}$, and since M_f is isometric to K^*, the t-equation is valid globally and shows that $t^{*\prime}$ and hence t^* are defined on all of $\mathbf{R}$.

As usual in the extreme and slow cases, horizons make matters more interesting. Note that a necessary condition for γ to cross a horizon H in K^* (or $^*K$) is that $E \neq 0$, for if $\gamma(s_1) \in H$, the axial r-equation reduces to $r'(s_1)^2 = E^2$.

Now let γ be a locally defined geodesic in the axis of M_e or M_s. Suppose first that γ has nonzero energy $E \neq 0$. Let s_1 be the first parameter value (if such exist) for which $r(s_1) = r_+, r_-$, or M. By the r-equation, $r'(s_1) = \pm E$.

Lemma 4.11.2 *With hypotheses as above, γ can be geodesically extended past s_1.*

(1) *If $r'(s_1) = -E$, then γ crosses a K^* horizon at the point $\gamma(s_1)$.*

(2) *If $r'(s_1) = E$, then γ crosses a $^*K$ horizon at $\gamma(s_1)$.*

Proof. Since we assume $E \neq 0$, the two cases are distinct.

(1) If $r'(s_1) = -E$, then $r'(s_1) - E \neq 0$. Writing the r-equation as $(r' + E)/\Delta = q(r^2 + a^2)/(r' - E)$ shows that $(r' + E)/\Delta$ is well behaved at $s = s_1$. Thus the t^*-equation can be integrated to give a smooth coordinate function $t^*(s)$ for γ up to and past s_1. Since $r(s)$ is already globally defined, this means that γ can be geodesically extended past s_1. Evidently it crosses the K^* horizon at $\gamma(s_1)$.

(2) If $r'(s_1) = E$, then $r'(s_1) + E \neq 0$. Now the r-equation shows that $t^{*'}$ blows up at s_1. In fact, since $|r'(s) + E|$ is bounded away from 0 near s_1, integration of this equation shows that, as $s \to s_1$, the geodesic γ races to infinity in Kerr-star coordinates. But, as in the regular case, switching to star-Kerr coordinates saves the situation. The function $r(s)$ is unchanged, but $^*t = t^* - 2T(r)$. Hence

$$t^{*'} = {}^*t' + 2\,dT/dr\,r' = {}^*t' + 2r'(r^2 + a^2)/\Delta.$$

Substituting this into the t^*-equation yields $^*t' = (E - r')(r^2 + a^2)/\Delta$. Now, as in case (1), $(E - r')/\Delta$ is well behaved near s_1, and hence $^*t$ is also. In short, γ extends across the horizon in $^*K$. $\square$

As usual, the extended r-coordinate of γ runs ahead of γ to see which horizons γ will cross.

Recall that all null geodesics in the axis A are principal. Lemma 4.11.2 shows that these axial principal nulls have the same behaviour as the regular ones. Specifically, the R-equation reduces to $r' = \pm E$, and nothing is lost by taking $E = 1$. The t^*-equation shows that if $r' = -1$, then $t^{*'} = 0$, so these are ingoers, defined for all $s \in \mathbf{R}$ (see Figure 2.8). But when $r' = +1$, this equation shows that t^* approaches infinity as $r(s)$ approaches a horizon radius; however, a shift to $^*K$ coordinates produces complete outgoers.

Corollary 4.11.3 *The axis in the extreme Kerr spacetime M_e is geodesically complete.*

Proof. The horizons of M_e are just the short horizons $r = $ M in its K^* and $^*K$ patches. We are supposing that γ starts in K^*. If $r(s)$ never equals M, then the t^*-equation can be integrated to give $t^*(s)$ on $\mathbf{R}$. If $E \neq 0$ and $r(s_1) = $ M for some s_1, Lemma 4.11.2 shows that γ crosses the horizon, possibly shifting to a $^*K$ patch. Continuing in this way, reversing K^* and $^*K$ as required, γ can be extended over $\mathbf{R}$. Finally, if $E = 0$, it follows from the t^*-equation that either γ is constant or it is a (complete) restphoton parametrizing $H \cap A$. $\square$

In the slow case M_s the previous arguments show that geodesics in A with energy $E \neq 0$ are complete. So suppose γ has $E = 0$. Horizon crossings are now possible but only at a crossing sphere $S(r_\pm)$.

If γ is timelike, the r-equation becomes $r'^2 = -\Delta/(r^2 + a^2)$. Hence $r_- \leq r(s) \leq r_+$, and $r(s)$ runs from one limit to the other. The t^*-equation yields

$$t^{*\prime} = r'(r^2 + a^2)/\Delta \quad \text{in } K^*, \qquad {}^*t' = -r'(r^2 + a^2)/\Delta \quad \text{in } {}^*K.$$

Thus the t coordinate is well defined while $r_- < r(s) < r_+$. To see what happens as $r(s)$ approaches r_- or r_+, we suppose for definiteness that γ is in block II of K^* and $r(s) \to r_+$. Then

$$t^*(s) - t^*(s_0) = \int_{s_0}^{s} \frac{(r^2 + a^2)r' \, ds}{\Delta} = \int_{r_0}^{r(s)} \frac{r^2 + a^2}{\Delta} \, dr$$

For r near r_+, $\Delta \approx r - r_+$, so t^* logarithmically approaches $+\infty$ as $r(s) \to r_+$. As in the regular case, it follows using KBL coordinates that γ meets a crossing sphere $S(r_+)$, and hence cuts through it into block II'. This behavior repeats itself, in the past as well as in the future. Thus the geodesic γ is a time traveller, as described in Section 4.10, passing through every crossing sphere of M_s. In particular, γ is complete.

If the $E = 0$ geodesic γ is null, then as before, the r-equation shows that $r(s)$ is constant; and t^*-equation then shows that it is a parametrization of $H \cap A$.

In the spacelike $E = 0$ case, the orbits are opposites of the block II time traveler orbits. The r-equation is $r'^2 = \Delta/(r^2 + a^2)$, hence either $r(s) \leq r_-$ or $r(s) \geq r_+$. For the latter, we can suppose γ starts in block I of K^* with $r'(s_0) < 0$. As s—and hence $r(s)$—decrease, $\Delta/(r^2 + a^2) \to 1$, so $r' \to -1$; thus $r(s)$ is defined for all $s \leq s_0$ and $r(s) \to \infty$ as $s \to -\infty$. As s and $r(s)$ increase, an argument like that in the timelike case shows that γ meets a crossing sphere $S(r_+)$ at some finite $s_1 < s_0$. Thus γ has a flyby orbit from $r = +\infty$ in block I, through $S(r_+)$ in block I' in the immediately past K^* patch—and on out to $r = +\infty$ in I'.

The $r(s) \leq r_-$ case is similar, with III replacing I. Evidently these spacelike $E = 0$ geodesics are complete, with orbits analogous to those in Figure 4.19. This finishes the proof of the following result.

Corollary 4.11.4 *The axis in the slow Kerr spacetime M_s is geodesically complete.*

The orbits of axial geodesics provide simple examples of general orbit types. They are readily found, since the axial r-equation can be written as an *energy equation*

$$r'^2 + \Phi(r) = E^2$$

by defining the *effective potential* $\Phi(r) = -q\Delta/(r^2 + a^2)$.

For timelike geodesics γ, we can read from Figure 4.21 (replacing E^2 by effective energy $e = E^2 - 1$) that γ has

- Transit type if $e > \mathrm{M}/a$,
- Flyby type if $0 \le e < \mathrm{M}/a$,
- Bound orbit if $-1 \le e < 0$.

As mentioned above, axial geodesics with $e > 0$ are vortical; these were already noticed in Section 4.9 (but are now known to be complete). The geodesic with $e = \mathrm{M}/a$ that separates transit and flyby types has r constant at $-a$, so it is a t-parameter curve. Hence flyby orbits from $r = -\infty$ and $r = +\infty$ turn before reaching $r = -a$ (the latter between $-a$ and 0). For axial geodesics, marginal escape energy $e = 0$ implies that escape actually occurs (not true in general), and as we saw, $e = -1$ gives the time traveler.

We can picture axial orbits as those of free-falling test particles ejected from a spacecraft hovering in, say, the north axis in the Kerr exterior

1. If a pebble is hurled downward toward the black hole with sizable energy $e > \mathrm{M}/a$, it will fall through the core on out to $r = -\infty$ (transit type). Hurled upward, it simply escapes to $r = +\infty$.

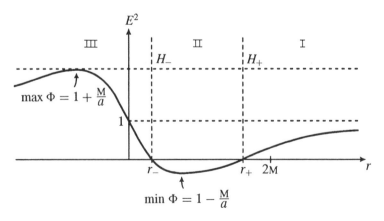

FIGURE 4.21.   Graph of the effective potential Φ of a timelike axial geodesic. (Note that the vertical axis is E^2 not E.)

2. If the pebble is thrown downward less energetically, $0 < e < \text{M}/a$, it crosses the throat $r = 0$ and then turns but does not rise back up to be recaptured at the spacecraft. Having already crossed both horizons in its initial patch K^*, it cannot recross them, so that when it meets $r = r_-$ after the turn, the pebble crosses into $*K$, thus achieving long flyby orbit (see Figure 4.19).

3. If the pebble is dropped downward with $e < 0$ (less than escape energy), it will cross the horizons but turn before reaching the throat. Then there is a second turn, giving a large bound orbit (see Figure 4.20).

Thus the pebble will return to the spacecraft only if it is tossed gently upward—a reasonable outcome. There is no way to produce a bound axial orbit in the Kerr exterior.

Since the axial null geodesics are principal, they all have transit orbits. The orbit types of spacelike axial geodesics can be read from a figure like Figure 4.21 with Φ replaced by $-\Phi$. Aside from an unstable orbit at (constant) $r = a$, the only orbit types are transit and long flyby, the latter turning in block II before reaching $r = a$.

4.12   Geodesics in Horizons

We examine the geodesics in the horizons of a maximal Kerr black hole M, showing that they are complete. We emphasize that these are the geodesics of M that are contained in the closed, totally geodesic hypersurfaces $\mathbb{H}_\pm$—being null, the horizons have no intrinsic geodesics of their own. We concentrate on the long horizons $\mathbb{H}_\pm$: $r = r_\pm$ in the slow Kerr black hole M_s, but the easier short horizons in M_e are covered by setting $r_\pm = a = \text{M}$.

For the horizons, various Kerr constants simplify considerably. First, since $\mathbb{H}$: $r = r_\pm$ is defined by $\Delta(r_\pm) = 0$, we have $r_\pm{}^2 + a^2 = 2\text{M}r_\pm$. This constant will be denoted by $b_\pm$

$$b_\pm = r_\pm{}^2 + a^2 = 2\text{M}r_\pm.$$

Then the first-integrals specialize as follows.

1. Except for restphotons *all geodesics γ in $\mathbb{H}$ are spacelike.* We saw in Chapter 3 that the restphotons are complete, so we can neglect them now, setting $q = +1$.

2. For every horizon geodesic, L and E are proportional: $b_\pm E = aL$. In fact, a geodesic in $r = r_\pm$ has $r' = 0$ and $\Delta(r_\pm) = 0$, hence the r-equation in Theorem 4.2.2 implies $\mathbb{P}(r_\pm) = 0$; that is, $(r_\pm{}^2 + a^2)E - aL = 0$. In particular, E and L are either both zero or both nonzero with the same sign.

3. $\mathcal{K} = r_{\pm}^2$. In fact, V is always in the principal plane Π and is null on $\mathbb{H}$, so by a formula following Lemma 4.2.7, $\langle \gamma_{\pi}', \gamma_{\pi}' \rangle = 0$. Then since $q = +1$, the lemma itself gives $\mathcal{K} = r_{\pm}^2$. (On horizons the $\mathcal{K}$ form of the Carter constant is usually more efficient than Q.)

Since the r-coordinate of a horizon geodesic is constant, the ϑ-equation becomes a first-order differential equation of degree 2.

$$(d\vartheta/ds)^2 = \Theta(\vartheta)/\rho_{\pm}^4, \quad \text{where} \quad \rho_{\pm}^2 = \rho_{\pm}(\vartheta)^2 = r_{\pm}^2 + a^2C^2.$$

Here $\rho_{\pm}^2 \geq r_{\pm}^2 > 0$, and the function Θ is always bounded above, so by Lemma 4.11.1, smooth solutions $\vartheta(s)$ exist, defined on all of **R**. Thus, for every ϑ_0 such that $\Theta(\vartheta_0) \geq 0$, we obtain, locally at least, a geodesic γ in $\mathbb{H}$. We must show that these goedesics are complete.

On a short horizon H, the first-order geodesic equations for φ^* and t^* (and ${}^*\varphi$ and ${}^*t$) collapse to a single equation, valid for both coordinate systems.

Lemma 4.12.1 *The Kerr-star coordinates of a geodesic γ in a short horizon $H : r = r_{\pm}$ are related by*

$$a t^{*\prime} - 2Mr_{\pm}\varphi^{*\prime} = -\mathbb{D}/S^2 = aE - L/S^2.$$

The same equation holds for ${}^*t'$ and ${}^*\varphi'$.

Proof. In general, for Kerr-star coordinates,

$$-E = \langle \gamma', \partial_t \rangle = +r' + g_{tt}t^{*\prime} + g_{t\varphi}\varphi^{*\prime},$$
$$L = \langle \gamma', \partial_\varphi \rangle = -aS^2 r' + g_{\varphi t}t^{*\prime} + g_{\varphi\varphi}\varphi^{*\prime}.$$

For star-Kerr coordinates, the signs of the r' terms reverse. Since $r' = 0$ on horizons the equations are valid also for ${}^*t$, ${}^*\varphi$. (The equations are not independent, since the determinant of coefficients is $g_{tt}g_{\varphi\varphi} - g_{t\varphi}^2 = -\Delta S^2 = 0$ on H.)

Now multiply the E equation by aS^2 and add this to the L equation. Then the metric identities (Section 2.1) give the result. $\square$

We now derive a second-order ordinary differential equation for the t^* (or ${}^*t$) coordinate of a geodesic γ in a Kerr horizon. Using it we can prove completeness and also determine when and how γ leaves a K^* patch for a ${}^*K$ patch, or the reverse. This equation could be gotten from the full geodesic equation for t^* (or ${}^*t$) by setting $r = r_{\pm}$, and using Lemma 4.12.1 to eliminate $\varphi^{*\prime}$ in favor of $t^{*\prime}$. We outline a less complicated derivation using Langrangian methods; details are omitted since the computation closely resembles that in Appendix C.

Two special notations are used.

1. To deal with Kerr-star and star-Kerr coordinates simultaneously, we write t for both t^* and $^*t$, and φ for φ^* and $^*\varphi$.

2. When the double signs $\pm$ and $\mp$ are used algebraically (on the line), the upper sign refers to K^*, the lower to $^*K$. This usage should not be confused with the *subscripts* such as $r_\pm$.

The line-element for K^* and $^*K$ gives the Langrangian

$$L = \frac{1}{2}\left[g_{tt}\dot{t}^2 + 2g_{t\varphi}\dot{t}\dot{\varphi} + g_{\varphi\varphi}\dot{\varphi}^2 + \rho^2\dot{\vartheta}^2 \pm 2\dot{r}\dot{t} \mp 2aS^2\dot{r}\dot{\varphi}\right].$$

For an arbitrary geodesic, since $\partial L/\partial \dot{r} = \pm\dot{t} \mp 2aS^2\dot{\varphi}$, the equation

$$(\partial L/\partial \dot{r})' = \partial L/\partial r$$

becomes

$$\pm t'' \mp aS^2\varphi'' \mp 2aSC\vartheta'\varphi' = \frac{1}{2}\left[\partial_r g_{tt}t'^2 + 2\partial_r g_{t\varphi}t'\varphi' + \partial_r g_{\varphi\varphi}\varphi'^2\right] + r\vartheta'^2. \quad (*)$$

For a horizon geodesic, $r \to r_\pm$, $\rho \to \rho_\pm$, and, by Lemma 4.12.1,

$$\varphi' = b_\pm^{-1}(at' + \mathbb{D}/S^2), \quad \text{hence} \quad \varphi'' = b_\pm^{-1}[at'' - 2L(C/S^3)\vartheta'].$$

Substituting these expression into the left side of equation $(*)$ yields

$$\pm\rho_\pm^2 b_\pm^{-1}t'' \mp 2a^2 b_\pm^{-1}SC\vartheta'(t' - E).$$

On the right side of equation $(*)$, we express the large first term as $\frac{1}{2}\mathcal{A}t' + \frac{1}{2}\mathcal{B}\varphi'$, where

$$\mathcal{A} = \partial_r g_{tt}t' + \partial_r g_{t\varphi}\varphi', \qquad \mathcal{B} = \partial_r g_{t\varphi}t' + \partial_r g_{\varphi\varphi}\varphi'.$$

Substituting for φ' gives

$$\mathcal{A} = [\partial_r g_{tt} + ab_\pm^{-1}\partial_r g_{t\varphi}]t' + b_\pm^{-1}\partial_r g_{t\varphi}\mathbb{D}/S^2$$

Differentiating the metric identities (as in Appendix C), and using Lemma 4.12.1 leads gives an expression for $\frac{1}{2}\mathcal{A}t' + \frac{1}{2}\mathcal{B}\varphi'$ of the form $ft'^2 + gt' + h$.

Since t'^2 is the only nonlinear part of equation $(*)$, the coefficient f of t'^2 here is crucial; it is

$$f = -b_\pm^{-2}\left[(r_\pm - \text{M})(b_\pm - a^2S^2) + r_\pm(bg_{tt} + ag_{t\varphi})\right].$$

But $b_\pm - a^2S^2 = \rho_\pm^2$, and on H, the metric identity (m4) becomes $b_\pm g_{tt} + ag_{t\varphi} = 0$. Thus the coefficient f is just $-b_\pm^{-2}(r_\pm - \text{M})\rho_\pm^2$.

Substitute into equation $(*)$ and solve for t''. Then simplification, using the fact that $\mathbb{D} = E\rho_\pm^2/a$, leads to the following result.

Proposition 4.12.2 *The t (= t^* or $^*t$) coordinate of a geodesic in the short horizon $r = r_\pm$ of K^* or $^*K$ satisfies the differential equation*

$$t'' = \mp b_\pm^{-1}(r_\pm - M)t'^2 + At' + B,$$

where $A = A(s) = 2a^2\rho_\pm^{-2}SC\vartheta' \pm 2b_\pm^{-1}\left[r_\pm - M + b_\pm r \pm \rho_\pm^{-2}\right]E$, and

$$B = B(s) = \pm r_\pm b_\pm \rho_\pm^{-2}\vartheta'^2 - 2a^2\rho_\pm^{-2}SCE\vartheta' + \frac{1}{2}N(\vartheta)E^2/(a^2 r \pm S^2 b_\pm \rho_\pm^2),$$

with $N(\vartheta) = 2r_\pm^6 + a^2(5C^2 - 1)r_\pm^4 + a^4(C^4 + 2C^2 - 1)r_\pm^2 + a^6C^2S^2$.
   (Recall that $b_\pm = 2Mr_\pm = r_\pm^2 + a^2$, and in the algebraic signs $\pm$ and $\mp$, the upper sign refers to K^*, the lower to $^*K$.)

Although these formulas are complicated, the differential equation itself is not. Since t does not appear explicitly, it is a first order equation for $v = t'$; in fact, a Ricatti equation. Furthermore, the v^2 coefficient is constant, and the other coefficients A and B are *bounded* functions of $\vartheta = vt(s)$, defined on the entire real line.

Corollary 4.12.3 *The geodesics in the horizons of the extreme Kerr spacetime M_e are complete.*

Proof. Horizons in M_e are the short horizons of K^* and $^*K$, and since $r_\pm = M$ the t'^2 term above vanishes, leaving a linear equation whose coefficients on defined on the entire real line **R**. Hence the solutions $t(s)$ are also defined on all of **R**. In view of Lemma 4.12.1, the same is true for the φ coordinate. We saw above that the same holds for $\vartheta(s)$. Since $r(s)$ is constant the geodesics of M_e are complete. $\square$

The proof in the slow case uses the following general observation.

Lemma 4.12.4 *Let $a(s)$ and $b(s)$ be smooth functions on **R** with absolute values bounded by k, and let v be a solution of the differential equation $v' = -v^2 + a(s)v + b(s)$, with maximal domain I.*
1. *If $v(s_0) \le -(k + 1)$ for some s_0, then $v'(s) < -1$ for $s > s_0$, and I has a (finite) right endpoint c, near which $v \sim 1/(s - c) < 0$.*
2. *If $v(s_0) \ge k + 1$ for some s_0, then $v'(s) < -1$ for $s < s_0$, and I has a (finite) left endpoint c, near which $v \sim 1/(s - c) > 0$.*

Proof. In both cases, $|v(s_0)| \geq k + 1$; hence

$$v^2(s_0) \geq (k + 1)|v(s_0)| \geq k|v(s_0)| + |v(s_0)| \geq |a(s_0)||v(s_0)| + |b(s_0)| + 1$$
$$\geq |a(s_0)v(s_0) + b(s_0)| + 1.$$

Thus $v'(s_0) < -1$. In the first case, where $v(s_0)$ is negative, it follows that $v'(s) < -1$ for all $s \geq s_0$. As v becomes arbitrarily large, the term v^2 dominates the $av + b$, so v resembles a solution of $v' = v^2$. In particular, the domain of v has a right endpoint c. The corresponding solution of $v' = v^2$ is $1/(s - c)$. The second case is similar. □

Replacing v by $-v$ gives corresponding results for the equation $v' = +v^2 + av + b$. In the applications below, the sign for v^2 depends on K^* vs $^*K$ and also on r_+ vs r_- since $r_+ - \text{M} > 0 > r_- - \text{M}$.

Corollary 4.12.5 *The geodesics in the horizons of slow Kerr spacetime M_s are complete.*

Proof. The crossing spheres, at which long horizons intersect, are compact, totally geodesic Riemannian submanifolds; hence the Kerr geodesics in them are their intrinsic geodesics—and these are complete, by the well-known Hopf–Rinow theorem.

In the general case we consider, for definiteness, a geodesic γ initially contained in the horizon $r = r_+$ of a K^* patch. This short horizon is contained in the long horizon $U^+ = 0$ of a *KBL* coordinate domain $\mathcal{D}(r_+)$ (see Section 3.4). The other cases are minor variants differing only in sign choices.

As in the extreme case, the only problem is to extend the t^* coordinate of γ, since its ϑ coordinate is already defined on **R**, φ^* will derive from t^*, and r is constant.

According to Proposition 4.12.2, $v = t^{*\prime}$ is a solution of

$$v' = -b_+^{-1}(r_+ - \text{M})v^2 + Av + B.$$

Let I be the largest possible domain of v (so I contains the given domain of γ).

First we consider the ordinary case where the solution v remains bounded on a (possibly infinite) subinterval J of I containing the domain of γ. Then

$$t^*(s) = t^*(s_0) + \int_{s_0}^{s} v(s) \, ds$$

is defined for all $s \in J$. Consider the *KBL* coordinates r_+, V^+, ϑ, φ^+ of γ. By definition, $V^+(s) = \exp(\kappa_+ t^*)$, where $\kappa_+ = (r_+ - \text{M})/(r_+^2 + a^2) = (r_+ - \text{M})/b_\pm$.

Thus $V^+(s)$ is defined for all $s \in J$. As before, $\vartheta(s)$ is already defined on $\mathbf{R}$. Finally, since $r = r_+$ the definition of φ^+ becomes

$$\varphi^+(s) = \varphi^*(s) - (a/b_\pm)t^*(s) + \left[A(r_+) - (a/b_\pm)T(r_+)\right].$$

According to Lemma 4.12.1, $\varphi^{*\prime} - (a/b_\pm)t^{*\prime} = (L/S^2 - aE)/b_\pm$. Here, as usual, $S^2 = \sin^2 \vartheta(s)$ is bounded away from 0, or else $L = 0$. Then integration of this equation shows that $\varphi^+(s)$ is defined for all $s \in J$. We conclude that γ can be extended over J.

Now we consider the two exceptional cases.

Case 1. The largest domain I has (finite) right endpoint c. Neglecting a positive factor, we apply Lemma 4.12.4 to find that, near c, $v \sim 1/(s - c) < 0$. Hence, near c, $t^*(s) = \ln(c - s) +$ (bounded function of s) $\sim \ln(c - s)$. Then $V^+(s) = \exp(\kappa_+ t^*) \sim (c - s)^{\kappa_+}$. Thus as $s \to c$, $V^+ \to 0$. The formula above shows that since $\vartheta(s)$ approaches a finite limit as $s \to c$, so does φ^+. We conclude that as $s \to c$, the *KBL* coordinates of γ approach the finite limits $U^+ = 0$, $V^+ = 0$, $\vartheta(c)$, $\varphi^+ = \varphi^+(\sin \vartheta(c))$.

In short, as $s \to c$, $\gamma(s)$ approaches a point of the crossing sphere S: $U^+ = 0$, $V^+ = 0$ in $\mathcal{D}(r_+)$. Thus $\gamma(s)$ is extendible to $s = c$, and past it. Since S is closed and totally geodesic, γ meets it transversally, so the extension cuts through S into the other half of the long horizon, that is, the short $r = r_+$ horizon between blocks II' and I' of $\mathcal{D}(r_+)$. (See Figure 3.13). The further continuation of γ on the other side of S follows by a similar argument—and shows that γ is complete.

Case 2. The domain I has finite left endpoint c. In this variant of the previous case, near c, $v \sim 1/(s - c) > 0$, and hence $t^*(s) \sim \ln(s - c)$. Then $V^+(s) = \exp(\kappa_+ t^*) \sim (s - c)^{\kappa_+}$. Thus as $s \to c$, $V^+ \to 0$. In this case the geodesic has emerged from S.

These cases cover all behaviors of v allowed by Lemma 4.12.4. We conclude that in its initial horizon, γ can be extended (in both directions) until either its domain stretches to infinity or, at a finite endpoint, γ meets S. In the latter case, γ crosses S into another short K^* horizon, and the situation repeats itself. Thus γ is ultimately extendible over $\mathbf{R}$.   □

The proof of completeness mod Σ (Theorem 4.3.1) is now itself complete.

We record a few properties of horizon geodesics. First, the following result shows that all have bounded energy and there are no vortical geodesics.

Lemma 4.12.6 *If γ is a spacelike geodesic in a horizon $r = r_\pm$ of a maximal Kerr black hole then $Q \geq 0$ and $E^2 \leq r_\mp/r_\pm$.*

Proof. In the ϑ-equation, the function Θ, when expressed in terms of $\mathcal{K}$ with $q = 1$, is $\mathcal{K} + a^2 C^2 - \mathbb{D}^2/S^2$, where

$$\mathbb{D} = L - aES^2 = b_{\pm}E/a - aES^2 = E(b_{\pm} - a^2 S^2)/a = E\rho_{\pm}^2/a,$$

Hence $\Theta = \rho_{\pm}^2\left[1 - E^2\rho_{\pm}^2/(a^2 S^2)\right]$. As mentioned above, the only criterion for existence of a horizon geodesic is $\Theta(\vartheta) \geq 0$, that is, $E^2 \leq a^2 S^2/\rho_{\pm}^2$. It follows that the largest value of E^2 occurs when $\vartheta = \pi/2$ where

$$a^2 S^2/\rho_{\pm}^2 = a^2/r_{\pm}^2 = r_{\mp}/r_{\pm}.$$

Substituting $\mathcal{K} = r_{\pm}^2$ and $L = b_{\pm}E/a$ into $Q = \mathcal{K} - (L - aE)^2$ yields the formula $Q = r_{\pm}[1 - (r_{\pm}/r_{\mp})E^2]$. Thus $Q > 0$ except that $Q = 0$ at the maximum value $E^2 = r_{\mp}/r_{\pm}$. □

These geodesics with $E^2 = r_{\mp}/r_{\pm}$ lie in the intersection of the horizon $r = r_{\pm}$ and the equatorial plane; they will reappear from the equatorial viewpoint in Section 4.14.

Of course, all geodesics in horizons are spherical so the question is: Which are stable? For γ in $\mathbb{H}$ an asymptotic approach with $r(s) \to r_0$ while $r(s) \neq r_0$ will be by a geodesic that travels far from $\mathbb{H}$, so for γ to be unstable is the same as being *unstably in* $\mathbb{H}$.

Lemma 4.12.7 *The spacelike geodesics in $\mathbb{H}_+$ are all unstable; those in $\mathbb{H}_-$ are unstable if and only if $E^2 \geq \frac{1}{2}(r_+/r_- - 1)$.*

Proof. We compute the function $R(r)$ for horizon geodesics using the formula in Theorem 4.2.2. Since $aL = (r_{\pm}^2 + a^2)E$,

$$\mathbb{P} = (r^2 + a^2)E - aL = E(r^2 - r_{\pm}^2).$$

Then since $\mathcal{K} = r_{\pm}^2$ and $\Delta = (r - r_-)(r - r_+)$,

$$R = \mathbb{P}^2 + \Delta(r^2 - r_{\pm}^2) = (r^2 - r_{\pm}^2)\left[E^2(r^2 - r_{\pm}^2) + (r - r_-)(r - r_+)\right]$$

Thus for $\mathbb{H}_+$ we get

$$R = (r - r_+)^2(r + r_+)\lambda(r), \quad \text{where} \quad \lambda(r) = (E^2 + 1)r + E^2 r_+ - r_-.$$

This has, as it must, a double zero at r_+. To see that these geodesics are all unstable it suffices to show $R(r_+ + \epsilon) \geq 0$ for sufficiently small $\epsilon > 0$. This positivity follows from the preceding formula, since $\lambda(r_+) = 2E^2 r_+ + r_+ - r_- > 0$.

For $\mathbb{H}_-$, reversing r_- and r_+, gives

$$R = (r - r_-)^2(r + r_-)\mu(r), \quad \text{where} \quad \mu(r) = (E^2 + 1)r + E^2 r_- - r_+.$$

As before the geodesic is unstable if and only if $\mu(r_-) \geq 0$. But now $\mu(r_-) = 2E^2 r_- + r_- - r_+$, so the result follows. $\square$

These spacelike horizon geodesics, though lacking physical significance, have nontrivial orbit structure. Those with $E = 0$, hence also $L = 0$, spiral helically along $\mathbb{H}$, some cutting through the central sphere in $\mathbb{H}$. If $E = 0$, then $\vartheta(s)$ oscillates around the equator $\pi/2$, with amplitude diminishing as energy increases, so that, as we have seen, at maximum E^2 the geodesic lies in $\mathbb{H} \cap Eq$.

4.13   Polar Orbits

We say that a Kerr geodesic γ has *polar orbit* if it passes through both the north and the south axes—and thus traverses all latitudes. We are interested principally in the bound case; there the geodesic passes through the axes infinitely many times, like an artificial Earth satellite. Meeting one axis would suffice to give angular momentum $L = 0$, meeting two eliminates various exceptional cases discussed Section 4.5. In Newtonian gravitation the number of meetings of a flyby (that is, hyperbolic) orbit with the axis is at most two, but here it turns out that there can be arbitrarily many meetings.

Our goal is to see what kind of Kerr polar orbits occur and how they depend on first-integrals and initial conditions. We concentrate on the timelike geodesics ($q = -1$) in the slow black hole.

Lemma 4.13.1   *For a polar geodesic γ the function $\vartheta(s)$ is strictly monotonic. If γ is r-bounded then ϑ' is bounded away from 0; otherwise the total turn angle $\Delta\vartheta$ is finite.*

Proof.   A check of the $L = 0$ cases in Section 4.5 shows that meeting both axes implies $Q > 0$ and if $E^2 < 1$, then $Q > -a^2(E^2-1)$. Hence $\Theta = Q+C^2a^2(E^2-1) > 0$. Thus the ϑ-equation $\rho^4\vartheta'^2 = \Theta$ implies that ϑ' is nonvanishing. Write this equation as $\vartheta' = \pm\sqrt{\Theta}\rho^2$. Evidently $\Theta > 0$ is bounded (above and below), so if $r(s)$ is bounded above, then so is $\rho^2 = r^2+a^2C^2$. Hence ϑ' is bounded away from 0. If γ is not r-bounded, then we know that $r(\tau)$ must approach infinity as $\tau \to +\infty$ or as $\tau \to -\infty$. Take the case $r(\tau) \to +\infty$ as $\tau \to +\infty$. It suffices to show that for some b, $\Delta\vartheta$ is finite on $[b,\infty)$. For b sufficiently large we obtain

$r' > 0$ on $[b, \infty)$, hence ϑ can be expressed as a function of r. Then by the r- and ϑ-equations, $d\vartheta/dr = \vartheta'/r' = \sqrt{\Theta}/\sqrt{R}$. Evidently, $\Theta = Q + C^2 a^2 (E^2 - 1) > 0$ is bounded above, say by B. In $R = (E^2 - 1)r^4 + 2Mr^3 + \ldots$ we have $E^2 - 1 \geq 0$ and $2M > 0$, so for b large enough, $R(r) > Br^3$ on $r \geq b$. Thus $d\vartheta/dr > 1/r^3$ there, and the integral $\Delta\vartheta$ of $d\vartheta/dr$ from b to ∞ is finite.   □

Note that since $L = 0$ implies $\mathcal{K} = Q + a^2 E^2$, the Q inequality above is equivalent to $\mathcal{K} > a^2$. In the special case $E = 0$, γ is a time traveler as described in Section 4.10.

The following account makes extensive use of a paper by E. Stoghianidis and D. Tsoubelis (1987). They observe that when $q = -1$ and $L = 0$, the function R in the r-equation $\rho^4 r'^2 = R$ becomes $R = (r^2 + a^2)^2 E^2 - \Delta(r^2 + \mathcal{K})$, so that division by $(r^2 + a^2)^2$ yields

$$(r^2 + a^2)^{-2}\rho^4 r'^2 + \Phi = E^2, \quad \text{where} \quad \Phi = \Phi_{\mathcal{K}}(r) = \Delta(r^2 + \mathcal{K})/(r^2 + a^2)^2.$$

Since the coefficient of r'^2 is always positive we can treat Φ as an effective potential, which we call the *polar potential*, and the r-intervals of a polar geodesic of energy E^2 are the maximal intervals on which $E^2 \geq \Phi$.

Evidently $\Phi(0) = \mathcal{K}/a^2 > 1$. For large r, $\Phi = (r^4 - 2r^3 + O(r^2))/(r^4 + O(r^2))$, so $\Phi(r) \to 1$ from below as $r \to +\infty$. Since Φ is a positive function times Δ, Φ is zero only at $r_{\pm}$ and negative only on $r_- < r < r_+$ (see Figure 4.22). The same holds for $\partial\Phi/\partial\mathcal{K}$, so except in $r_- \leq r \leq r_+$, *increasing $\mathcal{K}$ gives larger values for* $\Phi_{\mathcal{K}}(r)$.

A direct calculation of $d\Phi/dr$ shows that the critical points of Φ occur at the roots of the polynomial

$$N = N_{\mathcal{K}} = Mr^4 - (\mathcal{K} - a^2)r^3 + 3M(\mathcal{K} - a^2)r^2 - a^2(\mathcal{K} - a^2)r - Ma^2\mathcal{K}.$$

Also N can be written as $rW - \mathcal{K}Z$, where

$$W = Mr^3 + a^2 r^2 - 3Ma^2 r + a^4 \quad \text{and} \quad Z = r^3 - 3Mr^2 + a^2 r + Ma^2.$$

Evidently, N has a unique zero in $r < 0$, and the character of Φ shows there is another in $r_- < r < r_+$. Hence Φ has at most two critical points in $r > r_+$. When there are two, we denote them by $r_u < r_s$, and then as Figure 4.22 illustrates, r_u is a local maximum point (representing an *unstable* spherical orbit) and r_s is a local minimum (*stable* spherical orbit).

Lemma 4.13.2 *A bound polar geodesic γ with $E \neq 0$ in a slow Kerr black hole has either small bound orbit in the Kerr exterior or large bound orbit (crossing both horizons) in $r > 0$.*

Proof. By Proposition 4.8.2 all orbits in $r < 0$ are flyby, and a polar orbit cannot meet $r = 0$ since $Q > 0$. Thus we need only show that polar geodesics cannot be trapped in the core of the black hole by a knob, as discussed in Section 4.8. This will be true if Φ is decreasing in $0 < r < r_-$ since then this interval can contain no r-interval. Write $\Phi = fg$, with $f = \Delta/(r^2+a^2)$ and $g = (r^2+\mathcal{K})/(r^2+a^2)$. Now $\Phi' = fg' + gf'$. We have $g > 0$ and $g' < 0$ everywhere (the latter since $\mathcal{K} > a^2$), while $f > 0$ on $r_- < r < r_+$. Thus if $f' < 0$ on this interval we get $\Phi' < 0$ there. Now $(r^2 + a^2)^2 f' = 2M(r^2 - a^2)$. Thus $f' < 0$ out to $r = a$. But $a \geq r_-$, since $a/M < 1 \Rightarrow 1 - (a/M) \leq (1 - a^2/M^2)^{1/2} \Rightarrow a \geq r_- = M - (M^2 - a^2)^{1/2}$. $\square$

The existence of exterior polar orbits is governed by two critical values of the Carter constant $\mathcal{K}$; the following result gives the first.

Proposition 4.13.3 *For the polar potential $\Phi_{\mathcal{K}}$ of a slow Kerr black hole there are unique numbers $r_d > 3M$ and $\mathcal{K}_d > 0$, depending solely on the Kerr parameters M and a, such that*

1. *If $\mathcal{K} < \mathcal{K}_d$, then $\Phi_{\mathcal{K}}$ is strictly increasing on $r > r_+$.*
2. *If $\mathcal{K} = \mathcal{K}_d$, then Φ_d is strictly increasing on $r > r_+$, except at $r = r_d$, where $d\Phi_d/dr$ has a double zero.*
3. *If $\mathcal{K} > \mathcal{K}_d$, then $\Phi_{\mathcal{K}}$ has critical points r_u and r_s (notation as before) with $r_z < r_u < r_d < r_s$, where r_z is the largest root of the polynomial Z (hence $2M < r_z < 3M$). Furthermore, $r_u \to r_z$ and $r_s \to +\infty$ as $\mathcal{K} \to \infty$.*

Proof. Consider the zeros of the polynomials Z and W defined above. Descartes' rule asserts that Z has at most two positive roots. Writing Z as $r^2(r - 3M) + a^2(r + M)$ shows that $Z > 0$ on $r > 3M$. Now $Z(0) = Ma^2 > 0$, $Z(2M) = M(3a^2 - 4M^2) < 0$ (for slow Kerr), and $Z(3M) = 4Ma^2 > 0$. Thus Z has exactly two positive roots, say r_σ and r_z, with $0 < r_\sigma < 2M < r_z < 3M$.

As with Z, Descartes' rule shows that W has at most two positive roots. Again, $W(0) > 0$, $W(a) = 2a^3(a - M) < 0$, and $W(M) = (M - a^2)^2 > 0$. Hence, $W(r) > 0$ for $r > M$.

On $r > r_z$, where $Z(r) > 0$, consider the function $K(r) = rW/Z > 0$. Since $W > 0$ also here, $K > 0$. Regard the polynomial N defined above as a function of both r and $\mathcal{K}$. Then

$$\mathcal{K} = K(r) \Leftrightarrow N(r, \mathcal{K}) = 0 \Leftrightarrow r \text{ is a critical point of } \Phi_{\mathcal{K}}.$$

Since $W > 0$ at r_z we have $K(r) \to +\infty$ as $r \to r_z$. Both Z and W are cubic polynomials so $K(r) \to +\infty$ as $r \to \infty$. Thus K has an absolute minimum point

r_d, the minimum value $K(r_d)$ denoted by $\mathcal{K}_d$. Hence any horizontal line at height $\mathcal{K} > \mathcal{K}_d$ must meet the graph of K at least twice in $r > r_z$—in fact, exactly twice, since such a meeting represents a root of N. We know that Φ has at most two critical points in $r > r_+$ and that it is increasing at r_+ and for large r. Hence the assertions in the proposition easily follow.   □

Remark 4.13.4 *The critical values of $\mathcal{K}$.*

1. To find explicit formulas for $\mathcal{K}_d$ and r_d, let $Ar + B$ be the remainder when $N_{\mathcal{K}}(r)$ is divided by $(r - r_d)^2$; then the equations $A = 0$, $B = 0$ can be solved explicitly for r_d and $\mathcal{K}_d$ in terms of M and a. For example, if M $= 1$ and $a^2 = 0.84$, then $\mathcal{K}_d \approx 12.114$, $r_d \approx 5.411$, (and the largest root of Z is $r_z \approx 2.539$).

2. A second value of $\mathcal{K}$ governs the existence of flyby orbits in the Kerr exterior. Evidently these are present when $\Phi_{\mathcal{K}}(r_u) > 1$ (as in Figure 4.22) but not when $\Phi_{\mathcal{K}}(r_u) < 1$. The two cases are separated by a unique constant $\mathcal{K}_1$ such that $\Phi_{\mathcal{K}_1}(r_u) = 1$, that is, for which the spherical orbit has marginal escape energy. We write r_1 for the coresponding value of r_u. Explicit formulas for K_1, r_1 can be gotten by simultaneously solving $\Phi_{\mathcal{K}}(r) = 1$, $N = 0$. Thus for M $= 1$, $a^2 = 0.84$ we find $K_1 \approx 15.886$, $r_1 \approx 3.504$.

The various polar orbits can now be described.

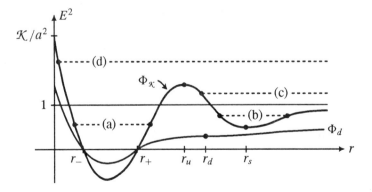

FIGURE 4.22.   Graphs of polar potentials (not to scale). The potential Φ_d has a double critical point at r_d. For $\mathcal{K} > \mathcal{K}_d$, $\Phi_{\mathcal{K}}$ has a maximum at r_u (unstable spherical orbit) and a minimum at r_s (stable spherical orbit). The dashed lines show the r-intervals for polar geodesics whose orbits are (a) large bound, (b) exterior bound, (c) exterior flyby, and (d) long flyby. Case (c) does not occur when $\Phi_{\mathcal{K}}(r_u) < 1$.

Case 1. $\mathcal{K} < \mathcal{K}_d$. There are no exterior orbits. All orbits with escape energy are long flyby; those with less than escape energy are large bound, with $0 < r_{min} < r_- < r_+ < r_{max}$. For the latter, $E = 0$ gives a time traveler, and as E^2 increases from 0 to 1 (limit values), r_{min} decreases from r_- and r_{max} increases from r_+ to $+\infty$.

Case 2. $\mathcal{K} = \mathcal{K}_d$. Same as Case 1 except that an unstable spherical orbit appears at $r = r_d$.

Case 3. $\mathcal{K}_d < \mathcal{K} < \mathcal{K}_1 \Leftrightarrow E_u^2 = \Phi_{\mathcal{K}}(r_u) < 1$. Similar to Case 4 except that there are no exterior flyby orbits, and two varieties of large bound orbits can be distinguished.

Case 4. $\mathcal{K}_1 < \mathcal{K} \Leftrightarrow E_u^2 = \Phi_{\mathcal{K}}(r_u) > 1$. The various orbit types can be recognized in Figure 4.22, where we find (a) large bound orbits if $E^2 < E_u^2$ and $r(0) < r_u$; (b) exterior bound orbits if $E_s^2 \leq E^2 < 1$ and $r(0) > r_u$; (c) exterior flyby orbits if $1 \leq E^2 < E_u^2$; and (d) long flyby orbits if $E^2 > E_u^2$. As usual, asymptotic orbits flank the unstable spherical orbit at $r = r_u$.

For polar geodesics, by contrast with the general case, it is easy to locate the spherical radii (as roots of the polynomial N) and find their first-integrals. In fact, the proof of Proposition 4.13.3 shows that for every $r > r_z$ there is a spherical polar orbit of radius r and constant $\mathcal{K} = K(r) = r W/Z$. The energy required for a spherical orbit at r satisfies $E^2 = \Phi_{\mathcal{K}}(r)$, and elimination of $\mathcal{K}$, using the definition of $\Phi_{\mathcal{K}}$, gives

$$E^2 = r\Delta^2/(r^2 + a^2)Z.$$

(see Stoghionidis, Tsoubelis 1987). Then Q can be inferred from $\mathcal{K} = Q + a^2 E^2$. Since $q = -1$ and $L = 0$ all first-integrals are thus determined.

In the $r-L$ plots of Section 4.8 the r-intervals of polar orbits, of course, lie on the r-axis. They can be specified in some detail by first locating the spherical ones, which signal change of orbit type. Spherical polar orbits are represented in an $r-L$ plot by a point at which the r-axis is tangent to the curve $R = 0$. By earlier results (namely, absence of knobs and Proposition 4.13.3) all are in the exterior $r > r_+$ and have constant $\mathcal{K} > \mathcal{K}_d$. There are just four possibilities described (for $E > 0$) by the point of tangency on the r-axis: (In the figures referenced below, the r-axis is not so tangent.)

- in unstable cases, (1) South tip of North continent $\mathcal{N}$ (Fig. 4.9), (2) North "headland" at entrance to the bay (Fig. 4.12).
- in stable cases, (3) North shore of the bay (Fig. 4.12), (4) North shore of lake (Figure 4.15).

Other horizontal tangents cannot reach the r-axis since knobs are excluded and southern tangents lie below the midline (hence below the r-axis). These cases can be calculated by using the e–Q chart, Figure 4.8, to compare Q with the functions Q_+ and Q_-. Consider first the unstable spherical polar orbits. As $\mathcal{K}$ increases from $\mathcal{K}_d$ the corresponding spherical radius r_u is initially at the N headland of the bay, but as $\mathcal{K}$ passes through $\mathcal{K}_1$, the orbit attains escape energy and is at the S tip of $\mathcal{N}$. The stable spherical orbits never have escape energy (see Figure 4.22). For $\mathcal{K}$ near $\mathcal{K}_d$ the stable radius r_s is initially at the N shore of the bay, but as $\mathcal{K}$ increases, the entrance to the bay closes, leaving r_u at the N shore of the resulting lake—where it remains.

The relative simplicity of polar geodesics also makes it feasible to consider their longitudinal motion—particularly for the exterior orbits, where Boyer–Lindquist coordinates suffice. Setting $L = 0$ in the equations of Proposition 4.1.5 yields $\rho^2 \varphi' = 2Mra E/\Delta$. In the Kerr exterior, both r and Δ are positive, so φ' is never zero: there are no longitudinal turning points. In fact, choosing $E > 0$ as usual, we have $0 < A \leq \varphi' \leq B <\infty$ for bound orbits, while for unbound orbits, $\varphi' \to 0^+$ as $r \to \infty$. Thus, as γ orbits from pole to pole around the horizon $r = r_+$ it is steadily advancing in the direction of rotation of the source of the hole.

Meanwhile the Boyer–Lindquist time t of γ increases according to

$$\rho^2 t' = EA(r, \vartheta)/\Delta, \quad \text{where} \quad A(r, \vartheta) = (r^2 + a^2)\rho^2 + 2Mra^2 S^2 > 0.$$

Thus the Boyer–Lindquist stationary observers regard the time-rate of longitudinal rotation as $d\varphi/dt = (\varphi'/t') = 2Mar/A(r, \vartheta)$.

To measure this dragging we combine the ϑ and φ first-integral equations to get

$$d\varphi/d\vartheta = \varphi'/\vartheta' = 2Ma Er/(\Delta\sqrt{\Theta}),$$

where $\Theta = Q + (E^2 - 1)a^2 C^2$.

For spherical orbits this equation can be solved in terms of elliptic integrals. Since $E^2 - 1 < 0$ for the stable spherical orbit, we replace the ϑ coordinate by *latitude* $\theta = \frac{1}{2}\pi - \vartheta$, which produces the reversals $S = \sin(\vartheta) = \cos(\theta)$ and $C = \cos(\vartheta) = \sin(\theta)$. Then $\Theta(\theta) = Q - a^2(1 - E^2)\sin^2\theta$, so for constant $r = r_0$, this equation has the solution

$$\varphi(\theta) = \varphi_0 + \frac{2Ma Er_0}{\Delta_0} \int_{\theta_0}^{\theta} \frac{d\theta}{\sqrt{\Theta(\theta)}}$$
$$= \varphi_0 - (2Ma Er_0/(\Delta_0\sqrt{Q}))\, F(\theta, k),$$

where F is the elliptic integral of the first kind, with modulus k such that $0 < k^2 = a^2(1 - E^2)/Q < 1$.

Thus the dragging of a spherical polar geodesic γ can be measured by the displacement $\Delta\varphi$ of Boyer–Lindquist longitude for a single polar "circuit" where ϑ increases from 0 to 2π. Then γ is invariant (but for a reparametrization $\tau \to \tau + \Delta\tau$) under the rotation $\varphi \to \varphi + \Delta\varphi$ (see Fig. 4.23).

Lemma 4.13.5 *Let γ be a timelike spherical polar geodesic of radius $r_0 = r_s$. If $E^2 < 1$, then the advance of longitude in one circuit is*

$$\Delta\varphi = (8MaEr_0)/(\Delta_0\sqrt{Q})\, K(k)$$

where $K(k) = F(\pi/2, k)$ is the complete elliptic integral of the first kind for $0 < k^2 = a^2(1 - E^2)/Q < 1$ (Stoghianidis, Tsoubelis 1987).

Figure 4.23 shows two circuits of such a spherical orbit. In it Boyer–Lindquist coordinates in the Kerr exterior are treated as Euclidean spherical coordinates (the version in Remark 1.7.2). The first-integrals are inferred, as described above, from the choice of radius r_0.

Such bound polar orbits invite comparison with (nonrelativistic) polar Earth orbits, the relativistic dragging of the former around the horizon $r = r_+$ mimicking the effect of the rotating Earth beneath the latter's planar Keplerian orbit.

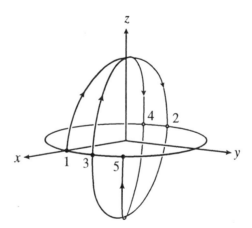

FIGURE 4.23.   A timelike spherical polar orbit of (constant) radius $r = 10$. Two circuits 1, 2, 3 and 3, 4, 5 are shown, and the dragging angle $\Delta\varphi \approx 21°$ appears as $< 1\,o\,3$ and $< 3\,o\,5$ (where $o =$ origin). With $M = 1$, $a = 0.84$, hence $r_d \approx 5.4 < 10$, this orbit is stable, with $E \approx .956 < 1$ and $\mathcal{K} \approx 14.939$, hence $Q \approx 14.126$. (Compare Fig. 3 of Stoghianidis, Tsoubelis 1987).

Figure 4.24 shows a nearby nonspherical orbit. The Keplerian analogue of this orbit is an ellipse in the xz-plane. In the Schwarzschild model, although the orbit precesses relativistically (e.g., advance of the perihelion of Mercury) it remains in the same plane. But finally, the Kerr rotation adds longitudinal dragging to this precession.

Let us locate the spherical orbit of Figure 4.23 in the e–Q chart, Figure 4.8. This orbit has $e = E^2 - 1 \approx -.0863 > -1/9$ and $Q \approx 14.171$. The formulas for $Q_{\pm}$ in Section 4.7 yield $Q_+(e) \approx 13.046$ and $Q_-(e) \approx 14.219$. Thus (e, Q) lies in the region $d > 0$ of the chart for which the r–L plot contains a lake—and the r-axis is necessarily tangent to its N shore. For the nonspherical orbit in Figure 4.24 the r-axis cuts through the lake in the stated r-interval.

The spherical orbit in Figure 4.23 has less than escape energy, $E^2 < 1$; hence it could be either stable or unstable. If $E^2 \geq 1$ (so the orbit is unstable) the switch to θ is not needed. If $E^2 = 1$, then $\varphi(\vartheta) = \varphi_0 + (2MaEr_0/(\Delta_0\sqrt{Q}))\vartheta$; and if $E^2 > 1$, then

$$\varphi(\vartheta) = \varphi_0 + 2MaEr_0 / \left[\Delta_0(\mathcal{K} - a^2)^{1/2}\right] F(\vartheta, k),$$

where now

$$k^2 = a^2(E^2 - 1)/(\mathcal{K} - a^2) < 1.$$

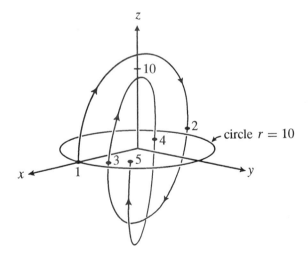

FIGURE 4.24.   A bound polar orbit in the Kerr exterior. The dots represent points at which the orbit meets the equatorial plane. The parameters are the same as in Figure 4.23, except that energy has been increased very slightly to $E = .957$. This is enough to replace the constant radius $r_s = 10$ by an r-interval $8.35 \leq r \leq 12.16$.

Evidently these orbits are qualitatively the same as the $E^2 < 1$ case in Figure 4.23. Now we consider flyby polar orbits in the Kerr exterior. These must have $\mathcal{K} > \mathcal{K}_1$; then any energy $1 < E^2 < \Phi_{\mathcal{K}}(r_u)$ suffices. Like all flyby orbits, these are nearly straight when $r \gg 1$, since there Kerr geometry is approximately Minkowskian. For smaller values of E^2 these orbits differ from a Keplerian hyperbola (through the axes) only in having the outgoing leg dragged away from the ingoing leg in the direction of Kerr rotation. Naturally, these mildly relativistic orbits meet the axis just twice. However, the number of axis crossings can be increased indefinitely. As Figure 4.22 indicates, at $E^2 = \Phi(r_u)$, a polar geodesic from $r = \infty$ asymptotically approaches the unstable spherical orbit at radius r_u (like that in Figure 4.23). Thus moving E^2 ever closer to $\Phi(r_u)$ produces orbits with arbitrarily many axis crossings. It is not easy *a priori* to imagine how a particle in free fall can manage to cross the axis as many as four times; this phenomenon is illustrated in Figure 4.25.

For polar geodesics with large bound orbits, longitudinal motion is not so easily understood, since we lack a single longitude function covering the entire orbit. Such a geodesic γ necessarily has less than escape energy, and we can assume that γ is a standard particle starting in the exterior (block I) of a K^* patch, with $r'(s_0) < 0$ and $0 < E < 1$. Since $\mathbb{P}(r_\pm) = (r^2 + a^2)E \neq 0$, the first horizons that γ crosses are the short horizons in K^*. Thus we begin with the Kerr-star longitude function φ^*. For $L = 0$ and these sign choices the first-integral equation for φ^* in Corollary 4.3.2 reduces to

$$\rho^2 \varphi^{*\prime} = (a/\Delta)[2\mathrm{M}Er - \sqrt{R}].$$

This differs from the exterior case only by the curvature term, which (by previous theory) ensures that φ^* is well-defined as γ crosses the horizons. After its r-turning point at $r_{\min}$ in block III, γ traverses a $*K$ patch, so both signs $\oplus$ and $\mathrm{sgn}(r')$ reverse, giving the same differential equation as above, but now for $*\varphi$. Duplications of these two cases cover the entire global trajectory of γ (see Figure 4.20).

Evidently, $\varphi^{*\prime} > 0$ at $r_{\min}$ and $r_{\max}$, since $R = 0$ there. But by constrast with the exterior case, where φ is strictly monotone, we shall see that φ^* can have turning points.

Lemma 4.13.6 *Let γ have large bound polar orbit. As γ meets the horizon $r = r_\pm$, $\rho^2 \varphi^{*\prime} = a[-E + k/(2E)]$, where $k = (r_\pm^2 + \mathcal{K})/(r_\pm^2 + a^2) > 1$.*

Proof. Since $L = 0$, $R = \mathbb{P}^2 - \Delta(\mathcal{K} + r^2) = (r^2 + a^2)^2 E^2 - \Delta(\mathcal{K} + r^2)$. Hence, as $\Delta \to 0$ the binomial theorem lets us replace $\sqrt{R}$ by

$$(r^2 + a^2)E - \Delta(\mathcal{K} + r^2)/[2(r^2 + a^2)E].$$

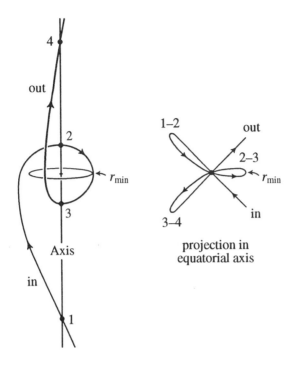

FIGURE 4.25. A timelike flyby polar orbit that meets the Axis four times. (Parameters: $M = 1$, $a^2 = .84$, $\mathcal{K} = 25 > K_1 \approx 15^+$, with $E^2 = 1.13456$, very close to $\Phi_{\mathcal{K}}(r_u) \approx 1.3484$.) The distant Axis crossings 1, 4 are at $r \approx 18.7$, the inner crossings 2, 3 at $r \approx 3.6$, with $r_{min} \approx 3.14$.

Thus as $r \to r_\pm$, $N/\Delta = [2MEr - \sqrt{R}]/\Delta$ approaches the limit

$$(N/\Delta)|_{r_\pm} = \left[2Mr_\pm E - (r_\pm{}^2 + a^2)E + \Delta(\mathcal{K} + r^2)/(2(r_\pm{}^2 + a^2)E)\right]/\Delta$$
$$= -E + (\mathcal{K} + r_\pm{}^2)/(2(r_\pm{}^2 + a^2)E)$$
$$= -E + k/(2E).$$

$\square$

Since $\rho^2\varphi^{*\prime}$ is positive at r_{min} and r_{max}, it is clear that suitable choices of first-integrals will produce turning points of φ^* variously in $r_{min} < r < r_-$, at r_-, in $r_- < r < r_+$ (block II), at r_+, and in $r_+ < r < r_{max}$. Note that $r^2\varphi^{*\prime}$ (and $\varphi^{*\prime}$) are always larger at r_- than at r_+.

When $\varphi^{*\prime} = 0$, the numerator $N = 2MEr - \sqrt{R}$ is zero; hence $R - 4M^2E^2r^2 = 0$, so there can be at most four turning points of $\varphi^{*\prime}$ as r travels from r_{max} to r_{min}.

Lemma 4.13.7 *Let γ be a timelike geodesic with large bound polar orbit. Then $\varphi^{*\prime} = d\varphi^*/d\tau$ changes sign at most twice as $r(s)$ moves from $r_{\max}$ to $r_{\min}$ (or the reverse).*

Proof. As before, write the φ^* equation as $\rho^2 \varphi^{*\prime} = aN/\Delta$, where $N = N(r) = 2\mathrm{M}Er - \sqrt{R}$ and

$$R = \mathbb{P}^2 - \Delta(\mathcal{K} + r^2) = (r^2 + a^2)^2 E^2 - \Delta(\mathcal{K} + r^2).$$

A necessary condition for $\varphi^{*\prime} = 0$ at radius r is that $N(r) = 0$, or equivalently, $R - 4\mathrm{M}^2 E^2 r^2 = 0$. This polynomial equation has degree 4, so N has at most four zeros. In fact, an appeal to Descartes' rule of signs shows that there are zero, two, or four. Here zero is impossible, because r_- and r_+ are always zeros of N. In fact, $N(r_\pm) = 2\mathrm{M}Er_\pm - (r_\pm{}^2 + a^2)E = \Delta(r_\pm)E = 0$.

Since $\Delta'(r_\pm) \neq 0$, L'Hospital's rule shows that if the zero of N at $r_\pm$ is simple, then $\varphi^{*\prime} = 0$ there, but if N has a double zero at $r_\pm$, then φ^* has a strict turning point; that is, $\varphi^{*\prime} \sim N/\Delta$ changes sign. (N has at most a triple zero at $r_\pm$; then $\varphi^{*\prime} = 0$ at $r_\pm$ but with no sign change.)

To examine the zeros of $\varphi^{*\prime}$, or equivalently of N/Δ, assume first that the zeros of N at $r_\pm$ are simple, so $N/\Delta \neq 0$ there. Thus, other than $r_\pm$, N has either no zeros or two zeros. Since Δ vanishes only at $r_\pm$, it follows that N/Δ has either no zeros or two zeros. When the latter are distinct, $\varphi^{*\prime}$ changes sign twice; otherwise φ^* pauses, but remains monotonic.

Next suppose N has a double zero at r_-, so $N/\Delta = 0$ has a simple zero there. It follows from Lemma 3.13.6 that N/Δ is smaller at r_+ than at r_-. Thus N/Δ is negative at r_+, so N has a simple zero there. Hence N has another (simple) zero, and we conclude that $\varphi^{*\prime}$ changes sign twice. The case where N has a double zero at r_+ is analogous.

If N has a triple zero at one of the horizon radii, say r_-, then the zero at r_+ is simple, so $\varphi^{*\prime}$ has a single repeated zero, hence φ^* remains monotonic.   □

To an inattentive observer, travel in a large polar orbit might seem to differ from travel in an exerior orbit only in that the oscillating radial motion carries him back and forth through horizons. But of course the horizons are always new ones, and there is no return to a past Boyer–Lindquist block.

4.14   Equatorial Geodesics

In the early study of Kerr spacetime the geodesics of the equatorial plane were emphasized since these compare directly with Schwarzschild geodesics, which are all, in effect, equatorial; thus the consequences of Kerr rotation stand out. The equator is still a good place to see certain aspects of Kerr geodesics in concrete terms. Taking advantage of the simplification produced by the equatorial property $Q = 0$ (Corollary 4.5.7), we consider geodesics of all three causal characters.

Recall that the equator Eq is disconnected by the ring singularity $r = 0$, and the $r > 0$ side, denoted by Eq^+, is connected, while the $r < 0$ side, Eq^-, has infinitely many components (see Fig. 3.16). Circles replace 2-spheres in the bundle structure of Eq, so its horizons are circles (enduring through time), and its geodesics with r constant are *circular*.

Geodesics in the equator are included in the broader class of geodesics with Carter constant $Q = 0$, and some of the general formulas below assume only $Q = 0$. However, orbital consequences are radically different for the equatorial and nonequatorial cases as we see at the end of this section.

Corollary 4.14.1 *For a geodesic γ with $Q = 0$, we have $R(r) = r\psi(r)$ where the function ψ is given by*

$$\psi(r) = (E^2 + q)r^3 - 2Mqr^2 + (a^2(E^2 + q) - L^2)r + 2M(L - aE)^2$$
$$= E^2r^3 + (a^2E^2 - L^2)r + 2M(L - aE)^2 + qr\Delta.$$

When $Q = 0$, r–L plots are determined solely by E, thus the e–Q chart (Figure 4.8) shows that timelike r–L plots have either the bay or continents configuration. The polynomial $R(r)$ is always zero at $r = 0$, otherwise ψ and R have the same zeros—and the same repeated zeros. As always, the r coordinate of a geodesic γ is limited to intervals on which $R(r) \geq 0$. Evidently the signs of $R = r\psi$ and ψ are the same when $r > 0$, but opposite when $r < 0$. Thus in r–L plots *regions with $\psi < 0$ are forbidden in $r > 0$, but allowed in $r < 0$*. As usual when a sign choice is required for $E \neq 0$ we take $E > 0$. For r–L plots the equatorial case is distinctive in that $r = 0$ represents the ring singularity (not a part of any spacetime) while elsewhere it is the throat $T: r = 0$, a hypersurface in Boyer–Lindquist block III.

The orbit types in Eq^- are easily described.

Lemma 4.14.2 *Every geodesic in Eq^- has flyby orbit except for those null or spacelike geodesics with $L = aE$. The latter meet the ring singularity and have crash-escape orbit type.*

Proof. According to Proposition 4.8.2 all timelike geodesics in Eq^- have flyby orbit, so suppose γ is either null or spacelike. Recall from Corollary 4.6.7 that geodesics in $r < 0$ have high rotational energy $a^2(E^2 + q) > L^2$ and hence escape energy $E^2 + q > 0$.

If $L \neq aE$, then the formulas above show that $R(0) = 0$ and $(dR/dr)(0) = 2M(L - aE)^2 > 0$. Thus $R(r) < 0$ for small $r < 0$. Also $R(r) > 0$ for $|r|$ large. A check of signs for Descartes' rule shows that $R(r)$ has a unique root with $r < 0$; hence these geodesics also have flyby type.

Now suppose $L = aE$. Recall that for geodesics in Eq—but not in general—this means that γ is principal (see proof at the end of Section 4.5). When $L = aE$ the formulas above for $R(r)$ reduce to

$$R(r) = (E^2 + q)r^4 - 2Mqr^3 + qa^2r^2 = r^2(E^2r^2 + q\Delta)$$

Since $q \geq 0$, we have $R(r) > 0$ for all $r < 0$. If γ is null we already know that principal null geodesics in Eq have crash-escape orbit type. If γ is spacelike ($q = +1$) then for $|r|$ small, $R(r)$ has order r^2. In Eq the r-equation becomes $r^4r'^2 = R(r)$, so near $r = 0$, $r^2r'^2 \approx 1$. Thus $r(s)$ reaches $r = 0$ on a finite parameter interval; in short, γ meets the ring singularity.   □

Thus in the "white hole" side Eq^- of the equator, of the geodesics aimed toward Σ, only the null and spacelike principal geodesics manage to hit it.

On the $r > 0$ side, Eq^+, the most striking changes from the general $Q \neq 0$ case occur in the strip $0 < r < r_-$. In an r–L plot for $Q \neq 0$ a "shore" protects particles from the throat $r = 0$—and hence from the ring singularity (see, for example, Figures 4.9 and 4.12). But for $Q = 0$ the shore largely disappears, since (as noted above) $R(0) = 0$ and $(dR/dr)(0) = 2M(L - aE)^2 \geq 0$, so the region in Eq^+ near $r = 0$ is allowed except possibly near the point $(0, aE)$. This point $(0, aE)$ is evidently a critical point of $R(r, L)$, so the pattern of allowed/forbidden regions near it is sensitive to small changes in first-integrals.

Since the degree 4 polynomial has in effect been replaced by the discriminant polynomial ψ, each horizontal line now meets Z in at most three points.

The early results from Section 4.7 can be simplified as follows.

Corollary 4.14.3 *Let γ be a $Q = 0$ geodesic, with ψ as in Corollary 4.14.1.*
(1) As a polynomial in L,

$$\psi(r, L) = (2M - r)L^2 - 4MaEL + \psi_0(r),$$

where $\psi_0(r) = \psi(r, L = 0) = E^2[r^3 + a^2r + 2Ma^2] + qr\Delta(r)$.
(2) If $r \neq 2M$, the discriminant of ψ as a polynomial in L is $4D(r)$, where $D(r) = r\Delta(r)[E^2r + qr - 2Mq]$.

(3) *The set Z of zeros of* $\psi(r, L)$ *(or of* $R(r, L)$*) is the union of the graphs of the two functions*

$$L_\pm(r) = \frac{2MaE \pm \sqrt{r\Delta[E^2r + qr - 2Mq]}}{2M - r}$$

defined on maximal r-intervals on which $D(r) \geq 0.$

Thus the reduction of R to ψ has replaced the discriminant polynomial ϕ of Section 4.7 by the linear expression $E^2r + qr - 2Mq$. As in Section 4.7 the discriminant is the key to determining the configuration of the set Z, since (for $r \neq 2M$) the sign $(-, 0, +)$ of $D(r)$ tells whether there are zero, one, or two points, respectively, of $R = 0$ on the vertical line $L \to (r, L)$.

For the exceptional line $r = 2M$, we find as before that the function $L \to \psi(2M, L)$ is linear with negative slope when $E > 0$. Thus as L increases, the line $r = 2M$ enters a forbidden region at $L_{2M} = \psi(r, 0)/(4MaE)$ and remains there. (When $E = 0$, ψ is constant on $r = 2M$.)

Proposition 4.14.4 *In a* $Q = 0$ *r–L plot with* $E^2 + q \neq 0$ *and* $E \neq 0$*, the only vertical lines that contain a unique point* (r, L_r) *of* Z *are* $r = 0$, r_-, r_+, $2M$, *or* $w = 2Mq/(E^2 + q)$. *The corresponding angular momenta are* L_{2M} *(as above) for* $r = 2M$, *and* $L_r = 2MaE/(2M - r)$ *for the others. Explicitly,*

$$L_0 = aE, \qquad L_\pm = 2MaE/r_\mp, \qquad L_w = a(E^2 + q)/E.$$

Proof. For $r \neq 2M$, there is exactly one such point if and only if the discriminant $D(r) = 4r\Delta(r)[E^2r + qr - 2Mq]$ vanishes at r. Hence $r = 0$, $r_\pm$, or $w = 2Mq/(E^2 + q)$. The (now trivial) quadratic formula gives the general expression for L_r. Then, in particular, $2M - w = 2ME^2/(E^2 + q)$; hence $L_w = a(E^2 + q)/E$. $\qquad \square$

Here w need not be different from the other points. At most of these uniqueness points, Z is smooth with a vertical tangent. In fact, it can be shown that the only exceptions occur when $w = 0$ (with γ null) and $w = r_\pm$ (with γ spacelike). For example, in Figure 4.26 it is clear that $(0, a)$ is a singular point. In the spacelike case the formulas above predict singularities at $r = r_\pm$, these occuring when the energy satisfies $r_\pm = 2M/(E_\pm{}^2 + 1)$ and hence $E_\pm{}^2 = r_\mp/r_\pm$. Then, unexpectedly, we get $L_\pm = 2M$ for both.

The asymptotic lines of the set Z follow from the general case, Lemma 4.8.3:

1. As $L \to \pm\infty$, Z is asymptotic to the vertical line $r = 2M$, approaching from the left as $L \to +\infty$ and from the right as $L \to -\infty$ (for $E > 0$).
2. When γ has escape energy, then (in all four quadrants) Z is asymptotic to the lines $A_\pm(r) = \mp(E^2 + q)^{1/2}(M + r)$.

NULL GEODESICS IN Eq

Geodesic reparametrization of any null geodesic γ changes L and E by the same constant factor, so L and E can be replaced (when $E \neq 0$) by the single first-integral $\lambda = L/E$, called the *impact parameter* of γ. In effect, λ is the angular momentum of γ when it is reparametrized to have unit energy. Thus a single r–λ plot is enough to describe the radial motion of all equatorial null geodesics.

Corollaries 4.14.1 and 4.14.3 give the following result.

Corollary 4.14.5 *For $Q = 0$ null geodesics with $E \neq 0$,*

$$\psi = R/r = r^3 + (a^2 - \lambda^2)r + 2M(\lambda - a)^2.$$

The zero set $Z: \psi = 0$ is the union of the graphs of the two functions

$$\lambda_\pm(r) = (2Ma \pm r\sqrt{\Delta})/(2M - r), \quad \text{where} \quad \Delta > 0.$$

Thus there are two points of Z on each vertical line in the strip $0 < r < r_-$ and in $r > 2M$, one on $r = r_+$ (and $r = 2M$), and none in $r_- < r < r_+$ (since there are no turns in II). Arguing as before, we can see that the resulting r–λ plot will have the two usual continental forbidden regions $\mathcal{N}$ and $\mathcal{S}$ in the exterior $r > r_+$, but in the strip $0 < r < r_-$ the forbidden region is a "tongue" $\mathcal{T}$, first observed by Boyer and Lindquist (1967). $\mathcal{T}$ emerges from $r = 0$ at the critical point $(0, a)$ and stretches to $r = r_-$, and the principal line $\lambda = a$ (corresponding to $L = aE$) is tangent to $\mathcal{T}$ from below (see Figure 4.26).

To determine the shape of $\mathcal{T}$ analytically, write

$$\lambda - a = (2Ma \pm r\sqrt{\Delta})/(2M - r) - a = [ar \pm r\sqrt{\Delta})]/(2M - r)$$

For r small, $\sqrt{\Delta} \approx a - Mr/a$ and $2M - r \approx 2M$, so

$$\lambda - a \approx [ar \pm \{ar - (M/a)r^2\}]/2M.$$

Thus, near $(0, a)$, Z is approximated by the union of the line $\lambda - a = (a/M)r$ (omitting the r^2 term) and the parabola $\lambda - a = r^2/(2a)$. This produces the result shown in Figure 4.26. Evidently $\mathcal{T}$ is the null analogue of the oval discussed in Section 4.8 (see Figure 4.10).

Since $E = 1$ the asymptotic lines reduce to $A_\pm = \mp(M + r)$ and are evident in Figure 4.26. In $r < 0$, a direct computation of $d\lambda/dr$ shows that the two branches of Z are strictly monotonic.

As Figure 4.26 suggests, a null geodesic γ coming in from $r = +\infty$ has flyby orbit if $|\lambda|$ is large (that is, $|L| \gg E$) and will probably hit the ring singularity if $|\lambda|$ is small. But the tongue provides a novelty: if λ is just larger than the principal level $\lambda = a$, it meet the tongue in an r-turning point, thus acquiring a long flyby orbit.

In contrast to the general $Q > 0$ case, it is now easy to locate the horizontal tangents to Z, that is, the circular photons—and hence to specify all null orbit types.

Proposition 4.14.6 *In the slow Kerr equator Eq there are, aside from restphotons, exactly three circular null orbits. For $E > 0$, these have radii and impact parameters (r_τ, λ_τ), (r_ρ, λ_ρ), $(r_\delta, \lambda_\delta)$ with*

$$0 < r_\tau < r_- < r_+ < r_\delta < 3M < r_\rho < 4M, \quad and$$

$$\lambda_\rho < -7a < 0 < a < \lambda_\tau < 2a < \lambda_\delta.$$

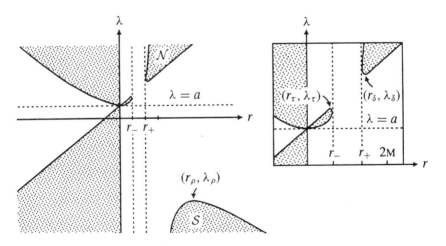

FIGURE 4.26.   The r–λ plot for equatorial null geodesics. As usual, forbidden regions $(R = r\psi < 0)$ are shaded. The inset shows a neighborhood of the tongue. The L-axis $r = 0$ represents the ring singularity, which severs Eq^- from Eq^+. Compare Fig. 8 of Boyer, Lindquist (1967).

The radii are the roots of the polynomial $p(r) = r(r - 3M)^2 - 4Ma^2$, *and the impact parameters are the roots of* $q(\lambda) = (\lambda + a)^3 - 27M^2(\lambda - a)$.

Proof. Circular orbits occur at simultaneous solutions (r, λ) of the equations $\psi = 0$, $\partial\psi/\partial r = 0$. The latter gives

(1) $$\lambda^2 = 3r^2 + a^2$$

Then substituting $r^2 = (1/3)(\lambda^2 - a^2)$ into $\psi(r, \lambda) = r^3 + (a^2 - \lambda^2)r + 2M(\lambda - a)^2 = 0$ leads to $r(a^2 - \lambda^2) + 3M(\lambda - a)^2 = 0$. Hence

(2) $$r = 3M(\lambda - a)/(\lambda + a).$$

Equations (1) and (2) are the basic relations between the radius and the impact parameter of a circular photon. Substituting equation (2) into $\partial\psi/\partial r = 3r^2 + a^2 - \lambda^2 = 0$ gives $27M^2(\lambda - a)^2/(\lambda + a)^2 + a^2 - \lambda^2 = 0$. By omitting the (wellknown) principal nulls $\lambda = a$, we can cancel $\lambda - a$ from the preceding equation, leaving

(3) $$q(\lambda) = (\lambda + a)^3 - 27M^2(\lambda - a) = 0.$$

Now solve equation (2) for $\lambda = -a(r + 3M)/(r - 3M)$, and substitute this into equation (3) to get

(4) $$p(r) = r(r - 3M)^2 - 4Ma^2.$$

Evidently this polynomial has no negative roots, and $p(0) < 0$, $p(3M) < 0$, $p(4M) = 4M^2 - 4Ma^2 > 0$. The placing of $r_\pm$ derives from

(5) $$p(r_\pm) = p(M \pm (M^2 - a^2)^{1/2}) = (M^2 - a^2)r_\pm > 0.$$

It follows at once that $p(r)$ has three roots obeying the stated inequalities. The function $\lambda(r) = a(3M + r)/(3M - r)$ is strictly increasing on $\{r < 3M\}$ and $\{r > 3M\}$, hence $\lambda(r_\rho) < \lambda(4M)$, which gives $\lambda_\rho < -7a$, and $\lambda(0) < \lambda(r_\tau) < \lambda(M) < \lambda(r_\delta)$, which gives $a < \lambda_\tau < 2a < \lambda_\delta$. $\square$

For a null geodesic (with $E^2 = 1$) the rotational energy $\tilde{E}M = a^2(E^2 + q)$ reduces to a^2. Thus the inequalities in this result show that these circular photons have low rotational energy (Definition 4.5.2) hence are *stably* in the equator; that is, a small change in first-integrals produces orbits with $|\vartheta(s) - \pi/2|$ small.

If a *circular photon* is defined to be a null geodesic with both $r(s)$ and $\vartheta(s)$ constant, then (except for restphotons) the three in Eq_+ are the only ones in the Kerr black hole . This can be seen by checking the cases in Section 4.6 for which ϑ is constant.

Remark 4.14.7 (Schwarzschild comparison.) Setting $a = 0$ in the Proposition 4.14.6 shows that in Schwarzschild spacetime, where $r = 0$ is the central singularity, there is a circular photon orbit only at $r = 3M$, with direct and retrograde impact parameters $\lambda = \pm 3\sqrt{3}M$. By spherical symmetry such orbits fill a spatial sphere outside the unique Schwarzschild horizon $r = 2M$. Kerr rotation has eliminated most of these, and in Eq^+ has given different radii $r_+ < r_\delta < 3M < r_\rho < 4M$ to the (direct and retrograde) circular photons in the exterior—and has created a new one at $r = r_\tau$ within the horizon $r = r_-$.

Once the circular photons are located in the null r–λ plot, the equatorial null orbit types can readily be described. There can obviously be no transit types, and we know that principal null geodesics hit the ring singularity. Lemma 4.14.2 showed that all nonprincipal geodesics in Eq^- have flyby orbits. In Eq^+, Proposition 4.14.6 shows that except for circular orbits, there are no bound null orbits. For $E \neq 0$ it also lets us characterize the remaining flyby and crash orbits in terms of impact parameter, as follows:

1. Flyby orbit (for r large) or crash orbit (for r small) $\Leftrightarrow$ if the impact parameter $\lambda = L/E$ is in one of the intervals: $(-\infty, \lambda_\rho)$, (a, λ_τ), (λ_δ, ∞);
2. Crash-escape orbit $\Leftrightarrow$ λ lies in the interval $(\lambda_\rho, a]$ or $(\lambda_\tau, \lambda_\delta)$;
3. Circular orbit (or associated asymptotic orbit) $\Leftrightarrow$ $\lambda = \lambda_\rho, \lambda_\tau$, or λ_δ.

In the remaining case $E = 0$, γ is a restphoton if $L = 0$, otherwise it has crash-crash orbit in $0 < r \leq 2M$. To see this note first that when $L = 0$, all the first-integrals of γ are zero, which is possible only for restphotons. If $L \neq 0$, then $\psi(r, L) = (2M - r)L^2$, hence γ lies in the strip $0 < r \leq 2M$. The r-equation is $r^3 r'^2 = L^2(2M - r)$, so γ hits Σ—at both ends, with a turn at $r = 2M$.

Note that null geodesics are *central* in the sense that, for large E, both the spacelike and the timelike sets Z resemble that of the null case. In fact, if $\psi^\pm$ denotes the function in Corollary 4.14.1 when $q = \pm 1$, then for $E^2 \gg 1$, $\psi^\pm(r, L)/E^2$ is close to $\psi(r, \lambda) = L/E$. But note that since $L = \lambda E$, r–L plots for $\psi^\pm$ are stretched in the L direction by a factor E.

TIMELIKE GEODESICS IN Eq

The main qualitative change from general $0 < Q < Q_+$ timelike r–L plots such as Figure 4.9 is that for all values of E the only forbidden region in the strip $0 < r < r_-$ is an oval tangent to the lines $r = 0$ and $r = r_-$ (see Lemma 4.8.5 and Figure 4.27).

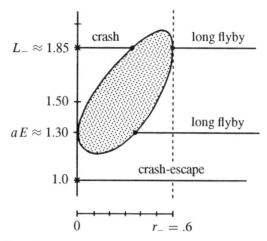

FIGURE 4.27.   The oval in a timelike equatorial r–L plot with escape energy. (Parameters: $M = 1, a^2 = .84, e = E^2 - 1 = 1$.) Some related orbital types are shown. There are circular orbits at the top ($r \approx .53, L \approx 1.96$) and bottom ($r \approx .12, L \approx 1.17$) of the oval, and these are flanked, as usual, by asymptotic orbits.

Corollary 4.14.8   *For timelike $Q = 0$ geodesics,*

$$\psi(r, L) = (E^2 - 1)r^3 + 2Mr^2 + [a^2(E^2 - 1) - L^2]r + 2M(L - aE)^2,$$

and the set $Z \colon \psi = 0$ is the union of the graphs of

$$L_\pm(r) = [2MaE \pm D(r)^{1/2}]/(2M - r),$$

defined on intervals where $D(r) = r\Delta(r)[(E^2 - 1)r + 2M]$ is non-negative.

Taking $E > 0$ we consider the equatorial timelike r–L plots as energy E decreases from ∞ to 0. For large E the oval approximates the shape of the tongue in the null case. As E decreases the oval sinks, since it meets the lines $r = 0$ and $r = r_-$ at $L_0 = aE$ and $L_- = 2MaE/r_+$, respectively.

Outside the strip $0 < r < r_-$, the r–L plots for timelike geodesics with escape energy resemble the null case in Figure 4.26, with continents $\mathcal{N}$ and $\mathcal{S}$ framed by asymptotic lines. The wandering vertical tangent $w(E) = -2M/(E^2 - 1)$ is negative and and marks the closest that a material particle of energy E in Eq^- can get to the ring singularity.

Thus most orbit types are as usual: In Eq^+, exterior flyby for $|L|$ large, and crash-escape for $|L|$ small—unless the oval is hit. But as Figure 4.27 illustrates, that produces crash orbits and long flyby orbits (the two–three rule shows in fact that the latter cannot meet $\mathcal{N}$ or $\mathcal{S}$).

When decreasing E reaches marginal escape energy $E = 1$ the orbits in Eq^- vanish, but those in Eq^+ are little changed. However, for $E < 1$ the change is radical, and all r–L plots have the bay configuration, like those with $0 < Q < Q_+$ discussed in Section 4.8. The wandering vertical tangent $w(E) = 2M/(1 - E^2)$ is positive, now marking the farthest a geodesic with energy $E < 1$ can get from the ring singularity.

When E is just less than 1 the bay is large, and there are plenty of exterior bound orbits, both direct ($L > 0$) and retrograde ($L < 0$). As E decreases, the bay shrinks, pushed by decreasing $w(E)$, the "end of the bay." At a critical energy E_r, specified below, the (stable) retrograde bound orbits have shrunk to a single (unstable) circular orbit, its radius denoted by c_r. (See Figure 4.28). This retrograde orbit vanishes as E passes below E_r, leaving only direct exterior bound orbits.

When decreasing energy reaches a second critical level $E_d < E_r$, the exterior direct orbits have been reduced to a single (unstable) circular orbit, of radius c_d. Then for $E < E_d$ there are no more exterior orbits, only crash orbits and other orbit types (notably large bound) produced by the oval. At $E = 0$, $w = 2M$, and the r–L plot, as always when $E = 0$, is symmetrical in L.

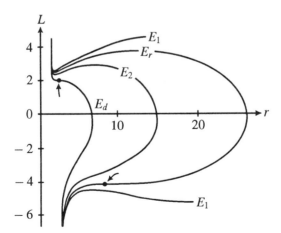

FIGURE 4.28. Some r–L plots for timelike equatorial geodesics with less than escape energies $1 > E_1 > E_r > E_2 > E_d > 0$. The minimum-energy direct and retrograde orbits are indicated by arrows.

The following result gives the radii of all timelike circular orbits in the equator; the corresponding values of angular momentum can then be computed from the formula in Corollary 4.14.8.

Lemma 4.14.9   *The radii of the circular timelike ($q = -1$) geodesics in Eq with effective energy $e = E^2 - 1$ are the real roots of the polynomial*

$$p(r) = c_6 r^6 + c_5 r^5 + c_4 r^4 + c_3 r^3 + c_2 r^2 + c_1 r + c_0,$$

where

$$c_6 = e^2, \quad c_5 = 2Me(1 - 3e), \quad c_4 = M^2(9e^2 - 14e + 1),$$
$$c_3 = 2M[-2a^2e^2 - a^2e + 12M^2e - 4M^2],$$
$$c_2 = 2M^2[-5a^2e - 2a^2 + 8M^2], \quad c_1 = -8M^3a^2, \quad c_0 = M^2a^4.$$

To find this polynomial, (1) take the derivative dL_+/dr, with L_+ as in Corollary 4.14.8; (2) in the numerator of dL_+/dr form the difference of the squares of the square-root terms and non square-root terms (thus including the L_- case); and (3) from the resulting degree 8 polynomial, divide out the factor $(r - 2M)^2$. The degree 6 cannot be reduced, since, as we have seen, for just less than escape energy (e small negative) there are six such circular geodesics.

We now find the critical radii c_r, c_d of the minimum-energy exterior bound orbits, retrograde and direct, respectively, along with their energies $E_r > E_d$ and angular momenta $L_r < 0 < L_d$.

Lemma 4.14.10   (1) *The critical radii $c_d < c_r$ in the Kerr equator are the two real roots of the Boyer–Lindquist polynomial (1967).*

$$\beta(r) = r^4 - 12Mr^3 + 6(6M^2 - a^2)r^2 - 28Ma^2r + 9a^4.$$

(2) *For $c = c_d$, c_r they occur at energy E_c such that $E_c^2 - 1 = -2M/(3c)$, and they have angular momentum L_c such that $(L_c)^2 = 2M(c + a^2/(3c))$, with $L_r < 0 < L_d$.*

Proof.   The radii c_r, c_d are limits of stable and unstable circular radii for which the signs of $L''_\pm(r)$ are opposite. Hence they occur at radii where $L_+(r)$ and $L_-(r)$ have a horizontal inflection points. These radii can be found by simultaneous solution of the equations $\psi = 0$, $\partial\psi/\partial r = 0$, $\partial^2\psi/\partial r^2 = 0$. The latter equation gives $3(E^2 - 1)r + 2M = 0$, that is, $3er + 2M = 0$. Using this to eliminate e from the polynomial $p(r)$ above yields the polynomial $\beta(r)$—and also the formula for E_c. And using $3(E^2 - 1)r + 2M = 0$ again in

$$\partial\psi/\partial r = 3(E^2 - 1)r^2 + 4Mr + (a^2(E^2 - 1) - L^2) = 0$$

gives the formula for L_c. (Note that the critical radius c is exactly $w_c/3$ for $w_c = -2M/(E_c^2 - 1) > 0$.) □

For example, with $M = 1$, $a^2 = 0.84$, we find

$$\text{direct:} \quad c_d \approx 2.21, \, E_d \approx 0.835, \, L_d = 2.16$$
$$\text{retrograde:} \quad c_r \approx 8.76, \, E_r \approx 0.961, \, L_r \approx -4.19$$

(These are the values used in Figure 4.28.) Applying this lemma one can read from the figure a reasonable characterization of the various orbit types.

Remark 4.14.11 (Schwarzschild comparison.) In the unextended Schwarzschild black hole a single horizon $r = 2M$ surrounds a central singularity at $r = 0$ (see Remark 2.5.8). Setting $a = 0$ in Corollary 4.14.8 gives

$$\psi_{SS} = (E^2 - 1)r^3 + 2Mr^2 + L^2(2M - r).$$

Thus r–L plots are symmetrical in L, with direct and retrograde orbits differing only in sign of L. The entire region $r < 2M$ is allowed, so once inside the horizon there is no protection from the singularity.

For geodesics with escape energy $E \geq 1$ the r–L plots have two symmetrical continents $\mathcal{N}$ and $\mathcal{S}$, but no oval. For $E < 1$ there is a symmetrical bay, initially containing both direct and retrograde exterior bound orbits. As E descends, setting $a = 0$ in Lemma 4.14.10 shows that there is a single critical energy $E_c = 2\sqrt{3}M$. This has direct and retrograde circular orbits at the same radius $c = 6M$ with angular momenta $\pm3\sqrt{2}M$. (Note that this energy E_c is equivalent to $e = -1/9$, as in the e–Q chart, Fig. 4.8.)

The equatorial plane is a particularly good place to look at the longitudinal motion of geodesics. In the equatorial plane the r- and φ-equations (Theorem 4.2.2, Proposition 4.1.5) reduce to

$$r' = \pm\sqrt{r\psi}/r^2, \qquad \varphi' = ((r - 2M)\,L + 2MaE)/(r\Delta).$$

(In the Schwarzschild limit, $a = 0$, the φ-equation further reduces to the equation $r^2\varphi' = L$ of Kepler's second law.) For Figures 4.29 and 4.30 these equations are integrated to give relativistic flyby and bound orbits different from Keplerian conic sections. In Figure 4.29, reducing the angular momentum tends toward a circular orbit (see Lemma 4.14.9) hence will yield orbits that circle the horizon arbitrarily many times. In Figure 4.30 the perihelion advance is huge compared to the historic, small advance for Mercury in our barely relativistic solar system.

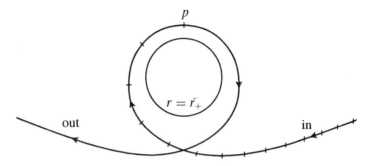

FIGURE 4.29.   A timelike flyby orbit in the equatorial plane, with parameters $M = 1$, $a^2 = .84$, $e = E^2 - 1 = 1$, $L = 4.2$. The circle represents the outer horizon $r = r_+ = 1.4$, so the orbit is in the Kerr exterior. The orbit is direct with $r_{min} \approx 2.16$. Analogous retrograde orbits would require $L < -8.5$.

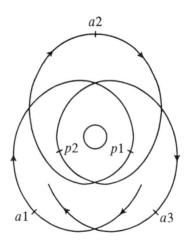

FIGURE 4.30.   A timelike bound exterior orbit in the equatorial plane. (Parameters: $M = 1$, $a^2 = .84$, $e = E^2 - 1 = -.1$, $L = 3.1$.) The circle represents the outer horizon $r = r_+ = 1.4$. Three apcenters and two pericenters are shown: $a1$, $p1$, $a2$, $p2$, $a3$, with $r_{min} \approx 5.35$ and $r_{max} \approx 13.25$. The pericenter advance from $p1$ to $p2$ is about $142°$.

SPACELIKE GEODESICS

Corollary 4.14.12 *For spacelike equatorial geodesics,*

$$\psi(r, L) = (E^2 + 1)r^3 + 2Mr^2 + [a^2(E^2 + 1) - L^2]r + 2M(L - aE)^2,$$

and the set $Z \colon \psi = 0$ *is the union of the graphs of*

$$L_\pm(r) = [2MaE \pm D(r)^{1/2}]/(2M - r),$$

defined on intervals where $D(r) = r\Delta(r)[(E^2 + 1)r - 2M]$ *is non-negative.*

From a purely geometric viewpoint it is the spacelike equatorial geodesics that are the most interesting. According to Lemma 4.14.2 we need only consider Eq^+. Outside the strip $0 < r < 2M$ the r–L plots generally resemble the null case (Figure 4.26). But the wandering vertical tangent $w(E) = 2M/(E^2+1)$ is positive, and its interaction with the horizon radii $r_\pm$ produces a considerable variety of orbits.

When E is large, the only significant difference from the null case is that the tongue has become a smooth island at the short distance $w(E) > 0$ from the L-axis. But as E decreases, this distance increases, while the right side of the island, as always, remains pinned at (r_-, L_-). Thus the island shrinks, until at $w = r_-$ it is reduced to the single point (r_-, L_-). As predicted following Proposition 4.14.4, this occurs when $E = (r_+/r_-)^{1/2}$, with $L_- = L_w = 2M$. Thus $(r_-, 2M)$ is an isolated maximum of $\psi(r, L)$. Evidently this remarkable point represents an unstable circular orbit in the intersection of the equator and the horizon $\mathbb{H}_-$, an orbit already distinguished in Section 4.12 as having maximum energy E^2 among all geodesics in $\mathbb{H}_-$.

When E decreases below $(r_+/r_-)^{1/2}$, $w(E)$ increases past r_- and the island reappears, now running from $r = r_-$ to $r = w(E) > r_-$. Thus, by contrast with null and timelike geodesics, spacelike ones have no difficulty turning in block II.

As w approaches r_+, the island stretches toward the western shore of $\mathcal{N}$. At $E^2 = r_-/r_+$, w reaches r_+, and the island meets $\mathcal{N}$ at the critical point $(r_+, 2M)$ (see Figure 4.31). This represents the unstable circular orbit of the maximum E^2 geodesic in $\mathbb{H}_+$.

Further decrease in E moves w past r_+ where it pushes $\mathcal{N}$ eastward. The island is left behind, stretching from r_- to r_+. Thereafter, it sinks until at $E = 0$ a symmetric plot is reached.

We now determine which equatorial geodesics hit the ring singularity.

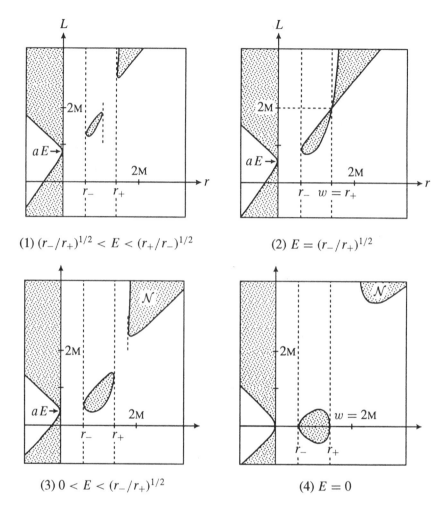

FIGURE 4.31.   Spacelike r–L plots for equatorial geodesics. As E decreases from $+\infty$, the island shrinks at first, until at $E = (r_+/r_-)^{1/2}$ it is reduced to a single point $(r_-, 2\mathrm{M})$. Thereafter, as the figures above illustrate, it moves through block II to a meeting with the north continent $\mathcal{N}$ when $E = (r_-/r_+)^{1/2}$ and then declines to symmetry at $E = 0$.

Remark 4.14.13   In the equator, if a geodesic has $r(s) \to 0$, with no intervening forbidden regions, then it will *hit* the ring singularity, that is, it will meet it on a finite parameter interval. This follows immediately from the r-equation $\rho^4 r'^2 = R(r)$ (see proof of Lemma 4.14.2). Now, since $R(r) = \cdots + (a^2(E^2 + q) - L^2)r^2 + 2\mathrm{M}(L - aE)^2 r$, follows that (1) If $L \neq aE$, such an approach is always possible from Eq^+, but never from Eq^-. (2) If $L = aE$, then we know principal nulls hit Σ, and for nonnull geodesics, $R(r) = \cdots + qa^2r^2$, so spacelike approach is possible from both sides $Eq^\pm$, timelike approach from neither.

GEODESICS WITH $Q = 0$

The class of geodesics with $Q = 0$ is important for the problems in the next section. Though all equatorial geodesics have $Q = 0$, very many $Q = 0$ geodesics are not equatorial—in fact, there is one through every point of a Kerr spacetime. This is a consequence of Proposition 4.2.6, since for any r_0, ϑ_0, one need only take E^2 sufficiently large to have $R(r_0) > 0$ and $\Theta(\vartheta_0) > 0$.

Remarks 4.14.14

1. If a geodesic has $Q > 0$ then its ϑ-motion is oscillation around $\pi/2$, so it cuts repeatedly through the equator. However, a nonequatorial $Q = 0$ geodesic cannot cross the equatorial plane. In fact, if a $Q = 0$ geodesic ever meets Eq it is entirely contained in Eq. *Proof:* If $\gamma(s_0) \in Eq$, so $\vartheta(s_0) = \pi/2$, then when the ϑ-equation $\rho^4\vartheta'^2 = \Theta = C^2(a^2(E^2 + q) - L^2/S^2)$ is evaluated at s_0 it reduces to $r_0^4\vartheta_0'^2 = 0$. But $C_0 = 0$ excludes $r_0 = 0$ (ring singularity), so $\vartheta_0' = 0$. This means that $\gamma'(s_0)$ is tangent to Eq. Since Eq is a closed totally geodesic hypersurface, it follows that γ lies entirely in Eq.

2. Nonequatorial geodesics with $Q = 0$ have high rotational energy $a^2(E^2 + q) > L^2$, and hence escape energy $E^2 + q > 0$. This consequence of Section 4.6 can be seen directly from $\Theta = C^2(a^2(E^2 + q) - L^2/S^2) \geq 0$, since if $C^2 > 0$ then $S^2 < 1$).

Since r–L plots do not involve ϑ, they cannot distinguish equatorial geodesics from nonequatorial geodesics that share the same first-integrals. Nevertheless, radial motions in the two cases are by no means the same. Of course, in the equatorial case, $r = 0$ represents the ring singularity (not in the spacetime), while for $\vartheta \neq \pi/2$, it represents the throat T, a respectable hypersurface in block III. But more than that, since nonequatorial $Q = 0$ geodesics have high rotational energy, a horizontal line in an r–L plot for energy E represents a *nonequatorial* geodesic orbit if and only if $L_0^2 < a^2(E^2 + q)$. For example, in the null r–λ plot (Figure 4.26), nonequatorial orbits exist only for $-a < \lambda < a$. In $r > 0$ this entire horizontal band meets no forbidden regions, so every line $\lambda = \lambda_0$ from $r = 0$ to $+\infty$ is an r-interval. In particular, all three momenta $\lambda_\tau, \lambda_\delta, \lambda_\rho$ in Proposition

4.14.6 exceed a in absolute value. Thus there are no $Q = 0$ spherical null orbits outside Eq.

If a nonequatorial $Q = 0$ geodesic γ is timelike, then the fact that γ has high rotational energy means that in

$$R(r) = (E^2 - 1)r^4 + 2Mr^3 + (a^2(E^2 - 1) - L^2)r^2 + 2(L - aE)^2 r$$

every coefficient is strictly positive. Thus as before, in $r > 0$ the entire band $L^2 < a^2(E^2 - 1)$ is allowed, so its geodesics, ingoing from $r = +\infty$, fall without obstruction toward $r = 0$. In the next section we will determine what happens to geodesics (of all causal characters) that approach $r = 0$.

For detailed studies of equatorial geodesics see, for example, Bardeen (1973) and de Felice (1968).

4.15   Approaching the Center

A natural question for any spacetime with a singularity is: Which geodesics hit it? Accordingly, for the Kerr black hole we want to know which geodesics hit the ring singularity Σ. Recall that by Corollary 4.5.1, such geodesics have Carter constant $Q = 0$.

The case of *equatorial* geodesics is settled. All have $Q = 0$, there is no throat in Eq, and Remark 4.14.13 shows which equatorial geodesics hit Σ. So a primary question now is: Which nonequatorial $Q = 0$ geodesics, if any, hit Σ? The following negative result, due to Brandon Carter (1968), is fundamental.

Proposition 4.15.1 *If a timelike or null Kerr geodesic γ hits the ring singularity, then γ lies entirely in the equatorial plane Eq.*

The causal restriction is necessary since it turns out that many spacelike geodesics hit Σ from outside Eq. To construct a proof of the proposition we consider the broader question of *what happens to geodesics that approach the center*, that is, whose radial coordinate $r(s)$ approaches 0. Here the critical case remains $Q = 0$ since we know that such an approach is impossible if $Q > 0$, while vortical ($Q < 0$) geodesics that approach $r = 0$ avoid the ring singularity Σ and pass unobstructed through the throat.

Although the ring singularity Σ is not part of Kerr spacetime it is natural to imagine it as the intersection of the equator Eq: $\vartheta = \pi/2$ and throat T: $r = 0$.

Lemma 4.15.2 *If a* $Q = 0$ *geodesic meets* T, *then either* (1) $r(s) \geq 0$ *holds for all* s, *and* γ *has an* r-*turning point at* $r = 0$, *or* (2) γ *lies entirely in* T. *In particular,* γ *cannot cut through* T.

Proof. Suppose $\gamma(s_0) \in T$, that is, $r(s_0) = 0$. But $\vartheta(s_0) \neq \pi/2$, so $\rho^2(s_0) \neq 0$. Then since $Q = 0$ the r-equation shows that $r'(s_0) = 0$, hence $\gamma'(s_0)$ is tangent to T. As we see below, T is not totally geodesic, so we cannot deduce that γ is contained in T. Now suppose γ has $L \neq aE$. Then $(dR/dr)(0) = 2\mathrm{M}(L - aE)^2 > 0$. It follows that $r = 0$ is an r-turning point and that $r(s) \geq 0$ holds for s near s_0. This is true for any other meeting of γ with T; hence (1) holds. But if $L = aE$, then function $R(r)$ has a repeated zero at $r = 0$. This implies that $r(s)$ is constant, so γ is entirely contained in T . $\qquad\square$

The hypersurfaces Eq and T channel the possible approaches to the ring singularity as follows.

Corollary 4.15.3 *Let* γ *be a* $Q = 0$ *geodesic defined on* $[s_0, s_1)$. *If* γ *approaches* Σ *as* $s \to s_1$, *then exactly one of the following holds:*
(1) γ *lies entirely in the equator* Eq.
(2) γ *lies entirely in the throat* T.
(3) $\Theta(\vartheta(s)) > 0$ *and* $R(r(s)) > 0$ *hold on some final subinterval* $[b, s_1)$; *hence* γ *does not meet either* Eq *or* T.

Proof. Exclude possibility (1). Then, according to Remark 4.14.14, γ never meets the equator, so $\vartheta(s) \to \pi/2$ through values $\neq \pi/2$. As Figure 4.3 illustrates, if $\Theta(\vartheta(s^*)) = 0$ for $s^* \neq \pi/2$, then this is a ϑ-turning point.

Also, if $L = 0$ there are no such ϑ-turns, and if $L \neq 0$ then past one there cannot be another . Hence $\Theta(\vartheta(s)) > 0$ on some final interval.

Now exclude both (1) and (2). To show that (3) holds, we consider how many r-turning points γ has.

Suppose γ has no r-turning points. Then $R(r(s)) > 0$ on $[s_0, s_1)$, and by Lemma 4.15.2 γ does not meet T, so (3) holds.

Suppose γ has exactly one turning point, say r_1. Now $r_1 \neq 0$, for if γ turned at $r = 0$, a second turn would be required, since $r(s) \to 0$. But having turned at $r_1 \neq 0$, $r(s)$ cannot turn again as it proceeds to $r = 0$. Since it does not turn, $R(r(s))$ is positive, so γ cannot meet T. Again, (3) holds.

Finally, suppose γ has two r-turning points. Since $r(s) \to 0$ and (according to Lemma 4.15.2) γ cannot cross T, these must occur at $r_2 > r_1 = 0$. Conceivably, $r(s)$ might oscillate between $T \colon r = 0$ and $r = r_2$ many times, and then, due to changes in $\vartheta(s)$, hit Σ. But $r(s) \to 0$ requires a final interval $[b, s_1)$ in $[s_0, s_1)$ on which $r = r_2$ is not taken on. Then $r = 0$ can also not be taken on—since after an $r = 0$ turn there would be no turn to enable $r(s) \to 0$. Thus on some $[b, s_1)$, $R(r(s))$ is never zero, so γ does not meet T. Hence (3) holds.   □

Remark 4.15.4 The ways in which Kerr geodesics with $Q = 0$ can approach $r = 0$ are severely restricted. As in the equatorial case (Remark 4.14.13), examination of the polynomial $R(r)$ shows that if $L \neq aE$, then, as noted earlier, γ can approach $r = 0$ only from $r > 0$. If $L = aE$,

$$R(r) = (E^2 + q)r^4 - 2\mathsf{M}qr^3 + qa^2r^2.$$

Then: (a) If γ is timelike it can approach $r = 0$ from neither side. (b) If γ is null it is a principal null geodesic in the equator. (*Proof:* In the ϑ-equation, $C^2 > 0 \Leftrightarrow S^2 < 1 \Leftrightarrow \Theta < 0$, so $\vartheta = \pi/2$; then a final remark in Section 4.5 shows that γ is principal.) Hence γ hits Σ. (c) If γ is spacelike it can approach $r = 0$ from either side.

The next three results show that if a $Q = 0$ geodesic is aimed toward $r = 0$ from outside Eq, with no forbidden regions in the way, it will hit the ring singularity if and only if it is spacelike with $L = aE$.

Proposition 4.15.5 *Let γ be a nonequatorial Kerr geodesic with $Q = 0$ such that $r_0 > 0$ and $r_0' < 0$, with $R(r) > 0$ on $(0, r_0]$. Then γ will hit the ring singularity if and only if it is spacelike with $L = aE$. Otherwise, γ has an r-turning point at $r = 0$.*

Proof. Since γ is not contained Eq, it never meets Eq, so $\vartheta_0 \neq \pi/2$. As shown earlier, $\vartheta(s)$ must, after at most one turn, approach $\pi/2$, so we may as well suppose this is true initially. For definiteness, take $\vartheta_0 < \pi/2$, $\vartheta_0' > 0$. Hence $\Theta(\vartheta) > 0$ on $[\vartheta_0, \pi/2)$.

As the parameter s of $\gamma(s)$ increases, $r(s) > 0$ decreases, and $\vartheta(s)$ increases toward $\pi/2$. But $\vartheta(s)$ can never take on the value $\pi/2$ since γ does not meet Eq. Because of geodesic completeness mod Σ, the hypothesis on $R(r)$ requires that γ is defined at least on an interval $[s_0, s_1)$ on which the following conditions hold: $r(s) > 0$, $\vartheta(s) \neq \pi/2$, $\lim_{s \to s_1} r(s) = 0$, and $\lim_{s \to s_1} r(s)$ is some number $\vartheta_1 \geq \pi/2$.

Applying equation (1) of Proposition 4.3.3 gives

$$\int_{r_0}^{0} \frac{dr}{\sqrt{R}} = -\int_{\vartheta_0}^{\vartheta_1} \frac{d\vartheta}{\sqrt{\Theta}}.$$

Since $Q = 0$, we have

$$R(r) = (E^2 + q)r^4 - 2Mqr^3 + (a^2(E^2 + q) - L^2)r^2 + 2M(L - aE)^2r,$$
$$\Theta(\vartheta) = C^2(a^2(E^2 + q) - L^2/S^2).$$

We assert that the ϑ-integral is finite if $\vartheta_1 < \pi/2$ and infinite if $\vartheta_1 = \pi/2$. In fact, if $\vartheta_1 < \pi/2$, then we have just the integral of a continuous positive function over a bounded interval. If $\vartheta_1 = \pi/2$, the integral is improper, and

$$\sqrt{\Theta} = |C|(a^2(E^2 + q) - L^2/S^2)^{1/2} < k|C|,$$

where $k = (a^2(E^2 + q) - L^2)^{1/2} > 0$. Hence

$$\int_{\vartheta_0}^{\pi/2} \frac{d\vartheta}{\sqrt{\Theta}} \geq \frac{1}{k} \int_{\vartheta_0}^{\pi/2} \frac{d\vartheta}{C} = \frac{1}{2k} \ln \left| \frac{1 + S}{1 - S} \right| \Big|_{\vartheta_0}^{\pi/2} = \infty.$$

Case 1. $L = aE$. Then

$$R(r) = (E^2 + q)r^4 - 2Mqr^3 + qr^2,$$
$$\Theta(\vartheta) = a^2C^2((E^2 + q) - E^2/S^2) = (a^2C^2/S^2)(qS^2 - E^2C^2)$$

The latter equation shows that $\Theta(\vartheta) < 0$ if $q \leq 0$; hence γ can only be spacelike. Thus for r small, $R(r)$ is of order r^2. But then the integrand of the r-integral above has order $1/r$, so this improper integral from r_0 to $r = 0$ is infinite. Hence the ϑ-integral is infinite. Thus, as noted above, $\vartheta_1 = \pi/2$, and this means that γ hits the ring singularity.

Case 2. $L \neq aE$. Now the formula for $R(r)$ shows that for r small, $R(r)$ has order r. Thus the integrand of the r-integral has order $1/\sqrt{r}$, so the integral from r_0 to (limit) $r = 0$ is finite. Thus $\vartheta_1 < \pi/2$. Hence γ misses the ring singularity, so it is extendible past s_1, meeting the throat at s_1. Lemma 4.15.2 shows that it turns there and reenters $r > 0$. □

As expected, reaching $r = 0$ from the "white hole" side $r < 0$ is more difficult.

Corollary 4.15.6 *Let γ be a nonequatorial Kerr geodesic with $Q = 0$. If $r_0 < 0$ and $r_0' > 0$, with $R(r) > 0$ on $[r_0, 0)$, then γ reaches $r = 0$ if and only if it is spacelike with $L = aE$, and in this case γ hits the ring singularity.*

Proof. Remark 4.15.4 shows that γ must be spacelike with $L = aE$, and in this case the previous proof that γ hits Σ remains valid. □

Of course many geodesics with $L \neq aE$ can successfully reach the throat from $r < 0$ (for example, principal null geodesics), but they must have $Q < 0$, and hence they cut through T into $r > 0$.

It remains to decide which Kerr geodesics, if any, can approach Σ from within the throat. Recall from Section 2.3 that T is an intrinsically flat hypersurface $S^2\langle 0 \rangle \times \mathbf{R}$ in block III representing the the central sphere $S^2\langle 0 \rangle = \{r = 0, t = 0\}$ enduring through time. $S^2\langle 0 \rangle$ is cut by ring singularity into two components, and each of the resulting components of T is isometric to the interior $D \times \mathbf{R}$ of a cylinder $S^1 \times \mathbf{R}^1$ of radius $r = a$ in Minkowski 3-space. An explicit isometry ψ is given by

$$x = aS\cos\varphi, \qquad y = aS\sin\varphi, \qquad t = t.$$

Extended to $\vartheta = \pi/2$, ψ carries the ring singularity to the boundary cylinder $S^1 \times \mathbf{R}^1$.

Lemma 4.15.7 *A geodesic β of T is a Kerr geodesic if and only if it is everywhere orthogonal to the vector field $\partial_\varphi + a\partial_t$. All such geodesics are spacelike with $L = aE$.*

Proof. First we compute the shape tensor S of the throat (Remark 1.7.13). For vector fields X, Y tangent to T, $S(X, Y)$ is the component $\mathrm{nor}\nabla_X Y$ of the Kerr covariant $\nabla_X Y$ that is normal to T. In the Boyer–Lindquist frame field (Section 2.6) the vector fields $E_2 = \partial_\vartheta/|\partial_\vartheta|$, $E_3 = W/|W|$, $E_0 = V/|V|$ are tangent to T, and $E_1 = \partial_r/|\partial_r|$ is normal to T. On T, $\rho^2 = a^2 C^2$ and $\Delta = a^2$, so the covariant derivative formulas in Section 2.6 show that all but one of the pairs E_i, E_j tangent to T have $S(E_i, E_j) = 0$. The exception is $S(E_0, E_0) = \mathrm{nor}\nabla_{E_0} E_0 = FE_1$, where in the case at hand, $F = -\mathrm{M}/(a^2 C) \neq 0$. Thus the throat is almost totally geodesic, bending principally in the E_0 direction, with magnitude approaching infinity as $\vartheta \to \pi/2$ (the ring singularity).

First suppose γ is a Kerr geodesic contained in T. In particular, $S(\gamma', \gamma') = \mathrm{nor}\,\gamma'' = 0$. Writing γ' as a linear combination of E_0, E_2, E_3 (by orthonormal expansion) shows that γ' has no E_0 component, that is, $\gamma' \perp E_0$. Thus γ is spacelike since both E_2 and E_3 are. Since γ' is orthogonal to $E_0 = V/|V|$ and $V = a(\partial_\varphi + a\partial_t)$ on T we get $L = aE$ since $\langle \gamma', a\partial_t + \partial_\varphi \rangle = 0 \Leftrightarrow -aE + L = 0$.

Now suppose that β is a geodesic of T with $\beta' \perp \partial_\varphi + a\partial_t$. Its Kerr acceleration β'' is normal to T, and since $\beta' \perp E_0$, $\beta'' = S(\beta', \beta') = 0$. Thus β is a Kerr geodesic, and as before, $\beta' \perp \partial_\varphi + a\partial_t$ implies β spacelike and $L = aE$. $\square$

The Kerr geodesics that lie in T can be determined explicitly by using the isometry ψ defined just before Lemma 4.15.7. Since $d\psi(\partial_\varphi + a\partial_t) = -y\partial_x +$

$x\partial_y + a\partial_t$, this lemma shows that Kerr geodesics in T correspond under ψ to straight lines in $D \times \mathbf{R}$ orthogonal to $-y\partial_x + x\partial_y + a\partial_t$. Such lines are easily found to be φ-rotations and t-translations of a line of the form $s \to (x_0 + As, Bs, aBx_0s)$, with $A^2 + B^2 = 1$. These straight lines all hit the boundary of the cylinder at each end; hence the (spacelike) Kerr geodesics in T all hit the ring singularity at both ends, and thus have crash-crash orbit type. In the simplest case, $B = 0$, they are the ϑ-parameter curves in any t-constant slice $S\langle 0 \rangle$ of T, and run from Σ—the "equator" of $S\langle 0 \rangle$—through either pole to Σ again.

Proof of Proposition 4.15.1. Let γ be a Kerr geodesic that hits Σ but does not lie in Eq. In view of Corollary 4.15.3 the three preceding results—for $r_0 > 0$, $r_0 < 0$, $r_0 = 0$, respectively—show that γ is spacelike. $\qquad \square$

Thus an astronaut falling into a Kerr black hole can expect to avoid the ring singularity—for even if he is initially in the equatorial plane, any small acceleration will get him out. (Nevertheless, the trip is not recommended.)

The preceding results on Kerr geodesics show collectively that although some hit Σ (finite parameter interval), none asymptotically approach Σ. This can be proved directly by using Proposition 4.3.3(2), which remains valid for geodesics with either constant $r = 0$ or $\vartheta = \pi/2$.

We can now tell what happens to a $Q = 0$ geodesic that approaches $r = 0$:

Proposition 4.15.8 *Let γ be a $Q = 0$ Kerr geodesic moving toward $r = 0$ with no intervening forbidden regions.*

(1) *If γ is timelike, then γ reaches $r = 0$ if and only if $L \neq aE$. The approach is only from $r > 0$. If γ lies in Eq it hits Σ; otherwise, as shown in Figure 4.32, it bounces off T, with long flyby orbit.*

(2) *If γ is null, the same conclusions hold, with one addition: if $L = aE$ (that is, $\lambda = a$) then γ is a principal null geodesic in Eq, and hence hits Σ.*

(3) *If γ is spacelike with $L \neq aE$, then γ, approaching only from $r > 0$, hits Σ if it lies in Eq, otherwise bounces off T. If $L = aE$, then γ, approaching from either side, must hit Σ.*

INEXTENDIBILITY

In Chapter 3 we saw that no Kerr spacetime containing Boyer–Lindquist block Ⅲ can be isometrically imbedded in a complete semi-Riemannian manifold $\widetilde{M}$. We can now prove the stronger property that the Kerr spacetimes M_f, M_e, M_s are

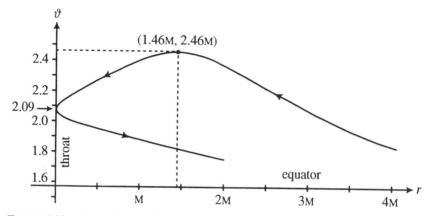

FIGURE 4.32.   Plot of $(r(\tau), \vartheta(\tau))$ for a typical nonequatorial timelike geodesic γ with $Q = 0$. (Parameters: $M = 1$, $a^2 = .84$, $L = 1$, $E = 2$.) The horizontal axis represents the equator $\vartheta = \pi/2$. As it comes in from $r = +\infty$, γ is moving away from the equator. After its ϑ-turn at $\vartheta \approx 51°$ (radians converted to North latitude) it glances off the throat $r = 0$ at $\vartheta \approx 30°$ and returns to $r = \infty$, asymptotically reapproaching the equator. The elapsed proper time of the pictured curve segment is $\Delta \tau \approx 3M$. Note that in reverse orientation, the curve still represents a geodesic.

inextendible, that is, cannot be isometrically imbedded in any larger spacetime whatsoever. Hence the same is true of all spacetimes covering or covered by these standard models. This justifies, at last, the term "maximal" applied to them in Chapter 3.

Proposition 4.15.9   *The Kerr spacetimes M_f, M_e, M_s are inextendible.*

Proof.   Assume that one of these, say M, is (isometric to) an open submanifold of a strictly larger spacetime $\widetilde{M}$. We deduce a contradiction.

(1) We show first that there is a timelike geodesic segment $\sigma: [0, b] \to \widetilde{M}$ with only $\sigma(b)$ not in M. Since $\widetilde{M}$ is connected, there is a point $p \in \widetilde{M}$ in the topological boundary of M in $\widetilde{M}$—hence p is not in M. Let $\mathcal{U}$ be a normal neighborhood of p, so there is a starshaped neighborhood $\widetilde{\mathcal{U}}$ of 0 in $T_p(\widetilde{M})$ mapped diffeomorphically onto $\mathcal{U}$ by the exponential map $\exp_p$ (see O'Neill 1983). Let $\widetilde{\mathfrak{T}}$ be the set of all timelike vectors in $\widetilde{\mathcal{U}}$

If the image $\mathfrak{T} = \exp_p(\widetilde{\mathfrak{T}})$ meets M, the proof is done, since $\exp_p$ carries timelike rays to timelike geodesics.

If not, we replace p by a better point as follows. Pick a sequence $\{p_n\}$ in M that converges to p, and a sequence of timelike tangent vectors z_n at p_n that converge to $z \in \widetilde{\mathfrak{T}} \subset T_p(\widetilde{M})$. Since $\widetilde{\mathfrak{T}}$ is open in $T_p(M)$, $\mathfrak{T}$ is open in $\mathcal{U}$, hence in $\widetilde{M}$. Thus for n sufficiently large, $\exp_p(z_n) \in \mathfrak{T}$. Since $\mathfrak{T}$ does not meet M, the timelike geodesic $\gamma(s) = \exp_{p_n}(sz_n)$ runs from $\gamma(0) = p_n \in M$ to a point $\gamma(1) \in \mathfrak{T}$, not in M. Thus an initial segment σ of γ will have the required property.

(2) Evidently the restriction $\sigma_1 = \sigma|[0, b)$ is a timelike geodesic of M that is inextendible in M. Since M is complete mod Σ, σ_1 hits the ring singularity. Hence by Proposition 4.15.1, σ_1 lies in Eq. But we saw in Section 2.7 that the curvature invariant $k(\sigma_1(s)) \to \infty$ as $s \to b$, so σ_1 is inextendible in any containing spacetime. Thus the extension σ of σ_1 provides the required contradiction. $\square$

PETROV TYPES

The curvature tensors of spacetimes admit a natural classification into a small number of *Petrov types*. The classification applies not to the full curvature tensor R but rather to its trace-free part, the *Weyl conformal tensor* C. (For Ricci-flat spacetimes such as Kerr's, $R = C$.) At different points of a spacetime M its Weyl tensor may have different types, but in the common case where the type is the same at all points we refer to the *Petrov type of M*.

This chapter is a general exposition of Petrov types with Kerr spacetime as the key example. The type of a spacetime gives considerable information, not merely about the character of its curvature, but also about other geometric invariants, notably null geodesics. For example, Kerr spacetime turns out to have Petrov type D, and this gives an independent derivation of its *principal null geodesics*.

Several descriptions of the Petrov classification are known. The simplest proceeds as follows: The curvature tensor R induces, at each point $p \in M$, a self-adjoint linear operator $\mathbb{R}$ called the *curvature transformation* on the second exterior product $\Lambda^2 = \Lambda^2(T_p M)$ of the tangent space at p. For a spacetime, Λ^2 is a real vector space of dimension 6 (see Appendix D). Similarly, the Weyl tensor C induces an operator $\mathbb{C}$ on Λ^2, and this case we can do better. The particular dimension and signature of a spacetime mean that the *Hodge star* $*$ provides a complex structure on Λ^2, which thus becomes a complex vector space of only three dimensions. But $\mathbb{C}$ commutes with $*$, and is thus a *complex* linear operator on Λ^2. The Petrov classification then appears naturally when we look at the (complex) eigenvalues and eigenvectors of $\mathbb{C}$ (see Figure 5.1).

The complex operator $\mathbb{C}$ determines real null directions at each point of M, and the resulting *principal null congruences* lead to a second characterization of Petrov type (see Figure 5.2).

To study the geometry of null congruences a natural weapon is the *Newman–Penrose formalism*, in which orthonormal frame fields are replaced by frame fields containing null vector fields. This formalism has proved its usefulness in many areas of relativity theory. After a brief exposition, we use it to relate the Petrov type of a spacetime to the properties of its principal null congruences. In particular, for type D spacetimes such as Kerr's, these congruences are not only geodesic but also "shearfree," roughly speaking, if a beam of such light initially has circular cross section, then it keeps this property as it propagates.

A final perspective on Kerr spacetime is provided by the Goldberg–Sachs theorem, which for arbitrary Ricci-flat spacetimes gives necessary and sufficient curvature conditions for the existence of shearfree geodesic null congruences.

5.1   Weyl Tensor

The definition and elementary properties of Petrov types involve only the curvature tensor on individual tangent spaces $T_p(M)$ of a spacetime M. Thus to clarify the role of the symmetries of curvature in the theory, we replace $T_p(M)$ by a scalar product space V, that is, an real vector space furnished with a scalar product $g = <, >$. For the generalities in this section V can be arbitrary; thereafter it is a four-dimensional Lorentz vector space.

Definition 5.1.1 *Let F be an (0,4) tensor on V, that is, a multilinear real-valued function on $V^4 = V \times V \times V \times V$. Then*

1. *F is* biskew *if $F(v, w, x, y)$ is skew-symmetric in v, w and in x, y;*
2. *F has* cyclic symmetry *if $F(v, x, y, z) + F(v, y, z, x) + F(v, z, x, y) = 0$ for all v, x, y, z in V.*

If F has both properties, it is called a curvature *on V.*

Evidently the Riemannian curvature tensor R of a manifold is, at each point p, a curvature on $T_p(M)$. The same proof as for R shows that every curvature F is *pair-symmetric*, that is, $F(v, w, x, y) = F(x, y, v, w)$ for all v, w, x, y in V.

Clearly a biskew tensor F has, but for sign, at most one nonzero contraction. To fix that sign, we agree to take CF to be the contraction $C_{13}(F)$—or, equivalently, $C_{24}(F)$. Then for an arbitrary basis $e_1, \ldots, e_n$,

$$(CF)(v, w) = \Sigma g^{ab} F(e_a, v, e_b, w) = \Sigma g^{ab} F(v, e_a, w, e_b),$$

where, as usual, g^{ab} is the inverse matrix of $g_{ab} = <e_a, e_b>$, and all indices run from 1 to $n = \dim V$. For an orthonormal basis, $(CF)(x, y)$ simplifies to $\Sigma \varepsilon_a F(e_a, x, e_a, y,)$, where $\varepsilon_a = <e_a, e_a> = \pm 1$. With the case of Riemannian curvature R in mind, we call CF the *Ricci tensor* of F.

In general, a tensor is said to be *tracefree* if all contractions of it are zero. Clearly, a biskew (0,4) tensor is tracefree if and only if its Ricci tensor is 0.

Remark 5.1.2 Informally, the second exterior product $\Lambda^2 = \Lambda^2(V)$ of V consists of all linear combinations of wedge products $v \wedge w$ of vectors $v, w \in V$. The wedge product operation is bilinear, associative, and alternate (or skew) on vectors, that is, $w \wedge v = -v \wedge w$. We can picture $v \wedge w$ as an oriented parallelogram with sides v and w. As discussed in Appendix D, Λ^2 has the following *universal property*:

If $A: V \times V \to W$ is a skew-symmetric, bilinear function into a vector space W, there is a unique linear function $\ell: \Lambda^2 \to W$ such that

$$\ell(v \wedge w) = A(v, w) \qquad \text{for all} \quad v, w \in V.$$

If F is a biskew (0,4) tensor, two applications of the universal property turn it into bilinear form on the second exterior product $\Lambda^2 = \Lambda^2(V)$, with

$$F(v \wedge w, x \wedge y) = F(v, w, x, y) \qquad \text{for all} \quad v, w, x, y.$$

(It should be safe to use the same notation for such closely related objects.) This transition can be used to extend the scalar product $<, >$ from V to $\Lambda^2 = \Lambda^2(V)$. Consider the following real-valued function on V^4:

$$(v, w, x, y) \to \begin{vmatrix} <v, x> & <v, y> \\ <w, x> & <w, y> \end{vmatrix}.$$

This is evidently multilinear and biskew; hence it determines a bilinear function $<, >$ on Λ^2 characterized by

$$(v \wedge w, x \wedge y) = \begin{vmatrix} <v, x> & <v, y> \\ <w, x> & <w, y> \end{vmatrix} \qquad \text{for all} \quad v, w, x, y \text{ in } V.$$

By construction $<,>$ is bilinear, and it is evidently symmetric. That $<,>$ is nondegenerate and hence is a scalar product follows from the fact, established below, that Λ^2 has an orthonormal basis, for expressing $\beta \perp \Lambda^2$ in terms of the basis implies $\beta = 0$.

If $e_1, \ldots, e_n$ is an orthonormal basis for V, then the canonical basis $\{e_a \wedge e_b : 0 \leq a < b \leq n\}$ for Λ^2 (Appendix D) is orthonormal relative to $<,>$, since, for example,

$$<e_1 \wedge e_2, e_1 \wedge e_3> = <e_1, e_1> <e_2, e_3> - <e_1, e_3> <e_2, e_1> = 0, \quad \text{but}$$

$$<e_1 \wedge e_2, e_1 \wedge e_2> = <e_1, e_1> <e_2, e_2> - <e_1, e_2> <e_2, e_1> = \varepsilon_1 \varepsilon_2 = \pm 1$$

On any scalar product space there is a natural one-one correspondence between bilinear forms and linear operators. Thus the bilinear form on Λ^2 determined by a biskew tensor F gives rise to a linear operator $\mathbb{F}$ on Λ^2 characterized by

$$<\mathbb{F}(v \wedge w), x \wedge y> = F(v \wedge w, x \wedge y) = F(v, w, x, y) \quad \text{for all } v, w, x, y \text{ in } V.$$

In tensor terms, $\mathbb{F}$ is the $(2, 2)$ tensor obtaining by type-changing of the $(0, 4)$ tensor F. On a semi-Riemannian manifold M, the $(2, 2)$ tensor $\mathbb{R}$ derived from Riemannian curvature tensor R is called the *curvature transformation* of M. Often $\mathbb{R}$ provides the easiest way to deal with curvature.

Now we consider some natural ways to construct abstract curvatures. These are used to decompose an arbitrary curvature into simpler pieces—one being the Weyl tensor.

Definition 5.1.3 *If A and B are symmetric bilinear forms on V, let $A \wedge B \colon V^4 \to R$ be the function given by*

$$(A \wedge B)(v, w, x, y) = \begin{vmatrix} A(v, x) & A(v, y) \\ B(w, x) & B(w, y) \end{vmatrix} \quad \text{for all } v, w, x, y \text{ in } V.$$

A direct computation shows that $A \wedge B$ has cyclic symmetry. It need not be biskew, but the special case $A \wedge A$ is. Thus, the identity

$$(A + B) \wedge (A + B) = A \wedge A + A \wedge B + B \wedge A + B \wedge B$$

shows that $A \wedge B + B \wedge A$ is biskew. Since it has cyclic symmetry it is a curvature.

In looking for simpler pieces of R, Definition 5.1.3 suggests $g \wedge g$ (where g is the metric tensor) and $g \wedge \text{Ric} + \text{Ric} \wedge g$ (where $\text{Ric} = C_{13}(R)$ is the Ricci tensor of R). Since the Weyl tensor C of R is to be the tracefree part of R, we need a computational lemma.

Lemma 5.1.4 (1) $C(g \wedge g) = (n-1)g$, and (2) $C(g \wedge Ric + Ric \wedge g) = (n-2)Ric + Sg$, where S is the scalar curvature $C(Ric)$.

Proof. Let $\{e_a\}$ be an orthonormal basis for V. Then for assertion (1),

$$
\begin{aligned}
C(g \wedge g)(v, w) &= \Sigma \varepsilon_a (g \wedge g)(e_a, v, e_a, w) \\
&= \Sigma \varepsilon_a \begin{vmatrix} <e_a, e_a> & <e_a, w> \\ <v, e_a> & <v, w> \end{vmatrix} \\
&= \Sigma \varepsilon_a [\varepsilon_a <v, w> - <v, e_a> <e_a, w>] \\
&= (n-1)<v, w>
\end{aligned}
$$

The other proof is similar.   □

We are now ready to derive the Weyl tensor of a curvature R. Write

$$ R = C + \alpha(g \wedge Ric + Ric \wedge g) + \beta g \wedge g $$

where the numbers α and β are to be determined so that $C(C) = 0$. By the preceding lemma, applying C gives

$$ Ric = C(C) + \alpha[(n-2)Ric + Sg] + \beta(n-1)g. $$

Clearly we must have $\alpha = 1/(n-2)$. (This imposes the definition: $C = 0$ if $n = 2$.) There remains

$$ 0 = C(C) + Sg/(n-2) + \beta(n-1)g. $$

Thus, if $C(C)$ is to be zero we must have $(n-1)(n-2)\beta = -S$.

Definition 5.1.5 *Let R be a curvature on a scalar product space V of dimension $n \geq 3$. The* Weyl *conformal curvature tensor C of R is characterized by*

$$ R = C + \frac{1}{n-2}(g \wedge Ric + Ric \wedge g) - \frac{S}{(n-1)(n-2)} g \wedge g. $$

Corollary 5.1.6 *The Weyl tensor C is a tracefree curvature.*

Proof. Since C is a linear combination of curvatures, it is itself a curvature. By construction, the contraction $C(C)$ is zero, and since C is biskew, all its contractions are zero.   □

To get a tensor component formula, note that for an arbitrary basis, Definition 5.1.3 gives, for example,

$$(g \wedge \text{Ric})(e_a, e_b, e_c, e_d) = \begin{vmatrix} g_{ac} & g_{ad} \\ R_{bc} & R_{bd} \end{vmatrix} = g_{ac} R_{bd} - g_{ad} R_{bc}.$$

Applying such expressions to the invariant definition of C gives the required formula. For simplicity we record only the $n = 4$ case:

$$R_{abcd} = C_{abcd} + \frac{1}{2}[g_{ac} R_{bd} - g_{ad} R_{bc} + R_{ac} g_{bd} - R_{ad} g_{bc}]$$
$$-(S/6)(g_{ac} g_{bd} - g_{ad} g_{bc})$$

In dimension $n = 2$ we have already defined C to be 0. Also $C = 0$ for $n = 3$ (an easy consequence of its biskew and tracefree properties).

For any semi-Riemannian manifold M the Weyl tensor of its Riemannian curvature tensor R is the *Weyl conformal curvature tensor of M*. As the name suggests, this tensor, discovered by Hermann Weyl, is invariant under conformal change of metric (Eisenhart, 1926). If C is identically zero, M is said to be *conformally flat*. Then, locally M is conformally equivalent to the Minkowski space of the same dimension and index.

Some perspective on the tensors in this section may be provided by counting dimensions when $n = \dim V = 4$. Then a tensor of type $(0, 4)$, such as curvature, is determined by $4^4 = 256$ numbers; that is, the vector space of all $(0, 4)$ tensors on V has dimension 256.

Biskew tensors correspond to bilinear forms on the six-dimensional space Λ^2, hence they have dimension 36. The pair-symmetry property further reduces the dimension to $6 \cdot 7/2 = 21$.

In the biskew case, cyclic symmetry is redundant unless all four vectors (or indices) are different. Thus it imposes only one new condition, say,

$$F_{0123} + F_{0231} + F_{0312} = 0.$$

Consequently, the space of curvatures on V has dimension 20.

Contraction C carries a curvature R to its Ricci curvature, a symmetric bilinear form on V. Furthermore, every such form is the Ricci curvature of some curvature, for instance, that given by the expression $R - C$ from Definition 5.1.5 (to see that this is a curvature, note the remark following Definition 5.1.3). Hence the Ricci curvatures have dimension 10, and C is a linear map of the 20-dimensional space of all curvatures onto a 10-dimensional space. The kernel of C is just the space of Weyl tensors, so it also has dimension 10.

In short, a curvature tensor on V is completely determined by 20 numbers: 10 for its Ricci curvature, 10 for its Weyl curvature.

5.2 Hodge Star

The Hodge star operator on an oriented scalar product space V is an algebraic version of the geometric operation that sends each subspace Π to its orthogonal complement $\Pi^{\perp}$. Formally, the Hodge star gives, for each $0 \leq p \leq n = \dim V$, a linear isomorphism $*$ from $\Lambda^p(V)$ to $\Lambda^{n-p}(V)$ that applied twice has $*^2 = \pm id$. The sign $\pm$ depends on *three* integers: n, p, and the index ν of V. For simplicity we concentrate on the spacetime bivector case $n = 4$, $p = 2$, $\nu = 1$. Here the special character of the Hodge star contributes to the simplicity of the Petrov classification.

Applying a general definition to a four-dimensional Lorentz vector space V, we say that an *orientation* of V is a choice, ω, of one of the two unit four-vectors in $\Lambda^4(V) \approx \mathbf{R}$ (see Appendix D). Then every orthonormal basis e_0, e_1, e_2, e_3 for V has $e_0 \wedge e_1 \wedge e_2 \wedge e_3 = \pm \omega$, and when this sign is positive, the basis is said to be *positively oriented*.

Remark 5.2.1 Let e_0, e_1, e_2, e_3 be an orthonormal basis for a four-dimensional Lorentz vector space V. As usual, e_0 is assumed to be timelike unless the contrary is explicitly mentioned. As noted in Section 5.1, the bivectors

$$e_0 \wedge e_1, e_0 \wedge e_2, e_0 \wedge e_3, e_1 \wedge e_2, e_1 \wedge e_3, e_2 \wedge e_3$$

form an orthonormal basis for the six-dimensional space $\Lambda^2 = V \wedge V$. However, future development will demonstrate that it is more convenient to readjust the last three bivectors to get a *Hodge basis*

$$e_0 \wedge e_1, e_0 \wedge e_2, e_0 \wedge e_3, e_2 \wedge e_3, e_3 \wedge e_1, e_1 \wedge e_2.$$

Now each of the first three "early" bivectors wedged with the corresponding "late" bivector gives the same orientation of V, since

$$e_0 \wedge e_1 \wedge e_2 \wedge e_3 = e_0 \wedge e_2 \wedge e_3 \wedge e_1 = e_0 \wedge e_3 \wedge e_1 \wedge e_2.$$

Because e_0 is timelike, the early bivectors are timelike; the late ones are spacelike (so Λ^2 has index 3).

In expressing the linear algebra of Λ^2 relative to such a basis, it is only reasonable to index the basis bivectors not by $i = 1, 2, \ldots, 6$ but by the *Hodge double indices* $01, 02, 03, 23, 31, 12$.

Since the Hodge star $*$ is to be the algebraic version of the operation $\Pi \to \Pi^\perp$ on 2-planes, we expect that it sends a decomposable bivector $v \wedge w \neq 0$ such that $\mathrm{span}\{v, w\} = \Pi$ to a bivector $x \wedge y$ such that $\mathrm{span}\{x, y\} = \Pi^\perp$. (As shown in Appendix D, $v \wedge w$ determines not only Π but also an orientation of Π, intuitively the one for which the rotation from v to w is positive.) If we require that $*$ preserve bivector scalar products *up to sign*, then for a Hodge basis, as above, $*$ should *up to sign* effect the exchanges

$$e_0 \wedge e_1 \leftrightarrow e_2 \wedge e_3, \quad e_0 \wedge e_2 \leftrightarrow e_3 \wedge e_1, \quad e_0 \wedge e_3 \leftrightarrow e_1 \wedge e_2.$$

In the positive definite case these exchanges determine $*$ without sign ambiguity; in the Lorentz case, causal character and orientation are harmonized by the following invariant characterization of $*$.

Proposition 5.2.2 *Let the 4-vector ω be an orientation of V. There exists a unique linear operator $* : \Lambda^2 \to \Lambda^2$, called the* Hodge star operator, *such that the* Hodge *identity $\xi \wedge *\eta = -{<}\xi, \eta{>}\omega$ holds for all $\xi, \eta \in \Lambda^2 = \Lambda^2(V)$.*

(With our conventions this minus sign gives best results; removing it gives a conjugate version of later complex-valued objects.)

Proof. (1) *Uniqueness.* Let $*$ and $\star$ be two such operators. Then

$$\xi \wedge *\eta = -{<}\xi, \eta{>}\omega = \xi \wedge \star\eta \qquad \text{for all} \quad \xi, \eta \in \Lambda^2.$$

Fix η. Then $\xi \wedge (*\eta - \star\eta) = 0$ for all ξ. By taking ξ successively as the bivectors of a basis $e_a \wedge e_b$ for Λ^2 we find that $*\eta - \star\eta = 0$. Hence, $* = \star$.

(2) *Existence.* Fix a positively oriented orthonormal basis e_0, e_1, e_2, e_3 for V; thus $e_0 \wedge e_1 \wedge e_2 \wedge e_3 = \omega$. Define $*$ by the equations in assertion (1) of Lemma 5.2.3. Then a direct computation shows that

$$e_a \wedge e_b \wedge *(e_c \wedge e_d) = -{<}e_a \wedge e_b, e_c \wedge e_d{>}\omega$$

for all bivectors in these equations. For example, suppose that $a = c = 0$ and $b = d = 2$. Then, since $*(e_0 \wedge e_2) = e_3 \wedge e_1$, the left side above is $e_0 \wedge e_2 \wedge e_3 \wedge e_1 = e_0 \wedge e_1 \wedge e_2 \wedge e_3 = \omega$. On the right, ${<}e_0 \wedge e_2, e_0 \wedge e_2{>} = -1$, yielding the same result.

We have thus proved the property $\xi \wedge *\eta = -<\xi, \eta>\omega$ for sufficiently many bivectors ξ, η to form a basis for Λ^2. Since the property is bilinear in ξ and η, it holds for arbitrary bivectors. $\square$

Lemma 5.2.3   *The Hodge star $*$ has the following properties:*
(1) *If e_0, e_1, e_2, e_3 is a positively oriented orthonormal basis for V, then*

$$*(e_0 \wedge e_1) = e_2 \wedge e_3, \quad *(e_2 \wedge e_3) = -e_0 \wedge e_1,$$

$$*(e_0 \wedge e_2) = e_3 \wedge e_1, \quad *(e_3 \wedge e_1) = -e_0 \wedge e_2,$$

$$*(e_0 \wedge e_3) = e_1 \wedge e_2, \quad *(e_1 \wedge e_2) = -e_0 \wedge e_3.$$

(2) $*^2 = -id.$
(3) $*$ *is self-adjoint, that is, $< *\xi, \eta> = <\xi, *\eta>$ for all $\xi, \eta \in \Lambda^2$.*

Proof.   (1) This follows from the proof of Proposition 5.2.1, which shows that if the basis given here is used to define a star operation by means of these six equations, then that operation is the Hodge star $*$.
(2) The equations show at once that $*^2 = -id$ on a basis for Λ^2.
(3) Since bivectors commute under the wedge product, the Hodge identity $\xi \wedge *\eta = -<\xi, \eta>\omega$ gives

$$< *\xi, \eta>\omega = - *\xi \wedge *\eta = - *\eta \wedge *\xi = < *\eta, \xi>\omega = <\xi, *\eta>\omega.$$

$\square$

Because the sign in assertion (2) of the preceding lemma is minus, we see that in the spacetime bivector case $n = 4$, $p = 2$, $\nu = 1$, *the Hodge star $*$ is a complex structure on Λ^2.* Explicitly, Λ^2 becomes a complex vector space when scalar multiplication by a complex scalar $a + ib$ is defined by:

$$(a + ib)\xi = a\xi + b(*\xi) \quad \text{for all } a, b \in \mathbf{R}.$$

In short, $i\xi = *\xi$. The dimension of Λ^2 as a complex vector space is 3; for example, $\{e_0 \wedge e_1, e_0 \wedge e_2, e_0 \wedge e_3\}$ is a basis. Note that scalar multiplication by i gives the basis $\{e_2 \wedge e_3, e_3 \wedge e_1, e_1 \wedge e_2\}$.

To emphasize cases where Λ^2 is considered as a complex vector space we write Λ^2_{cx}. The scalar product $<, >$ on Λ^2 gives rise to a complex scalar product on Λ^2_{cx} as follows.

Definition 5.2.4   *Let $g(\xi, \eta) = <\xi, \eta> - i<\xi, *\eta>$ for all $\xi, \eta \in \Lambda^2$.*

The minus sign here is unrelated to the choice of sign for $*$ and is needed to prove the following convenient result.

Lemma 5.2.5 g *is a complex scalar product on* Λ^2_{cx}.

Proof. (1) *Symmetry.* Since $*$ is self-adjoint with respect to $<,>$—which is, of course, symmetric—we have

$$g(\xi, \eta) = <\xi, \eta> - i<\xi, *\eta> = <\eta, \xi> - i<\eta, *\xi> = g(\eta, \xi).$$

(Thus g is not a *Hermitian* product.)

(2) *Complex bilinearity.* Clearly g is bilinear for real scalars; thus, it suffices, by the symmetry of g to show that i can be factored out of, say, the first slot of g. Since $i\xi$ is just $*\xi$, the properties of $*$ give

$$g(i\xi, \eta) = <*\xi, \eta> - i<*\xi, *\eta> = <\xi, *\eta> - i<\xi, **\eta>$$
$$= -i^2<\xi, *\eta> + i<\xi, \eta> = i(-i<\xi, *\eta> + <\xi, \eta>)$$
$$= i(<\xi, \eta> - i<\xi, *\eta>) = ig(\xi, \eta).$$

(3) *Nondegeneracy.* Given ξ, suppose $g(\xi, \eta) = 0$ for all $\eta \in \Lambda^2$. The real part of $g(\xi, \eta)$ is $<\xi, \eta>$, so the nondegeneracy of $<,>$ already implies $\xi = 0$.   □

Since the Hodge star $*$ on Λ^2 is just scalar multiplication by i, a real linear operator on Λ^2 is complex linear if and only if it commutes with $*$.

Remark 5.2.6 If a linear operator T on Λ^2 is self-adjoint with respect to $<,>$ and commutes with $*$, then T is self-adjoint with respect to g. In fact,

$$g(T\xi, \eta) = <T\xi, \eta> - i<T\xi, *\eta> = <\xi, T\eta> - i<\xi, T(*\eta)>$$
$$= <\xi, T\eta> - i<\xi, *T\eta> = g(\xi, T\eta).$$

Finally we consider the effect of the Hodge star on the causal character of bivectors. Recall that a bivector is *decomposable* if it can be expressed as a wedge product of two vectors (Appendix D). Null decomposable bivectors have the key property that they determine a null directions in V.

Lemma 5.2.7 *A bivector* $\delta \in \Lambda^2(V)$ *is null and decomposable if and only if it can be expressed as* $v \wedge x$, *where* $v \in V$ *is null and* x *is spacelike with* $0 \neq x \perp v$. *Here* v *is unique up to nonzero scalar multiplication, so* δ *invariantly determines a null line* $N(\delta) = \mathbf{R}v$ *in* V.

Proof. For v and x as specified, $v \wedge x$ is clearly null and decomposable. Conversely, suppose δ is null (hence, by definition, nonzero) and can be written as a product $y \wedge z$. Then

$$0 = <y \wedge z, y \wedge z> = <y, y><z, z> - <y, z>^2.$$

This is exactly the condition that the plane $\Pi = \text{span}\{y, z\}$ have null causal character (Section 1.5). Hence by Lemma 1.5.11, Π contains a null vector v, unique up to scalar multiplication. (Geometrically, $\mathbf{R}v$ is the line $\Pi \cap \Lambda$ along which Π is tangent to the nullcone Λ of V.) Pick x in Π independent of v; then x is spacelike and $v \wedge x \neq 0$. Since $\dim \Pi = 2$, δ and $v \wedge x$ are dependent, so stretching x if necessary gives $\delta = v \wedge x$. □

Proposition 5.2.8 *The Hodge star on* $\Lambda^2 = V \wedge V$ *has the following properties:*

(1) $*$ *reverses causal character (that is, reverses spacelike and timelike, but carries null to null.)*

(2) *A bivector* β *is decomposable* $\Leftrightarrow \beta \wedge \beta = 0 \Leftrightarrow *\beta$ *is decomposable* $\Leftrightarrow *\beta \perp \beta$.

(3) *For a null decomposable bivector* $\delta = v \wedge x$ *expressed as in Lemma 5.2.7,* $*(v \wedge x) = v \wedge y$, *where* x, y, v *are mutually orthogonal and* $|y| = |x| > 0$. *(Intuitively,* $*$ *rotates the flag* δ *by* $90°$ *around the flagstaff* $N(\delta) = \mathbf{R}v$.*)*

Proof. (1) Since $*$ is self-adjoint and $*^2 = -id$, $< * \beta, *\beta> = <\beta, * * \beta> = -<\beta, \beta>$.

(2) A standard criterion (proved in Appendix D) for a bivector β to be decomposable is $\beta \wedge \beta = 0$. This is equivalent to $*\beta \perp \beta$, since by the Hodge identity,

$$<\beta, *\beta> = 0 \Leftrightarrow \beta \wedge * * \beta = 0 \Leftrightarrow \beta \wedge \beta = 0.$$

Then

$$\beta \text{ decomposable } \Rightarrow < * \beta, \beta> = 0 \Rightarrow *\beta \wedge *\beta = 0 \Rightarrow *\beta \text{ decomposable.}$$

Conversely, if $*\beta$ is decomposable so is $* * \beta = -\beta$.

(3) Since x is spacelike, $x^\perp$ is a three-dimensional Lorentz vector space. Choose an orthonormal basis e_0, e_1, e_2 for it such that $v = c(e_0 + e_1)$. Then $e_0, e_1, e_2, e_3 = x/|x|$ is an orthonormal basis for V with $x = de_3$. By Lemma 5.2.3,

$$*(v \wedge x) = *cd(e_0 \wedge e_3 + e_1 \wedge e_3)$$
$$= cd(e_1 \wedge e_2 + e_0 \wedge e_2)$$
$$= c(e_0 + e_1) \wedge (de_2) = v \wedge y,$$

where $y = de_2$. Evidently this vector y has the required properties. □

The transition $\delta \to N(\delta)$ described in Lemma 5.2.7 is important in Section 5.5 Note that if δ is scalar multiplied by a complex number $z = a + ib \neq 0$, then $z\delta$ is still null decomposable and $N(z\delta) = N(\delta)$. This follows from property (3) in the preceding proposition, since

$$z\delta = (a+ib)(v \wedge x) = a(v \wedge x) + b*(v \wedge x) = v \wedge ax + v \wedge by = v \wedge (ax+by),$$

where v, x, y are mutually orthogonal. It follows that the null decomposable bivectors that determine $N(\delta)$ are the nonzero elements of the complex line through δ in Λ^2_{cx}.

5.3   Commutativity

Our goal is to show that, in the spacetime bivector case, the Weyl curvature operator $\mathbb{C}: \Lambda^2 \to \Lambda^2$ commutes with the Hodge star $*$ and is thus a *complex* linear operator. To do so it suffices to show that their matrices commute, where these matrices are taken relative to the obvious choice of basis for Λ^2, namely a Hodge basis $e_0 \wedge e_1, e_0 \wedge e_2, e_0 \wedge e_3; e_2 \wedge e_3, e_3 \wedge e_1, e_1 \wedge e_2$ (see Remark 5.2.1). For the proof somewhat weaker hypothesis will suffice.

An *Einstein manifold* is a semi-Riemannian manifold whose Ricci curvature tensor is a constant times the metric tensor g. Thus we say that a biskew, pair-symmetric (0,4) tensor F on V is *Einstein* provided its Ricci tensor $\mathbb{C}F$ is a constant times the scalar product of V; explicitly, $\mathbb{C}_{13}F = k<,>$. (The cyclic symmetry of curvature is superfluous here.)

Proposition 5.3.1 *Let F be a biskew, pair-symmetric (0,4) tensor on V.*

(1) F is Einstein if and only if, relative to any Hodge basis for Λ^2, the matrix of F (as a symmetric bilinear form on Λ^2) has the form

$$\langle F \rangle = \begin{bmatrix} A & B \\ B & -A \end{bmatrix}, \tag{5.1}$$

where A and B are 3×3 symmetric matrices.

(2) Furthermore, if F is Einstein, then F has cyclic symmetry $\Leftrightarrow$ trace $B = 0$, and F is tracefree $\Leftrightarrow$ trace $A = 0$.

Proof. (1) Fix an orthonormal basis e_0, e_1, e_2, e_3 for V. Then $\mathbb{C}_{13}F = k<,>$ is equivalent to $\Sigma \varepsilon_m F_{mamb} = k\delta_{ab}\varepsilon_b$, where, as usual, $\varepsilon_b = <e_b, e_b>$ and $F_{abcd} =$

$F(e_a, e_b, e_c, e_d)$. By pair symmetry only $10 = 4 \cdot 5/2$ of these equations are needed. We list them, labeled by (ab).

$$\begin{array}{ll}
(00) & F_{1010} + F_{2020} + F_{3030} = -k \\
(11) & -F_{0101} + F_{2121} + F_{3131} = k \\
(22) & -F_{0202} + F_{1212} + F_{3232} = k \\
(33) & -F_{0303} + F_{1313} + F_{2323} = k
\end{array}$$

$$\begin{array}{ll}
(01) & F_{2021} + F_{3031} = 0 \\
(02) & F_{1012} + F_{3032} = 0 \\
(03) & F_{1013} + F_{2023} = 0
\end{array}$$

$$\begin{array}{ll}
(23) & -F_{0203} + F_{1213} = 0 \\
(31) & -F_{0301} + F_{2321} = 0 \\
(12) & -F_{0102} + F_{3132} = 0
\end{array}$$

It is easy to see that the first four equations above are equivalent to

$$k = -F_{0101} - F_{0202} - F_{0303} = F_{2323} + F_{3131} + F_{1212} \text{ and}$$

$$\alpha \begin{cases}
F_{0101} + F_{2323} = 0 \\
F_{0202} + F_{3131} = 0 \\
F_{0303} + F_{1212} = 0
\end{cases}$$

Converting the remaining two sets of equations to the Hodge double indices 01,02,03,23,31,12 gives

$$\beta \begin{cases}
F_{0212} = F_{3103} \\
F_{0112} = F_{2303} \\
F_{0131} = F_{2302}
\end{cases} \qquad
\gamma \begin{cases}
F_{0203} + F_{3112} = 0 \\
F_{0103} + F_{2312} = 0 \\
F_{0102} + F_{2331} = 0
\end{cases}$$

As in the previous section, F is regarded as a symmetric bilinear form on Λ^2, so its matrix $\langle F \rangle$ relative to the Hodge basis $\{e_a \wedge e_b\}$ is the 6×6 matrix with entries

$$F(e_a \wedge e_b, e_c \wedge e_d) = F(e_a, e_b, e_c, e_d) = F_{ab,cd},$$

where ab and cd are Hodge double indices. We write $\langle F \rangle$ in terms of 3×3 matrices A, B, C, D as

$$\begin{bmatrix} A & B \\ C & D \end{bmatrix}.$$

Since F is pair-symmetric, $\langle F \rangle$ is symmetric; hence A and D are symmetric, and $^tB = C$. Omitting subdiagonal entries, $\langle F \rangle$ is

	01	02	03	23	31	12
01	F_{0101}	F_{0102}	F_{0103}	F_{0123}	F_{0131}	F_{0112}
02		F_{0202}	F_{0203}	F_{0223}	F_{0231}	F_{0212}
03			F_{0303}	F_{0323}	F_{0331}	F_{0312}
23				F_{2323}	F_{2331}	F_{2312}
31					F_{3131}	F_{3112}
12						F_{1212}

Now we can see that the equations α assert that corresponding diagonal entries of A and D are negatives of each other, and the equations γ say their corresponding off-diagonal entries are negatives. Thus $D = -A$.

Since $\langle F \rangle$ is symmetric, the equations β assert that the matrix B is symmetric. But $^tB = C$, so $C = B$. We conclude that for F to be Einstein is equivalent to $\langle F \rangle$ having the specified form.

(2) As mentioned earlier the cyclic symmetry of F (Definition 5.1.1) is equivalent to

$$F_{0123} + F_{0231} + F_{0312} = 0,$$

and this sum is exactly trace B.

Finally, since $C_{13} F = k\langle\, ,\, \rangle$ is assumed, F is tracefree if and only if $k = 0$, but equation (00) above asserts that trace $A = -k$.   □

Remark 5.3.2 A more concise notation for double indices will simplify calculations. Let e_0, e_1, e_2, e_3 be a positively oriented orthonormal basis for V. If $\lambda = ab$ is any double index $01, 02, 03, 23, 31, 12$, assign the bivector e_λ and its sign ε_λ their expected meanings: $e_\lambda = e_a \wedge e_b$, and $\varepsilon_\lambda = \varepsilon_a \varepsilon_b$. In fact, ε_λ *is* the sign of e_λ, since

$$\langle e_\lambda, e_\lambda \rangle = \langle e_a \wedge e_b, e_a \wedge e_b \rangle = \langle e_a, e_a \rangle \langle e_b, e_b \rangle - \langle e_a, e_b \rangle^2 = \varepsilon_a \varepsilon_b = \varepsilon_\lambda.$$

Finally, early Greek letters $\alpha, \beta, \ldots$ are reserved for the "early" double indices $01, 02, 03$, and a prime on an early index turns it into the corresponding "late" double index. For example, if $\alpha = 01$, then $\alpha' = 23$.

In this notation a Hodge basis for Λ^2 can be written as $\{e_a, e_{\alpha'}\}$, the effect of the Hodge star on it (Lemma 5.2.2) is just $*(e_\alpha) = e_{\alpha'}$, $*(e_{\alpha'}) = -e_\alpha$, and the form of the matrix $\langle F \rangle$ in assertion (1) of Proposition 5.3.1 is given by $F_{\alpha\beta} + F_{\alpha'\beta'} = 0$, $F_{\alpha\beta'} = F_{\alpha'\beta}$.

In relating matrices to linear operators on Λ^2, we use the following nearly standard convention: the matrix of $T: X \rightarrow X$ relative to a basis $x_1, \ldots, x_n$ for X is $(a^i{}_j)$, where $T(x_j) = \Sigma a^i{}_j x_i$ for all j. (Only in abstract linear algebra or in the positive definite case, can the index i can be lowered without tensor consequences.)

The rules for raising and lowering indices given in Section 1.2 remain valid for double indices. Explicitly, let F be a biskew pair-symmetric $(0,4)$ tensor on V, and let $\mathbb{F}: \Lambda^2 \rightarrow \Lambda^2$ be the corresponding (self-adjoint) linear operator. Then the matrix $[\mathbb{F}]$ of $\mathbb{F}$ relative to a Hodge basis $\{e_\lambda\}$ for Λ^2 is $(F^\lambda{}_\mu)$, where $F^\lambda{}_\mu = \varepsilon_\lambda F_{\lambda\mu}$. Here, if $\lambda = ab$, $\mu = cd$, then $F_{\lambda\mu} = F_{abcd} = F(e_a, e_b, e_c, e_d)$.

Corollary 5.3.3 *Let F be a biskew pair-symmetric $(0,4)$ tensor on V. Then F is Einstein if and only if the corresponding linear operator $\mathbb{F}$ on Λ^2 commutes with the Hodge star $*$ on Λ^2.*

Proof. As usual, let $\langle F \rangle$ and $[\mathbb{F}]$ be the matrices of F and $\mathbb{F}$ relative to a Hodge basis $\{e_\alpha, e_{\alpha'}\}$ for Λ^2. Proposition 5.3.1 asserts that

$$F \text{ is Einstein} \iff \langle F \rangle \text{ has the form } \begin{bmatrix} A & B \\ B & -A \end{bmatrix}.$$

The remarks preceding this corollary show that $\langle F \rangle$ has this form if and only if $[\mathbb{F}]$ has the form

$$\begin{bmatrix} -A & -B \\ B & -A \end{bmatrix}.$$

It is easy to check that $[\mathbb{F}]$ has this form (the minus signs on A are irrelevant) if and only if it commutes with the matrix of $*$, namely

$$\begin{bmatrix} 0 & -I \\ I & 0 \end{bmatrix}.$$

Equivalently, $\mathbb{F}$ and $*$ commute. $\square$

Consequently, a spacetime is an Einstein manifold if and only if its curvature tensor commutes with the Hodge star at each point. There is a more direct criterion in terms of sectional curvature (see Thorpe, 1969).

Corollary 5.3.4 *A spacetime M is Einstein if and only if its sectional curvature K is the same on (nondegenerate) orthogonal tangent 2-planes: $K(\Pi) = K(\Pi^\perp)$.*

Finally, we record the relation between the real and complex matrices of a linear operator on Λ^2_{cx}.

Lemma 5.3.5   *Let $\mathbb{F}$ be a linear operator on Λ^2_{cx}. If the (real) matrix $[\mathbb{F}]$ of $\mathbb{F}$ relative to a Hodge basis $\{e_\alpha, e_{\alpha'}\}$ for Λ^2 is*

$$
\begin{bmatrix}
F^\beta{}_\alpha & F^\beta{}_{\alpha'} \\
F^{\beta'}{}_\alpha & F^{\beta'}{}_{\alpha'}
\end{bmatrix},
$$

then the (complex) matrix $[\mathbb{F}]_{cx}$ is $[F^\beta{}_\alpha + iF^{\beta'}{}_\alpha]$ relative to both bases $\{e_\alpha\}$ and $\{e_{\alpha'}\}$ for Λ^2_{cx}.

Proof.   Since $e_{\beta'} = *(e_\beta) = ie_\beta$ on Λ^2_{cx},

$$
\mathbb{F}(e_\alpha) = \Sigma_\lambda F^\lambda{}_\alpha e_\lambda = \Sigma F^\beta{}_\alpha e_\beta + \Sigma F^{\beta'}{}_\alpha e_{\beta'}
$$
$$
= \Sigma F^\beta{}_\alpha e_\beta + i\Sigma F^{\beta'}{}_\alpha e_\beta = \Sigma_\beta (F^\beta{}_\alpha + iF^{\beta'}{}_\alpha)\, e_\beta.
$$

The other basis $\{e_{\alpha'}\}$ differs only by scalar multiplication by i, hence yields the same matrix. $\square$

5.4   Petrov Classification

We describe the Petrov classification of tracefree curvatures on a four-dimensional oriented Lorentz vector space V. The principal application is to the Weyl conformal tensor at each point of a spacetime.

Let C be a tracefree curvature on V, with $\mathbb{C}$ the linear operator it induces on Λ^2. Evidently, tracefree curvatures are Einstein (with $k = 0$) so Corollary 5.18 shows that $\mathbb{C}$ commutes with the Hodge star and is a complex linear operator on Λ^2_{cx}. The following remark shows that the (complex) trace of $\mathbb{C}$ is zero.

Remark 5.4.1   Let F be a biskew, pair-symmetric (0,4) Einstein tensor on V. Then we have trace $_{cx}\mathbb{F} = 0$ if and only if F is tracefree and has cyclic symmetry (hence is a tracefree curvature). This follows at once from Proposition 5.3.1 since, in the notation of Section 5.3,

$$
\text{if } \langle F \rangle =
\begin{bmatrix}
A & B \\
B & -A
\end{bmatrix},
\text{ then } [\mathbb{F}] =
\begin{bmatrix}
-A & -B \\
B & -A
\end{bmatrix}
\text{ and } [\mathbb{F}]_{cx} = -A + iB.
$$

Linear operators with trace 0 on a three-dimensional complex vector space can be classified naturally by their eigenstructure. Petrov's original classification (1954) applied, pointwise, to the Riemannian curvature transformation $\mathbb{R}$ of Einstein manifolds (Ric $=$ kg) and had three types according to the number $\{3, 2, 1\}$ of independent eigenvectors of $\mathbb{R}$ (see John Thorpe's succinct account, 1969). Restricting the classification to the *Weyl curvature* makes it applicable, pointwise, to all spacetimes. Furthermore, each of the first two cases now splits in two. (Recall that the geometric multiplicity of an eigenvalue λ is the dimension of its eigenspace E_λ, the subspace spanned by the eigenvectors belonging to λ.)

Lemma 5.4.2 *Each tracefree curvature $C \neq 0$ on a four-dimensional Lorentz vector space V is in exactly one of the following five Petrov types, expressed in terms of the induced linear operator $\mathbb{C}$ on Λ^2_{cx}:*

Type I. $\mathbb{C}$ *has three distinct eigenvalues.*
Type D. $\mathbb{C}$ *is diagonalizable with exactly two distinct eigenvalues.*
Type II. $\mathbb{C}$ *has exactly two distinct eigenvalues, each of geometric multiplicity 1.*
Type N. $\mathbb{C}$ *has a unique eigenvalue, necessarily 0, and it has geometric multiplicity 2.*
Type III. $\mathbb{C}$ *has a unique eigenvalue, necessarily 0, and it has geometric multiplicity 1.*

(These types are represented graphically in Figure 5.1.)

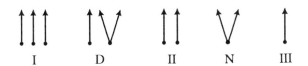

FIGURE 5.1.   Petrov types in terms of the eigenstructure of $\mathbb{C}$. Note how these "Roman numerals" collapse with each step to the right.

Proof. Consider the sum σ of the geometric multiplicities of the three (not necessarily distinct) eigenvalues of $\mathbb{C}$.

If $\sigma = 3$, then $\mathbb{C}$ is diagonalizable. In the ordinary case, I, the eigenvalues of $\mathbb{C}$ are distinct; in the "degenerate" case, D, only two are different. All three eigenvalues cannot be equal, for then trace $\mathbb{C} = 0$ would imply $\mathbb{C} = 0$, which has been ruled out.

If $\sigma = 2$, in the ordinary case, II, there are two distinct eigenvalues, each necessarily of geometric multiplicity 1; in the degenerate case, N, the eigenvalues are equal, hence are necessarily 0, again since trace $\mathbb{C} = 0$.

Finally, if $\sigma = 1$, then all three eigenvalues are equal, hence are 0, and this is type III.   □

Petrov type I—three distinct eigenvalues–is the generic case. The other types are said to be *algebraically special*.

For types D and II the eigenvalues can be written $-2\lambda, \lambda, \lambda$, with λ having geometric multiplicity 2 for type D and 1 for type II. For types N and III, since the eigenspaces of 0 are nonzero, $\mathbb{C}$ is not invertible (this can also occur for type I if an eigenvalue is zero.)

The case $C = 0$ is sometimes called Petrov type O. Thus, for a spacetime, type O means *conformally flat*.

By the theory of canonical forms (for example, see Mal'cev, 1963) there exist bases for Λ^2_{cx} that produce the matrices of g and $\mathbb{C}$ shown in the following table.

TABLE 5.1.   Canonical forms for $\langle g \rangle$ and $[\mathbb{C}]$ in the various Petrov types. Note that alternatively, *each matrix $\langle g \rangle$ can be multiplied by* -1, *leaving* $[\mathbb{C}]$ *unchanged.* (Simply scalar-multiply the selected basis for Λ^2_{cx} by i.)

Types	Matrix $\langle g \rangle$	Matrix $[\mathbb{C}]$	Conditions
I, D	$\begin{bmatrix} 1 & 0 & 0 \\ 0 & 1 & 0 \\ 0 & 0 & 1 \end{bmatrix}$	$\begin{bmatrix} \lambda_1 & 0 & 0 \\ 0 & \lambda_2 & 0 \\ 0 & 0 & \lambda_3 \end{bmatrix}$	$\Sigma\lambda_j = 0$ I: λ_j distinct D: $\lambda_2 = \lambda_3 = \lambda, \lambda_1 = -2\lambda$
II, N	$\begin{bmatrix} 1 & 0 & 0 \\ 0 & 0 & 1 \\ 0 & 1 & 0 \end{bmatrix}$	$\begin{bmatrix} -2\lambda & 0 & 0 \\ 0 & \lambda & 1 \\ 0 & 0 & \lambda \end{bmatrix}$	II: $\lambda \neq 0$ N: $\lambda = 0$
III	$\begin{bmatrix} 0 & 0 & 1 \\ 0 & 1 & 0 \\ 1 & 0 & 0 \end{bmatrix}$	$\begin{bmatrix} 0 & 1 & 0 \\ 0 & 0 & 1 \\ 0 & 0 & 0 \end{bmatrix}$	

The eigenstructure of $\mathbb{C}$ can easily be computed *explicitly* in terms of any matrix P of $\mathbb{C}$, since P is only 3×3 and has trace 0. General criteria for the type of $\mathbb{C}$ can be given, as follows, in terms of its characteristic polynomial

$$f(\lambda) = \det(P - \lambda I) = -\lambda^3 + c(P)\lambda + \det P.$$

In the cases below note that whether a 3×3 matrix has rank 1 is clear by inspection.

Lemma 5.4.3 *Let P be a matrix of $\mathbb{C} \neq 0$, and let $c = c(P)$, $d = \det(P)$. Then $\mathbb{C}$ has*
   (1) type I $\Leftrightarrow 27d^2 \neq 4c^3$,
   (2) type N $\Leftrightarrow$ *rank* $P = 1 \Leftrightarrow P^2 = 0$,
   (3) type III $\Leftrightarrow c = d = 0$ *and rank* $P \neq 1 \Leftrightarrow P^3 = 0$, *but* $P^2 \neq 0$,
   (4) either type D *or type* II $\Leftrightarrow 27d^2 = 4c^3 \neq 0$.
The repeated eigenvalue is $\lambda = -3d/2c$, and $\mathbb{C}$ has type D $\Leftrightarrow$ *rank* $(P - \lambda I) = 1$.

Proof. (1) The eigenvalues of $\mathbb{C}$, that is, the roots of $f(\lambda)$, are distinct if and only if f and its derivative f' are relative prime. By the Euclidean algorithm this is equivalent to $d - 4c^3/27d \neq 0$.

(2,3) Type N or III $\Leftrightarrow$ the only eigenvalue is $0 \Leftrightarrow f(\lambda) = -\lambda^3 \Leftrightarrow c = d = 0$. Type N has two independent eigenvectors so rank $(P - 0I) = $ rank $(P) = 1$. For type III this rank is 2. Only types N and III have all eigenvalues 0; for them the matrix P has Jordan canonical forms

$$\begin{bmatrix} 0 & 0 & 0 \\ 0 & 0 & 1 \\ 0 & 0 & 0 \end{bmatrix} \quad \text{and} \quad \begin{bmatrix} 0 & 1 & 0 \\ 0 & 0 & 1 \\ 0 & 0 & 0 \end{bmatrix}, \quad \text{respectively.}$$

The criteria in terms of powers of P follow.

(4) By the preceding cases, type D or II $\Leftrightarrow 27d^2 = 4c^3 \neq 0$. The euclidean algorithm shows that the greatest common divisor (f, f') is $(9d/2c)\lambda + c$. Thus $\lambda = -2c^2/9d$ is the repeated eigenvalue, but $27d^2 = 4c^3$ so $\lambda = -3d/2c$. Type D is then distinguished by the fact that for this λ, rank $(P - \lambda I) = 1$. $\qquad \square$

For other criteria for Petrov type, see Section 4.4 of Kramer et al. (1980).

Corollary 5.4.4 *Kerr spacetime has type* D.

Proof. Since any Kerr spacetime K is Ricci-flat, its Weyl tensor is just its curvature tensor R. Let $\{E_a\}$ be the Boyer–Lindquist frame field (Definition 2.6.1). Consider first blocks I and III, where E_0 is the timelike vector and the Hodge double indices are 01,02,03,23,31,12. With $\varepsilon = \mathrm{sgn}\Delta = +1$, Theorem 2.7.2 gives the curvature forms

$$
\begin{array}{ll}
\Omega^0{}_1 = 2\mathbf{I}\omega^0 \wedge \omega^1 - 2\mathbf{J}\omega^2 \wedge \omega^3 & \Omega^2{}_3 = 2\mathbf{J}\omega^0 \wedge \omega^1 + 2\mathbf{I}\omega^2 \wedge \omega^3 \\
\Omega^0{}_2 = -\mathbf{I}\omega^0 \wedge \omega^2 + \mathbf{J}\omega^3 \wedge \omega^1 & \Omega^3{}_1 = -\mathbf{J}\omega^0 \wedge \omega^2 - \mathbf{I}\omega^3 \wedge \omega^1 \\
\Omega^0{}_3 = -\mathbf{I}\omega^0 \wedge \omega^3 + \mathbf{J}\omega^1 \wedge \omega^2 & \Omega^1{}_2 = -\mathbf{J}\omega^0 \wedge \omega^3 - \mathbf{I}\omega^1 \wedge \omega^2
\end{array}
$$

where $\mathbf{I} = \mathrm{M}r\rho^{-6}(r^2 - 3a^2C^2)$ and $\mathbf{J} = \mathrm{M}aC\rho^{-6}(3r^2 - a^2C^2)$.

Evaluating the form $\Omega^a{}_b$ on E_c, E_d gives the matrix $R^a{}_{bcd}$, but in the cases above, the index b is never 0, so it may be raised without sign change. Hence the matrix of $\mathbb{R}: \Lambda^2 \to \Lambda^2$ relative to the Hodge basis associated with $\{E_a\}$ has

$$
R^{ab}{}_{cd} = R^a{}_{bcd} = \Omega^a{}_b(E_c, E_d)
$$

for paired indices ab, cd. Omitting the zero entries, we find

$$[\mathbb{R}] =$$

	01	02	03	23	31	12
01	2**I**			−2**J**		
02		−**I**			**J**	
03			−**I**			**J**
23	2**J**			2**I**		
31		−**J**			−**I**	
12			−**J**			−**I**

Lowering the index a for the entries of this array reverses the signs of the first three rows, giving for the (symmetric) matrix $(R_{ab\,cd})$ of R as a bilinear form on Λ^2

$$\langle \mathbb{R} \rangle =$$

	01	02	03	23	31	12
01	−2**I**			2**J**		
02		**I**			−**J**	
03			**I**			−**J**
23	2**J**			2**I**		
31		−**J**			−**I**	
12			−**J**			−**I**

This matrix exhibits in simplest form the structure predicted in Proposition 5.3.1.

Applying Lemma 5.3.5 to the earlier matrix $[\mathbb{R}]$ gives the complex matrix

$$P = [\mathbb{R}]_{cx} = \begin{bmatrix} 2(\mathsf{I}+i\mathsf{J}) & 0 & 0 \\ 0 & -\mathsf{I}-i\mathsf{J} & 0 \\ 0 & 0 & -\mathsf{I}-i\mathsf{J} \end{bmatrix}.$$

for $\mathbb{R}$ on Λ^2_{cx}. This matrix is the same for both the timelike basis $\{e_\alpha\}$ and the spacelike basis $\{e_{\alpha'}\}$ for Λ^2_{cx} since (as noted earlier) the bases differ only by scalar multiplication by i.

Evidently P displays the eigenvalue pattern $-2\lambda, \lambda, \lambda$ of type D *provided* $\lambda = -\mathsf{I}-i\mathsf{J}$ *is never zero*. To check on this, write $\rho^2 = r^2 + a^2 C^2$ as $(r+iaC)(r-iaC)$. Then

$$\mathsf{I}+i\mathsf{J} = \mathsf{M}\rho^{-6}\big[r(r^2 - 3a^2C^2) + iaC(3r^2 - a^2C^2)\big]$$
$$= \mathsf{M}\rho^{-6}(r+iaC)^3 = \mathsf{M}(r-iaC)^{-3}.$$

Thus *the repeated eigenvalue of* $\mathbb{C}$ *for Kerr spacetime is*

$$\lambda = -\mathsf{I}-i\mathsf{J} = -\mathsf{M}(r-iaC)^{-3}.$$

This function is never zero on any Kerr spacetime K since the functions r and $C = \cos\vartheta$ vanish simultaneously only on the ring singularity, not part of K.

The proof is not yet complete, since the computation of $\langle R \rangle$ excluded block II. There it is E_1, not E_0, that is timelike. But if we set $\varepsilon = -1$ in the curvature formulas (Theorem 2.7.2) and impose the Hodge ordering 10,12,13; 32,03,20, the matrix $\langle R \rangle$ for block II turns out to be exactly the same as $\langle R \rangle$ above. Thus $[\mathbb{R}]$ is the same, so block II has type D.

By continuity the same eigenvalue pattern persists on horizons since λ cannot approach 0 there. We conclude that K has type D globally. $\square$

A mathematical hazard throughout this chapter is the large number of notational choices that can affect the form of the final results. For example, our value for λ above agrees with Stewart, Walker (1973) but not with Chandrasekhar (1983).

5.5   Principal Null Directions

Petrov types can also be described in terms of null vectors. Although this is less direct than the complex eigenvalue approach, it is more valuable geometrically. For example, it yields a curvature characterization of the Kerr principal null geodesics.

Again let C be a tracefree curvature on V with $\mathbb{C}$ its induced linear operator on Λ^2_{cx}, and let $<$, $>$ be the real scalar product on $\Lambda^2 = \Lambda^2(V)$, g the complex scalar product (Definition 5.2.4) on Λ^2_{cx}. (Of course, we anticipate C being the Weyl tensor of a spacetime.)

Applying a general definition, a bivector β in Λ^2_{cx} is g-null if $g(\beta, \beta) = 0$ and $\beta \neq 0$. This is the same as saying that β is null and decomposable, since $g(\beta, \beta) = 0$ if and only if both $<\beta, \beta> = 0$ and $<\beta, *\beta> = 0$; but the latter is a criterion for decomposability (see Proposition 5.2.8).

Recall from Section 5.2 that a null decomposable bivector δ determines a unique null line $N(\delta) = (v)$, and that any nonzero bivector in the same complex line as δ determines the same null direction.

Definition 5.5.1 *A g-null bivector $\delta \in \Lambda^2_{cx}$ such that $g(\mathbb{C}(\delta), \delta) = 0$ is called a* principal null bivector of C, *the null line $N(\delta)$ it determines in V is a* principal null direction, *and any nonzero vector v in this line is a* principal null vector *(Debever, 1958).*

In particular, a g-null eigenvector δ of $\mathbb{C}$ is principal, since $\mathbb{C}(\delta) = \lambda\delta$ implies $g(\mathbb{C}\delta, \delta) = \lambda g(\delta, \delta) = 0$. This leads to a notion of multiplicity m for the principal null vectors v.

Definition 5.5.2 *A principal null vector $v \in N(\delta)$ of C has* multiplicity
$m = 1$ if δ is not an eigenvector of $\mathbb{C}$
$m = 2$ if $\mathbb{C}(\delta) = \lambda\delta$ with $\lambda \neq 0$,
$m = 3$ if $\mathbb{C}(\delta) = 0$ and dim Ker $\mathbb{C} = 1$,
$m = 4$ if $\mathbb{C}(\delta) = 0$ and dim Ker $\mathbb{C} = 2$

These are the only possiblities if $C \neq 0$. If v has multiplicity m, it is called m-*principal*. Also, 1-principal null vectors are said to be *simple*, and those with $m \geq 2$ are *repeated*. This terminology for v applies correspondingly to the bivector δ and the line $N(\delta)$ containing v.

Petrov types can now be characterized as follows.

Proposition 5.5.3 *The principal null directions of a tracefree curvature $C \neq 0$ are related to its Petrov type as follows.*

Type I:	*Four simple principal nulls*	(1,1,1,1)
Type II:	*One 2-principal and two simple nulls*	(2,1,1)
Type D:	*Two 2-principal nulls*	(2,2)

Type III: *One 3-principal and one simple*   (3,1)

Type N: *One 4-principal null*   (4).

(These types are represented graphically by Figure 5.2)

I   II   D   III   N

<small>FIGURE 5.2.   Petrov types in terms of principal null vectors. (Compare with Figure 5.1, where the ordering is different.)</small>

Proof.   First, suppose that C has type D. Let $-2\lambda, \lambda, \lambda$ be the (complex) eigenvalues of $\mathbb{C}$, with $\xi \in \Lambda^2$ the (essentially unique) eigenvector belonging to -2λ. A g-orthonormal basis of eigenvectors, as in Table 5.1 is readily found. Remark 5.2.6 shows that $\mathbb{C}$ is self-adjoint with respect to g, so a standard argument implies that ξ *is g-orthogonal to the eigenspace E_λ of* λ. Because g is nondegenerate, neither ξ nor the eigenspace E_λ are g-null; thus the latter has an g-orthogonal basis η_1, η_2. Using scalar multiplication we can arrange that the bivectors ξ, η_1, η_2 all have $g(\zeta, \zeta) = +1$. (Causal character in a complex vector space reduces to null vs nonnull.) Thus ξ, η_1, η_2 is a g-orthonormal basis for Λ^2.

If $\beta \in \Lambda^2$, write $\beta = a\xi + b\eta_1 + c\eta_2$. Then β is principal null if and only if

(1) $0 = g(\beta, \beta) = a^2 + b^2 + c^2,$

(2) $0 = g(\mathbb{C}\beta, \beta) = -2\lambda a^2 + \lambda b^2 + \lambda c^2.$

Now multiply equation (1) by λ and subtract it from (2) to get $0 = -3\lambda a^2$; hence $a = 0$. Then both (1) and (2) reduce to

(3) $b^2 + c^2 = 0.$

Since β is determined only up to scalar multiplication, we can take $b = 1$. Then $c = \pm i$, giving the principal null bivectors $\eta_1 \pm i\eta_2$. Evidently these are eigenvectors of $\lambda \neq 0$ for $\mathbb{C}$ (since η_1 and η_2 are); hence by definition, they have multiplicity 2.

Type I is a mildly more complicated variant, so consider type II. Its only difference from type D is that the repeated eigenvalue has just a single eigenvector. From Table 5.1 (or by an ad hoc argument) we get a basis ξ, v_1, v_2 for Λ^2 such that g and $\mathbb{C}$ have matrices

$$
\begin{bmatrix} 1 & 0 & 0 \\ 0 & 0 & 1 \\ 0 & 1 & 0 \end{bmatrix} \quad \text{and} \quad \begin{bmatrix} -2\lambda & 0 & 0 \\ 0 & \lambda & 1 \\ 0 & 0 & \lambda \end{bmatrix}, \text{ respectively.}
$$

Then, writing $\beta = a\xi + bv_1 + cv_2$ gives $\mathbb{C}(\beta) = -2\lambda a\xi + (c + \lambda b)v_1 + (\lambda c)v_2$. Consequently, a bivector $\beta \in \Lambda^2$ is principal null if and only if

(1) $$0 = g(\beta, \beta) = a^2 + 2bc,$$
(2) $$0 = g(\mathbb{C}(\beta), \beta) = 2\lambda(-a^2 + bc) + c^2.$$

Using equation (1) to simplify (2) gives

(3) $$0 = c(c + 6b\lambda).$$

Case 1. $c = 0$. Then according to equation (1), $a = 0$, so β is essentially just v_1. The formula for $\mathbb{C}$ shows that v_1 is an eigenvector, but because $\lambda \neq 0$ for type II, it is not in the kernel of $\mathbb{C}$. So v_1 has multiplicity 2.

Case 2. $c \neq 0$. If we take $b = -1/2$, then from equation (3), $c = 3\lambda$, and from equation (1), $a^2 = 3\lambda$. Thus, two principal null bivectors appear,

$$\beta = \pm\sqrt{3\lambda}\,\xi - \frac{1}{2}v_1 + 3\lambda v_2.$$

These are not eigenvectors, hence have multiplicity 1.

Type N is the simpler variant of the above with $\lambda = 0$. The argument for type III is a simpler variant of that for type II. $\square$

The indices, totaling four, in the proposition could be anticipated from Bezout's theorem in algebraic geometry since the principal null directions derive from the intersection of the curves $g(\delta, \delta) = 0$ and $g(\mathbb{C}\delta, \delta) = 0$ in the complex projective plane $\mathbb{CP}^2$.

The *Sachs criteria* in the proposition below directly relate a tracefree curvature and its principal null vectors without the intervention of bivectors or complex numbers. As is often advantagous when null vectors are important, the notion of orthonormal basis is replaced by:

Definition 5.5.4 *A real null tetrad in a four-dimensional Lorentz vector space V is a basis k, ℓ, x, y such that k, ℓ are null vectors with $\langle k, \ell \rangle = -1$, and the vectors x, y are orthonormal and orthogonal to k and ℓ.*

For such a tetrad, $\Pi = \text{span}\{k, \ell\}$ is timelike and $\Pi^{\perp} = \text{span}\{x, y\}$ is spacelike. All scalar products are 0 except $\langle k, \ell \rangle = -1$ and $\langle x, x \rangle = \langle y, y \rangle = +1$.

If e_0, e_1, e_2, e_3 is an orthonormal basis for V, then

$$k = (e_0 + e_1)/\sqrt{2}, \quad \ell = (e_0 - e_1)/\sqrt{2}, \quad x = e_2, \quad y = e_3$$

is a null tetrad said to be *associated* with $\{e_a\}$. (This indexing is convenient for Kerr spacetime, but permuted indices $1 \to 3 \to 2 \to 1$ are often used.) Such a tetrad gives the following basis for $\Lambda^2 = \Lambda^2(V)$:

$$k \wedge \ell, k \wedge x, k \wedge y, \ell \wedge x, \ell \wedge y, x \wedge y.$$

For these bivectors all scalar products are zero except

$$\langle k \wedge \ell, k \wedge \ell \rangle = \langle k \wedge x, \ell \wedge x \rangle = \langle k \wedge y, \ell \wedge y \rangle = -1, \quad \text{and}$$
$$\langle x \wedge y, x \wedge y \rangle = +1.$$

Proposition 5.5.5 *Let $C \neq 0$ be a tracefree curvature on V, and let k be a null vector in V. Then*
(1) *k is principal $\Leftrightarrow$ $C(k, x, k, y) = 0$ for all $x, y \perp k$.*
(2) *k is m-principal with $m \geq 2$ $\Leftrightarrow$ $C(k, x, k, s) = 0$ for all $x \perp k$ and $s \in V$.*
(3) *k is m-principal with $m \geq 3$ $\Leftrightarrow$ $C_{kx} = 0$ for all $x \perp k$.*
(4) *k is 4-principal $\Leftrightarrow$ $C_{ky} = 0$ for all $y \in V$, that is, $C(k, \cdot, \cdot, \cdot) = 0$.*

Proof. (1) Suppose k is principal. Then by definition there exists an $x_0 \perp k$ such that $\delta = k \wedge x_0$ is a principal null bivector of $\mathbb{C}$. We saw in Section 5.2 that for every spacelike vector $x \perp k$ the null bivector $k \wedge x$ also determines k. Since $k \wedge x$ is in the same complex line as $k \wedge x_0$, the condition $g(\mathbb{C}(k \wedge x), k \wedge x) = 0$ remains valid. Thus, $C(k, x, k, x) = 0$ holds for all $x \perp k$ (including multiples of k). By polarization, $C(k, x, k, y) = 0$ for all x, y in $k^{\perp}$. The converse is evident, so assertion (1) is proved.

(2) Note that since $i(k \wedge x) = *(k \wedge x) = k \wedge y$, the bivector $k \wedge x$ is an eigenvector of the complex linear operator $\mathbb{C}$ if and only if both $\mathbb{C}(k \wedge x)$ and $\mathbb{C}(k \wedge y)$ are in the complex line $P = \text{span}_R\{k \wedge x, k \wedge y\}$. Equivalently, both

$\mathbb{C}(k \wedge x)$ and $\mathbb{C}(k \wedge y)$ are orthogonal to $P^\perp$. There exists an ℓ such that k, ℓ, x, y is a null tetrad. Scalar products of the resulting bivectors show that

$$P^\perp = \text{span}\{k \wedge \ell, k \wedge x, k \wedge y, x \wedge y\}.$$

Thus k is m-principal with $m \geq 2$ if and only if both $\mathbb{C}(k \wedge x)$ and $\mathbb{C}(k \wedge y)$ are orthogonal to $P^\perp$.

Suppose k is m-principal with $m \geq 2$. Since $C(k, x, s, t) = <\mathbb{C}(k \wedge x), s \wedge t>$, by taking $s \wedge t$ successively to be $k \wedge \ell, k \wedge x, k \wedge y$ we find

$$C(k, x, k, \ell) = C(k, x, k, x) = C(k, x, k, y) = 0.$$

Since $C(k, x, k, k) = 0$, we conclude that $C(k, x, k, s) = 0$ for all $s \in V$. Replacing $k \wedge x$ by $k \wedge y$ gives the same result with x replaced by y. Thus the required condition on C holds.

Conversely, the hypothesized condition implies the triple equation displayed immediately above. If it is also true that $C(k, x, x, y) = 0$, then $\mathbb{C}(k \wedge x)$ is orthogonal to $P^\perp$. Hence $k \wedge x$ is an eigenvector of $\mathbb{C}$, so k is m-principal with $m \geq 2$. To prove $C(k, x, x, y) = 0$, we contract C using the null tetrad $\{t_\alpha\} = k, \ell, x, y$:

$$0 = (CC)(k, y) = \Sigma g^{ab} C(k, t_a, y, t_b) = -C(k, \ell, y, k) + C(k, x, y, x).$$

The condition on C implies $C(k, y, k, \ell) = 0$. But then $C(k, \ell, y, k) = 0$, so $C(k, x, x, y) = 0$.

The proofs of assertions (3) and (4) are similar exercises in multilinear algebra. □

5.6   Type D Curvature

We consider the Weyl curvature tensor of type D, relating it to the generic type, I, and considering some consequences for Kerr spacetime. Of course all the algebraically special types are degenerations of type I, but type D is the only one for which the Weyl operator $\mathbb{C}$ on Λ^2_{cx} remains diagonalizable.

Lemma 5.6.1 *Given a g-orthogonal basis ξ_1, ξ_2, ξ_3 for Λ^2_{cx} there exists a orthonormal basis e_0, e_1, e_2, e_3 for V such that*

$$\xi_1 = e_2 \wedge e_3, \quad \xi_2 = e_3 \wedge e_1, \quad \xi_3 = \pm e_1 \wedge e_2.$$

The proof shows that these vectors are unique except for relabeling and sign reversal. Scalar multiplication by $-i$ changes this (spacelike) basis to the timelike basis $e_0 \wedge e_1, e_0 \wedge e_2, e_0 \wedge e_3$, which acccounts for the index arrangements.

Proof. (Thorpe, 1969). Since ξ_1, ξ_2, ξ_3 are unit bivectors relative to g they are unit bivectors relative to $<, >$ and are decomposable. (Recall that $<\xi, *\xi> = 0$ implies ξ decomposable). The g-orthogonality of ξ_1 and ξ_2 means that $<\xi_1, \xi_2> = 0$ and $<\xi_1, *\xi_2> = 0$. The latter implies $\xi_1 \wedge \xi_2 = 0$ since by the Hodge identity

$$\xi_1 \wedge \xi_2 = -\xi_1 \wedge * * \xi_2 = <\xi_1, *\xi_2>\omega = 0.$$

(We now use some properties of bivectors given in Appendix D.) Since $\xi_1 \wedge \xi_2 = 0$ the decomposable bivectors ξ_1 and ξ_2 share a vector, that is, there are vectors v, w, e_3 in $V = T_p(M)$ such that

$$\xi_1 = v \wedge e_3, \quad \xi_2 = e_3 \wedge w.$$

We can modify these vectors, if necessary, to make them orthonormal. First, adjust e_3 to be a unit vector. Then in the spacelike 2-planes, P_1 and P_2 in V determined by ξ_1 and ξ_2, replace $v \in P_1$ and $w \in P_2$ by their components e_1 and e_2 orthogonal to e_3. The bivectors ξ_1 and ξ_2 are unchanged, and e_1 and e_2 necessarily have unit length. Note that $<\xi_1, \xi_2> = 0$ implies $<e_1, e_2> = 0$.

Since $e_1 \wedge e_2$ is g-orthogonal to both ξ_1 and ξ_2, we have $e_1 \wedge e_2 = z\xi_3$ for some complex z. Also $e_1 \wedge e_2$ has unit length relative to g, so $z^2 = 1$, hence $z = \pm 1$. $\qquad\qquad\square$

We call an orthonormal basis e_0, e_1, e_2, e_3 for V a *canonical frame* if the bivectors $e_2 \wedge e_3, e_3 \wedge e_1, e_1 \wedge e_2$ are eigenvectors of the Weyl operator $\mathbb{C}$ on Λ^2_{cx}. Such a basis produces the canonical forms given in Table 5.1 for types I and D (for the latter we always take $\lambda_2 = \lambda_3$). Also the timelike bivectors $e_0 \wedge e_1, e_0 \wedge e_2, e_0 \wedge e_3$ would serve just as well. Canonical frames always exist if $\mathbb{C}$ is diagonalizable since in the preceding lemma the sign of ξ_3 can be reversed without harm. The corresponding notion for the other Petrov types is *canonical null tetrad* $k, \ell, m, \overline{m}$ (see Section 5.8) for which $k \wedge \ell, k \wedge x, \ell \wedge y$ (suitably ordered) provide bases for the other canonical forms in Table 5.1. The idea is to produce simple, geometrically significant components for the Weyl tensor.

Lemma 5.6.2 *Suppose the Weyl operator* $\mathbb{C}$ *on* Λ^2_{cx} *is diagonalizable with eigenvalues* $\lambda_j = a_i + ib_j$ $(j = 1, 2, 3)$. *Then relative to an canonical basis the matrix* $\langle C \rangle$ *of C on* Λ^2, *zeros omitted, is*

	01	02	03	23	31	12
01	$-a_1$			$-b_1$		
02		$-a_2$			$-b_2$	
03			$-a_3$			$-b_3$
23	$-b_1$			a_1		
31		$-b_2$			a_2	
12			$-b_3$			a_3

Proof. By hypothesis,

$$\mathbb{C}(e_2 \wedge e_3) = \lambda_1(e_2 \wedge e_3) = a_1 e_2 \wedge e_3 - b_1 e_0 \wedge e_1,$$
$$\mathbb{C}(e_3 \wedge e_1) = \lambda_2(e_3 \wedge e_1) = a_2 e_3 \wedge e_1 - b_2 e_0 \wedge e_2,$$
$$\mathbb{C}(e_1 \wedge e_2) = \lambda_3(e_1 \wedge e_2) = a_3 e_1 \wedge e_2 - b_3 e_0 \wedge e_3.$$

As usual, $[\mathbb{C}]$ denotes the matrix of $\mathbb{C}$ on Λ^2 relative to the Hodge basis $\{e_0 \wedge e_1, \ldots, e_1 \wedge e_2\}$. Thus the last three rows of $[\mathbb{C}]$, zeros omitted, are

	01	02	03	23	31	12
23	$-b_1$			a_1		
31		$-b_2$			a_2	
12			$-b_3$			a_3

These rows of $[\mathbb{C}]$ are the same as the last three rows of $\langle C \rangle$, hence by Proposition 5.3.1, the full matrix $\langle C \rangle$ is as stated.   □

For type D, with its repeated eigenvalues $-2\lambda, \lambda, \lambda$, the matrix $\langle C \rangle$ for any canonical frame can be read from the preceding lemma: If $\lambda = a + ib$, set $a_1 = -2a$, $a_2 = a_3 = a$ and $b_1 = -2b$, $b_2 = b_3 = b$.

In the case of (Ricci-flat) Kerr spacetime the matrices in Section 5.4 show that the Boyer–Lindquist frames E_0, E_1, E_2, E_3 are canonical—a tribute to the simplicity of Boyer–Lindquist coordinates. We recover the matrix $\langle R \rangle$ in Section 5.4 by setting $\lambda = -\mathbf{I} - i\,\mathbf{J}$ in the preceding type D formulas.

We now consider some global properties of a type D spacetime M. The definition of Hodge star depends on an orientation of $V = T_p(M)$, so for simplicity we assume that M is orientable and orient it by a smooth field $p \to \omega_p$ of unit 4-vectors (see Appendix D). Thus a consistent complex structure is determined on every $\Lambda^2(T_pM)$. At each point p of M the Weyl curvature operator $\mathbb{C}$ is diagonalizable, with nonzero eigenvalues $-2\lambda(p)$, $\lambda(p)$, $\lambda(p)$. Because this eigenstructure is uniform over M it follows that the resulting complex-valued function λ on M is smooth.

Let Π_p be the tangent 2-plane at $p \in M$ spanned by the two principal null directions at p. Using terminology from Kerr spacetime we call Π_p the *principal plane* at $p \in M$. Then $p \to \Pi_p$ is a (smooth) distribution of timelike planes on M.

Lemma 5.6.3 *Let e_0, e_1, e_2, e_3 be an orthonormal frame on a type D spacetime. The following are equivalent: (1) the frame is canonical, (2) $e_0 \wedge e_1$ (or equivalently, $e_2 \wedge e_3$) is an eigenvector of the nonrepeated eigenvalue -2λ of $\mathbb{C}$, (3) $span\{e_0, e_1\}$ is the principal plane, (4) $e_0 \pm e_1$ are the principal null vectors.*

Proof. The implications (1) $\Rightarrow$ (2) and (3) $\Leftrightarrow$ (4) are obvious. We saw in Section 5.5 that the g-orthogonal complement of a -2λ eigenvector in Λ^2_{cx} is the λ eigenspace. Thus (2) $\Rightarrow$ (1). Assume (1) holds, so (by convention) $\eta_1 = e_1 \wedge e_2$ and $\eta_2 = e_3 \wedge e_1$ are λ eigenvectors of $\mathbb{C}$. The proof of Proposition 5.5.3 shows that the principal null directions are $N(\eta_1 \pm i\,\eta_2)$. But here

$$\eta_1 \pm i\eta_2 = e_1 \wedge e_2 \pm i e_3 \wedge e_1 = e_1 \wedge e_2 \pm *(e_3 \wedge e_1) = e_1 \wedge e_2 \pm e_0 \wedge e_2 = (e_1 \pm e_0) \wedge e_2.$$

Hence choosing familiar signs, the principal null directions are given by $e_0 \pm e_1$. Thus (1) $\Rightarrow$ (4).

Finally, suppose (4) holds. We know that there is a canonical frame, say u_0, u_1, u_2, u_3, for which $u_0 \pm u_1$ point in the principal directions. Thus $u_0 + u_1 = r(e_0 + e_1)$, $u_0 - u_1 = s(e_0 - e_1)$. Taking wedge products shows that $u_0 \wedge u_1$ and $e_0 \wedge e_1$ are collinear, hence $e_0 \wedge e_1$ is also an eigenvector of $\mathbb{C}$ belonging to -2λ. Thus (4) $\Rightarrow$ (2). $\square$

Corollary 5.6.4 *On Kerr spacetime the abstract curvature-induced principal null directions are exactly those given in Section 2.5 as determined by $\pm\partial_r + \Delta^{-1} V$.*

Proof. Since Boyer–Lindquist frames are canonical, the preceding lemma shows that the abstract principal null directions are determined by $E_0 \pm E_1$. To express these in coordinate terms, recall that

$$E_0 = \frac{1}{\rho\sqrt{\varepsilon\Delta}}[(r^2 + a^2)\partial_t + a\partial_\phi], \qquad E_1 = \frac{\sqrt{\varepsilon\Delta}}{\rho}\partial_r.$$

The directions determined by $E_0 \pm E_1$ are unchanged by nonzero scalar multiplication, and in this context ε times $\pm$ is still just $\pm$. So we multiply both expressions above by $\rho/\sqrt{\varepsilon\Delta}$ and rearrange signs to conclude that the null directions are given by

$$\partial_r \pm (\varepsilon/\Delta)\big[(r^2 + a^2)\partial_t + a\partial_\phi\big] \approx \pm\partial_r + (1/\Delta)\big[(r^2 + a^2)\partial_t + a\partial_\phi\big]$$
$$= \pm\partial_r + \Delta^{-1}V.$$

$\square$

In the Ricci-flat case the type D eigenvalues are related to sectional curvature as follows.

Corollary 5.6.5   *Let Π be the principal plane at a point of a Ricci-flat spacetime of type D, and let λ be the repeated eigenvalue of $\mathbb{R}$. Then*
(1) $K(\Pi) = K(\Pi^\perp) = -2\,\mathrm{Re}\,\lambda$, where K denotes sectional curvature, and
(2) if $u \in \Pi$ and $x \in \Pi^\perp$ are unit vectors (x necessarily spacelike), then $K(\mathrm{span}\{u, x\}) = \mathrm{Re}\,\lambda$.

Proof.   Let e_0, e_1, e_2, e_3 be a canonical frame at the point.
(1) Then e_2, e_3 is a basis for $\Pi^\perp$, with $<e_2 \wedge e_3, e_2 \wedge e_3> = +1$. Since $e_2 \wedge e_3$ is a -2λ eigenvector of $\mathbb{R}$ on Λ^2_{cx}, Lemma 5.6.2 yields

$$K(\Pi) = R(e_2, e_3, e_2, e_3) = <\mathbb{R}(e_2 \wedge e_3), e_2 \wedge e_3> = a_1 = -2a = -2\,\mathrm{Re}\lambda.$$

(For Kerr, where $\mathrm{Re}\,\lambda = -\mathsf{I}$, this gives $K(\Pi) = 2\mathsf{I}$ as in Chapter 2.) By Corollary 5.3.4 or a direct computation, $K(\Pi^\perp)$ is the same.
(2) If u is timelike, we can suppose $u = e_0$ and $x = e_2$. Then for $P = \mathrm{span}\{u, x\}$, $K(P) = -<\mathbb{R}(e_0 \wedge e_2), e_0 \wedge e_2> = -(-a_2) = a = \mathrm{Re}\,\lambda$. For u spacelike we can use e_1 and e_2, for which $K(P) = <\mathbb{R}(e_1 \wedge e_2), e_1 \wedge e_2> = a_3 = a = \mathrm{Re}\,\lambda$. $\square$

The Petrov view of Kerr spacetime has an important byproduct

Corollary 5.6.6   *Isometries of Kerr spacetime preserve the functions r and $C = \cos\vartheta$.*

Proof.   Let $\psi\colon M \to M$ be an isometry. For any smooth map the universal property of exterior products shows that for $m = 2, 3, 4$, the differential map $d\psi$ of ψ induces a linear map $\widehat{\psi} = \Lambda^m d\psi\colon \Lambda^m(p) \to \Lambda^m(\psi p)$, where $\Lambda^m(q) = \Lambda^m(T_q(M))$. Assume as usual that M is oriented by a field ω of unit four-vectors. Then $\widehat{\psi} = \pm\omega$, where the sign is constant on M.

We can suppose that ψ is orientation-preserving, that is, $\widehat{\psi}(\omega) = +\omega$, for otherwise, following ψ by the (orientation-reversing) equatorial isometry ϵ makes $\epsilon \circ \psi$ orientation-preserving. Then since ϵ preserves r and C, if the result holds for $\epsilon \circ \psi$, it holds for ψ.

The definition of Hodge star now shows that for all $p \in M$ the induced map $\widehat{\psi} = d\psi \wedge d\psi \colon \Lambda^2(p) \to \Lambda^2(\psi p)$ commutes with $*$, hence is complex linear.

Isometries preserve Riemannian curvature operators in the sense that

$$d\psi(R_{vw}(x)) = R_{d\psi v, d\psi w}(d\psi x) \quad \text{for all} \quad v, w, x \in T_p(M), \, p \in M.$$

It follows mechanically that $\widehat{\psi}$ commutes with the curvature transformations, that is, $\widehat{\psi} \circ \mathbb{R} = \mathbb{R} \circ \widehat{\psi} \colon \Lambda^2(p) \to \Lambda^2(\psi p)$. Similarly, $\widehat{\psi}$ preserves the Weyl tensor and commutes with $\mathbb{C}$ (obvious in the present case since $\mathbb{R} = \mathbb{C}$). Consequently, $\widehat{\psi}$ preserves the eigenvectors and ψ the eigenvalues of $\mathbb{C}$. In particular, since M has type D, ψ preserves the repeated eigenvalue λ (a complex-valued function on M). Thus, $\lambda(\psi p) = \lambda(p)$ for all $p \in M$, where $\lambda = -\mathrm{M}/(r + iaC)^3$.

Write $\tilde{r}$ for $r \circ \psi$, and $\tilde{C}$ for $C \circ \psi$, where $C = \cos \vartheta$. Then $(r + iaC)^3 = (\tilde{r} + ia\tilde{C})^3$. Hence $\tilde{r} + ia\tilde{C} = k(r + iaC)$, where k (constant by continuity) is a cube root of 1, namely, 1 or $(-1 + \sqrt{3}i)/2$. The complex roots are impossible since C is bounded but r is not. Thus $\tilde{r} + ia\tilde{C} = r + iaC$, that is, $\tilde{r} = r$ and $\tilde{C} = C$. $\qquad\square$

The Petrov approach has produced principal null directions in the individual vector spaces $V = T_p(M)$ of a spacetime M, and in favorable cases these assemble into one-dimensional distributions on M. The family of integral curves of such a distribution is a foliation of M (Section 1.7) called a *principal null congruence*. The rest of this chapter relates the Petrov type of M to the geometric properties of these congruences.

5.7   The Optical Scalars

The optical scalars are functions that describe geometric properties of a null geodesic congruence, say, a flow of light through a spacetime. They were introduced by R. K. Sachs (1961) as lightlike analogues of the invariants of a steady-state fluid flow whose "molecules" are material particles. (For their initial physical application to the study of gravitational radiation; see Pirani, 1962, and Sachs, 1964.) Such a geodesic null congruence can be described, locally at least, by a future-pointing null vector field k for which $\nabla_k k = 0$.

Since $<k,k> = 0$ we have $\nabla_v k \perp k$ for all vector fields v, so (at each point) the covariant derivatives of k all lie in the three-dimensional subspace $k^\perp$. Since $\nabla_k k = 0$, the covariant differential ∇k is completely determined by its restriction to $k^\perp$, that is, by the map $v \to \nabla_v k$ from $k^\perp$ to $k^\perp$. In fact, since $\nabla_{v+ck} k = \nabla_v k$, this map can be considered as a linear operator on the quotient vector space $k^\perp/k$ in which vectors are identified if they differ by a multiple of k. The metric tensor is also well-defined on $k^\perp/k$, since $<x + ak, y + bk> = <x,y>$ for $x, y \perp k$. Thus, at each point, $k^\perp/k$ is a two-dimensional inner product space, regarded as a "screen" showing cross sections of a beam of light from the congruence given by k.

The covariant differential ∇k is a quadratic approximation of the flow of k in the sense that given a point p, the value of k at a nearby point, say, $p + \varepsilon x$, is approximately $k + \varepsilon \nabla_x k$. Denote by D the linear operator on $k^\perp/k$ determined by $v \to \nabla_v k$. We use D to describe the infinitesimal change in a small circular disk $C = \{v : |v| \leq \varepsilon\}$ in $k^\perp/k$ under the flow of k. We can imagine that the screen will show how this disk evolves through time. (The screens at different point of a ray of k can be canonically identified using parallel translation along γ.)

This evolution is described by three near-invariants as follows. Let x, y be an orthonormal basis for $k^\perp/k$, and write, for example, $k_{y,x} = <\nabla_x k, y>$. Then the operator $D \approx \nabla k$ on $k^\perp/k$ has matrix

$$
\begin{bmatrix}
k_{x,x} & k_{y,x} \\
k_{x,y} & k_{y,y}
\end{bmatrix}.
$$

The optical scalars are expressed in terms of this matrix.

Definition 5.7.1 *Let k be a null geodesic vector field on a spacetime. Relative to an orthonormal basis x, y for $k^\perp/k$, the* optical scalars *of k are*

1. expansion $\theta = \frac{1}{2}(k_{x,x} + k_{y,y})$
2. rotation *(or* twist*)* $\omega = \frac{1}{2}(k_{y,x} - k_{x,y})$
3. complex shear $= (1/2)(k_{y,y} - k_{x,x}) + (i/2)(k_{x,y} + k_{y,x})$.

Here θ is the infinitesimal measure of increase (decrease) of the area of the disk C under the flow (see Figure 5.3). The rotation ω describes the skew-symmetric part of D, namely, $A = \frac{1}{2}(D - D^t)$, hence represents an infinitesimal rotation of the disk. Thus $\theta = \frac{1}{2}\text{trace}\, D$ is fully invariant, and ω is invariant up to sign since $\omega^2 = \det A$.

Since k is null, θ is actually half the divergence of k, while ω is a close relative of the curl of k. In fact, k is *hypersurface-orthogonal* (that is, $k^\perp$ is integrable) *if and only if $\omega = 0$*. To see this, note first that

$$\omega = 0 \Leftrightarrow <\nabla_y k, x> = <\nabla_x k, y> \Leftrightarrow <k, \nabla_y x> = <k, \nabla_x y>$$
$$\Leftrightarrow <k, [x, y]> = 0.$$

This suffices since $k^\perp = \mathrm{span}\{k, x, y\}$ and

$$<k, [k, x]> = <k, \nabla_k x> - <k, \nabla_x k> = <\nabla_k k, x> - \frac{1}{2}x<k, k> = 0.$$

The complex shear σ of the null congruence describes the tracefree symmetric part of D, that is, $T = \frac{1}{2}(D + D^t) - \theta I$, which has matrix

$$\frac{1}{2}\begin{bmatrix} k_{x,x} - k_{y,y} & k_{y,x} + k_{x,y} \\ k_{x,y} - k_{y,x} & k_{y,y} + k_{x,x} \end{bmatrix}.$$

Since T is a symmetric operator with trace zero, it represents a distortion of the circular disk C into an elliptical region of the same area (see Figure 5.3). Thus

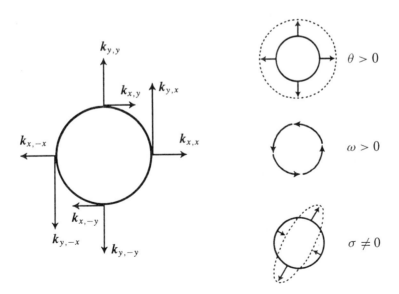

FIGURE 5.3. Infinitesimal deformations of the circular disk C, produced by ∇k.

shear σ measures "astigmatic focusing." Although σ itself it not invariant, its absolute value is, since $|\sigma|^2 = -\det T$.

The *Sachs equations* in the following result describe the rate of change of the optical scalars along null geodesics of the congruence.

Proposition 5.7.2   *If k is a geodesic null vector field, then*
  (1a) $k[\omega] = -2\theta\omega$.
  (1b) $k[\theta] = \omega^2 - \theta^2 - |\sigma|^2 - \frac{1}{2}Ric(k,k)$.
  (S2) $k[\sigma] = -2\theta\sigma + \frac{1}{2}(R_{kxkx} - R_{kyky}) - iR_{kxky}$, *where x and y are parallel orthonormal vector fields $\perp k$ and parallel in the k direction.*

(We can omit the proof since more general results are proved in the next section.)

The Sachs equations are rich in geometric information. Equation (1a) shows that (positive) expansion decreases rotation—as might be expected from conservation of angular momentum. Furthermore, if the rotation is ever zero along a ray it is identically zero. Equation (1b) shows that expansion is increased by rotation ("centrifugal force"), but decreased by expansion itself, shear, and matter—the latter represented relativistically by Ricci curvature (see Section 1.6).

Finally, equation (S2) shows that (positive) expansion tends to reduce shear. The curvature term here will be clarified in the next section.

The following example gives an intuitive base for comparison with the principle null congruences of Kerr spacetime.

Example 5.7.3   In Euclidean 3-space $\mathbf{R}^3$ a light source at the origin emits rays that, in terms of spherical coordinates, are the integral curves of ∂_r. In terms of special relativity, the rays γ are the integral curves of the geodesic null vector field $k = \partial_r + \partial_t$ on Minkowski spacetime $\mathbf{R}_1^4 = \mathbf{R}^3 \times \mathbf{R}_1^1$. A vector field x orthogonal to k and parallel along rays represents a point of the screen along γ. Under the projection π of $\mathbf{R}_1^4$ onto $\mathbf{R}^3$ the screen projects isomorphically onto the family of (canonically identified) planes in $\mathbf{R}^3$ orthogonal to the Euclidean ray $\pi \circ \gamma$.

A simple computation shows that $\nabla_x k = x/r$ for all such x. Hence the optical scalars are $\theta = 1/r$, $\omega = 0$, $\sigma = 0$. Thus, as expected, a beam of light with cross sections initially circular maintains this property. At radius r the cross-sectional area of such a beam is proportional to r^2 (for some fixed solid angle). As a divergence, θ gives rate of change per unit area, thus the rate of expansion of cross-sectional area is proportional not to $1/r$ but to r, again as expected.

The only nontrivial Sachs equation is (1a): $k[\theta] = -\theta^2$, verified by $k[\theta] = (\partial_r + \partial_t)[1/r] = \partial_r(1/r) = -1/r^2 = -\theta^2$. The integral hypersurfaces of $k^\perp$ (guaranteed by $\omega = 0$) are just the lightcones with vertices on the t-axis.

Now we consider Kerr spacetime. The principal null congruences are given (off the horizons) by the null vector fields $k_\pm = \pm\partial_r + \Delta^{-1}V$, which we know are geodesic. In terms of the Boyer–Lindquist frame field E_0, E_1, E_2, E_3 (Section 2.6), $k_\pm = (\rho/\sqrt{\varepsilon\Delta})(E_0 \pm E_1)$. Choosing $x = E_2$, $y = E_3$ as basis vectors for the screen gives

$$\nabla_x(k_\pm) = E_2[\rho/\sqrt{\varepsilon\Delta}](E_0 \pm E_1) + (\rho/\sqrt{\varepsilon\Delta})(\nabla_{E_2}E_0 \pm \nabla_{E_2}E_1)$$

where $\varepsilon = \mathrm{sgn}\Delta$. The first summand here can be ignored since only the $x = E_2$ and $y = E_3$ components are needed. Then the covariant derivative formulas in Corollary 2.6.7 give

$$\nabla_x(k_\pm) = \rho^{-2}(\pm rE_2 + \varepsilon aCE_3) + \text{(terms in } E_0 \text{ and } E_1).$$

Similarly,

$$\nabla_y(k_\pm) = \rho^{-2}(-\varepsilon aCE_2 \pm rE_3) + \text{(terms in } E_0 \text{ and } E_1).$$

Thus the matrix of $D = \nabla k$ on $k^\perp/k$ relative to the orthonormal basis x, y is

$$\begin{bmatrix} k_{x,x} & k_{y,x} \\ k_{x,y} & k_{y,y} \end{bmatrix} = \begin{bmatrix} \pm r/\rho^2 & \varepsilon aC/\rho^2 \\ -\varepsilon aC/\rho^2 & \pm r/\rho^2 \end{bmatrix}.$$

Hence $k_\pm$ has

$$\text{expansion } \theta_\pm = \pm r/\rho^2, \qquad \text{rotation } \omega_\pm = \varepsilon aC/\rho^2, \qquad \text{shear } \sigma = 0.$$

Thus the rotation of the gravitational source of Kerr spacetime produces rotation in the principle null geodesics, except for those in the equator. The sense of rotation is *opposite* in the northern and southern hemispheres. The nonequatorial principal nulls have θ change sign as they go through $r = 0$. For example, principal nulls from $r = -\infty$ have $\theta < 0$ as they converge toward the core of the black hole, but thereafter, for $r > 0$, they deserve their name "outgoers" and have $\theta > 0$ as in the corresponding Minkowski case. As $r \to \infty$ the optical scalars approach their Minkowski values.

Note that in Schwarzschild spacetime, invoked by $a = 0$, the principal nulls have optical scalars $\theta = 1/r$, $\omega = \sigma = 0$, just as in the Newton/Minkowki analogue in Example 5.7.3.

Since the Kerr principal nulls are shearfree—and Kerr spacetime is Ricci-flat—the first pair of Sachs equations become $k[\omega] = -2\theta\omega$, $k[\theta] = \omega^2 - \theta^2$, which may be tested on the preceding expressions for θ, ω.

The fact that equation (S2) is complex suggests the possiblity of combining (1a) and (1b) into a single complex equation. In fact, setting $\rho = -\theta + i\omega$ gives

$$k[\rho] = -k[\theta] + ik[\omega] = \theta^2 - 2i\theta\omega - \omega^2 + |\sigma|^2 + \tfrac{1}{2}\mathrm{Ric}(k, k).$$

Hence we obtain the complex equation (S1), namely

$$k[\rho] = \rho^2 + |\sigma|^2 + \tfrac{1}{2}\mathrm{Ric}(k, k).$$

The complex Sachs equations (S1) and (S2) serve to introduce the extensive generalization in the next section.

5.8   Newman–Penrose Formalism

In the study of null vectors on a spacetime it is natural to make use of *real tetrad fields* k, ℓ, x, y (Definition 5.5.4) where k and ℓ are null with $<k, \ell> = -1$, and x and y are unit vector fields orthogonal to each other and to k and ℓ. One further step produces a more concise notation: Define complex vector fields

$$m = (x - iy)/\sqrt{2} \qquad \overline{m} = (x + iy)/\sqrt{2}.$$

Then $k, \ell, m, \overline{m}$ is called a *complex null tetrad*. In fact, since $\{x, y\}$ is orthonormal, m and $\overline{m}$ are null, that is, $<m, m> = <\overline{m}, \overline{m}> = 0$. Also, $<m, \overline{m}> = +1$, so the metric components of such a tetrad are

$$(g_{ij}) = \begin{bmatrix} 0 & -1 & 0 & 0 \\ -1 & 0 & 0 & 0 \\ 0 & 0 & 0 & 1 \\ 0 & 0 & 1 & 0 \end{bmatrix}.$$

Ordinary (real) tensors now have complex tetrad components; for example, if A has type (0,2), then

$$A(k, \overline{m}) = A(k, (x + iy)/\sqrt{2}) = (1/\sqrt{2})\big[A(k, x) + iA(k, y)\big].$$

The usual symmetry properties of various curvature tensors remain valid. Contraction is given by the usual formula since

$$CA = \Sigma g^{ab} A_{ab} = -A(k, \ell) - A(\ell, k) + A(m, \overline{m}) + A(\overline{m}, m)$$

and the last two summands are

$$\tfrac{1}{2}\big[A(x - iy, x + iy) + A(x + iy, x - iy)\big] = A(x,x) + A(y,y).$$

The *Newman–Penrose formalism* is powerful notational system for expressing the geometry of a spacetime in terms of complex null tetrad fields. Ordinary notations such as covariant derivative ∇ and curvature tensor R are no longer appear as such: ∇ is replaced by analogues of the Christoffel symbols, and all tensors are replaced by their tetrad components. Furthermore, index notation is not used; equations are written out in full. Consequently, when the definition of curvature, its Ricci/Weyl decomposition, the Bianchi identities, etc., are so expressed, large arrays of equations necessarily result (Chandrasekhar 1983; Kramer et al. 1980; Newman, Penrose 1962). Nevertheless, this approach has certain substantial advantages over classical descriptions. For example, it provides a simple characterization of the multiplicity m of a principal null vector, making it easy to apply the differential equations of the formalism to the study of Petrov types. Furthermore, these differential equations, though numerous, involve only *first* derivatives $v[f]$ of complex-valued functions f.

We present a substantial part of the Newman–Penrose formalism, retaining some classical notation. The first step is to describe tetrad covariant derivatives by analogues of Christoffel symbols.

Definition 5.8.1 *The* spin coefficients *of a null tetrad field* $k, \ell, m, \overline{m}$ *are the complex-valued functions*

$$\kappa = -<\nabla_k k, m> \qquad \nu = <\nabla_\ell \ell, \overline{m}> \qquad \varepsilon = \tfrac{1}{2}\big[-<\nabla_k k, \ell> + <\nabla_k m, \overline{m}>\big]$$

$$\rho = -<\nabla_{\overline{m}} k, m> \qquad \mu = <\nabla_m \ell, \overline{m}> \qquad \gamma = \tfrac{1}{2}\big[<\nabla_\ell \ell, k> - <\nabla_\ell \overline{m}, m>\big]$$

$$\sigma = -<\nabla_m k, m> \qquad \lambda = <\nabla_{\overline{m}} \ell, \overline{m}> \qquad \beta = \tfrac{1}{2}\big[-<\nabla_m k, \ell> + <\nabla_m m, \overline{m}>\big]$$

$$\tau = -<\nabla_\ell k, m> \qquad \pi = <\nabla_k \ell, \overline{m}> \qquad \alpha = \tfrac{1}{2}\big[<\nabla_{\overline{m}} \ell, k> - <\nabla_{\overline{m}} \overline{m}, m>\big]$$

Context will distinguish this rare use of λ from its general use as an eigenvalue. The alphabetically irregular arrangement above displays the fact that building a new tetrad by the *tetrad reversal* $k \leftrightarrow \ell, m \leftrightarrow \overline{m}$ produces the exchanges

$$\kappa \leftrightarrow -\nu, \rho \leftrightarrow -\mu, \sigma \leftrightarrow -\lambda, \tau \leftrightarrow -\pi \quad \text{and} \quad \varepsilon \leftrightarrow -\gamma, \beta \leftrightarrow -\alpha$$

(This reversal is further developed in Geroch et al. 1973.) Note that reversing m and $\overline{m}$ only is the same taking complex conjugates.

For any vector fields v, w in a tetrad, $<v, w>$ is constant, hence the expressions $<\nabla_u v, w>$ are skew-symmetric in v and w. It follows, using complex conjugacy, that every $<\nabla_u v, w>$ can be expressed in terms of spin-coefficients. For example, we find $<\nabla_m \ell, k> = \bar{\alpha} + \beta$.

Then systematic use of *tetrad expansion*,

$$v = -<v, \ell>k - <v, k>\ell + <v, \bar{m}>m + <v, m>\bar{m},$$

leads to the following description of the covariant derivative.

Corollary 5.8.2 *A tetrad field k, ℓ, m, $\bar{m}$ has covariant derivatives*

$$\nabla_k k = (\varepsilon + \bar{\varepsilon})k - \bar{\kappa}m - \kappa\bar{m} \qquad \nabla_k \ell = -(\varepsilon + \bar{\varepsilon})\ell + \pi m + \bar{\pi}\bar{m}$$
$$\nabla_\ell k = (\gamma + \bar{\gamma})k - \bar{\tau}m - \tau\bar{m} \qquad \nabla_\ell \ell = -(\gamma + \bar{\gamma})\ell + vm + \bar{v}\bar{m}$$
$$\nabla_m k = (\bar{\alpha} + \beta)k - \bar{\rho}m - \sigma\bar{m} \qquad \nabla_m \ell = -(\bar{\alpha} + \beta)\ell + \mu m + \bar{\lambda}\bar{m}$$

$$\nabla_k m = \bar{\pi}k - \kappa\ell + (\varepsilon - \bar{\varepsilon})m \qquad \nabla_m m = \bar{\lambda}k - \sigma\ell + (-\bar{\alpha} + \beta)m$$
$$\nabla_\ell m = \bar{v}k - \tau\ell + (\gamma - \bar{\gamma})m \qquad \nabla_m \bar{m} = \mu k - \bar{\rho}\ell + (\bar{\alpha} - \beta)\bar{m}$$

Taking conjugates gives six more equations for the requisite total of sixteen. Actually even the ten equations above are redundant since the effect of the tetrad reversal $k \leftrightarrow \ell$, $m \leftrightarrow \bar{m}$ lets us read the formula for $\nabla_\ell \ell$ from that for $\nabla_k k$, and so on. Note that the four covariant derivatives involving only the real vector fields k and ℓ are themselves necessarily real.

The formula $[X, Y] = \nabla_X Y - \nabla_Y X$ then gives expressions for the brackets of the tetrad vector fields.

$$[k, \ell] = -(\gamma + \bar{\gamma})k - (\varepsilon + \bar{\varepsilon})\ell + (\pi + \bar{\tau})m + (\bar{\pi} + \tau)\bar{m} \quad \text{(real)}$$
$$[k, m] = (\bar{\pi} - \bar{\alpha} - \beta)k - \kappa\ell + (\varepsilon - \bar{\varepsilon} + \bar{\rho})m + \sigma\bar{m}$$
$$[\ell, m] = \bar{v}k + (\bar{\alpha} + \beta - \tau)\ell + (\gamma - \bar{\gamma} - \mu)m - \bar{\lambda}\bar{m}$$
$$[m, \bar{m}] = (\mu - \bar{\mu})k + (\rho - \bar{\rho})\ell + (\bar{\beta} - \alpha)m + (\bar{\alpha} - \beta)\bar{m} \quad \text{(imaginary)}$$

The equation $\nabla_k k = (\varepsilon + \bar{\varepsilon})k - \bar{\kappa}m - \kappa\bar{m}$ shows at once that k is

$$\text{pregeodesic} \Leftrightarrow \kappa = 0, \qquad \text{geodesic} \Leftrightarrow \kappa = \varepsilon + \bar{\varepsilon} = 0.$$

The spin coefficients include the optical scalars θ, ω, σ in the following way. Suppose that $\mathbf{m} = (1/\sqrt{2})(\mathbf{x} - i\mathbf{y})$ and $\overline{\mathbf{m}} = (1/\sqrt{2})(\mathbf{x} + i\mathbf{y})$, as above. Then

$$
\begin{aligned}
\rho &= -<\nabla_{\overline{m}}\mathbf{k}, \mathbf{m}> = -\tfrac{1}{2}<\nabla_{(x+iy)}\mathbf{k}, (\mathbf{x} - i\mathbf{y})> \\
&= -\tfrac{1}{2}\big[<\nabla_x\mathbf{k}, \mathbf{x}> + <\nabla_y\mathbf{k}, \mathbf{y}> + i(-<\nabla_x\mathbf{k}, \mathbf{y}> + <\nabla_y\mathbf{k}, \mathbf{x}>)\big] \\
&= -(1/2)(k_{x,x} + k_{y,y}) + (i/2)(k_{y,x} - k_{x,y}).
\end{aligned}
$$

The formulas for ω and θ in Definition 5.7.1 show that this is exactly $-\theta + i\omega$. Expanding the present definition of σ gives

$$
\begin{aligned}
\sigma &= -<\nabla_m\mathbf{k}, \mathbf{m}> = -\tfrac{1}{2}<\nabla_{(x-iy)}\mathbf{k}, (\mathbf{x} - i\mathbf{y})> \\
&= (1/2)\big[-<\nabla_x\mathbf{k}, \mathbf{x}> + <\nabla_y\mathbf{k}, \mathbf{y}>\big] + (i/2)\big[(<\nabla_x\mathbf{k}, \mathbf{y}> + <\nabla_y\mathbf{k}, \mathbf{x}>)\big].
\end{aligned}
$$

This is the complex shear as previously defined.

In working with tetrads we naturally want to fit them to the geometry of the problem. A common case is the study of a null vector field $\mathbf{k}$. We want to choose the remaining vectors $\boldsymbol{\ell}, \mathbf{m}, \overline{\mathbf{m}}$ so as to simplify spin coefficients. Often this is done by modifying some initial tetrad, so we must see how spin coefficients change under a change in tetrads. Consider first the change in $\mathbf{k}, \boldsymbol{\ell}, \mathbf{m}, \overline{\mathbf{m}}$ that leaves $\mathbf{k}$ unchanged and in real terms merely adds a scalar multiple of $\mathbf{k}$ to $\mathbf{x}$ and $\mathbf{y}$.

Lemma 5.8.3 *Let $\mathbf{k}, \boldsymbol{\ell}, \mathbf{m}, \overline{\mathbf{m}}$ be a tetrad field on M. Given any complex-valued function f on M, there is a unique tetrad field with $\mathbf{k}_1 = \mathbf{k}$ and $\mathbf{m}_1 = \mathbf{m} + f\mathbf{k}$. Conversely, given any null vector $\boldsymbol{\ell}_1$ with $<\mathbf{k}, \boldsymbol{\ell}_1> = -1$ there is a unique function f such that $\mathbf{k}, \boldsymbol{\ell}_1, \mathbf{m} + f\mathbf{k}, \overline{\mathbf{m}} + \overline{f}\mathbf{k}$ is a tetrad field, and then*

$$
\boldsymbol{\ell}_1 = |f|^2\mathbf{k} + \boldsymbol{\ell} + \overline{f}\mathbf{m} + f\overline{\mathbf{m}}.
$$

Proof. Given f, conjugation of $\mathbf{m}_1$ as above gives $\overline{\mathbf{m}}_1 = \overline{\mathbf{m}} + \overline{f}\mathbf{k}$. Then $\mathbf{m}_1$ and $\overline{\mathbf{m}}_1$ are null and have scalar product $+1$. Now write $\boldsymbol{\ell}_1 = a\mathbf{k} + b\boldsymbol{\ell} + c\mathbf{m} + d\overline{\mathbf{m}}$. Then

$$
<\boldsymbol{\ell}_1, \mathbf{k}> = -1 \Rightarrow b = 1, \qquad <\boldsymbol{\ell}_1, \mathbf{m}_1> = 0 \Rightarrow d = f,
$$
$$
<\boldsymbol{\ell}_1, \overline{\mathbf{m}}_1> = 0 \Rightarrow c = \overline{f}.
$$

Finally, $<\boldsymbol{\ell}_1, \boldsymbol{\ell}_1> = 0$ implies $a = f\overline{f} = |f|^2$, hence $\boldsymbol{\ell}_1 = |f|^2\mathbf{k} + \boldsymbol{\ell} + \overline{f}\mathbf{m} + f\overline{\mathbf{m}}$. Now $\mathbf{k}_1, \boldsymbol{\ell}_1, \mathbf{m}_1, \overline{\mathbf{m}}_1$ is the required tetrad field.

Conversely, for ℓ_1 as specified, the required complex-valued function f such that $0 = <\ell_1, m + fk> = <\ell_1, m> - f$ is evidently just $f = <\ell_1, m>$.   □

The spin coefficients of the new tetrad can readily be computed. For example, in the next section we need to know how $\tau = <\nabla_\ell k, m>$ changes. Consider

$$\tau_1 = -<\nabla_{\ell_1} k, m_1> = -<|f|^2 \nabla_k k + \nabla_\ell k + \bar{f}\nabla_m k + f\nabla_{\bar{m}} k, m + fk>.$$

On the right, fk can be discarded since the scalar products with k produce only zeros, so the new tetrad has

$$\tau_1 = |f|^2 \kappa + \tau + \bar{f}\sigma + f\rho.$$

The change in Lemma 5.8.3 is said to be of Class I. Class II is the corresponding change with ℓ in the role of k.

Finally, Class III changes k and ℓ by a *boost*: $k_1 = gk$, $\ell_1 = (1/g)\ell$ and changes m and $\bar{m}$ by a *rotation*: $m_1 = e^{i\vartheta}m$, $\bar{m}_1 = e^{-i\vartheta}\bar{m}_1$, where $g \neq 0$ and ϑ are real-valued functions on M.

It is easy to check that any orientation-preserving change of tetrads can be expressed in terms of these three classes.

Remark 5.8.4 Effects of tetrad change on spin coefficients:

CLASS I. It is immediate from their definitions that κ is invariant, and $\rho, \sigma, \varepsilon$ change only by addition of a multiple of κ. Thus, if $\kappa = 0$, the spin coefficients $\rho, \sigma, \varepsilon$ are also invariant. The changes in $\tau, \alpha, \beta, \gamma$ are algebraic and generally follow the pattern of the formula for τ. However, the changes in π, λ, μ, ν involve derivatives (see Newman, Penrose 1962, or Chandrasekhar, 1983).

CLASS II. Analogous to Class I, with k and ℓ interchanged.

CLASS III. $\kappa, \rho, \sigma, \tau$ and π, λ, μ, ν change only by multiplication by nonzero functions. In particular, changing only by a boost $k_1 = gk$, $\ell_1 = (1/g)\ell$ leaves π and τ invariant, while

$$\kappa_1 = g^2\kappa, \rho_1 = g\rho, \sigma_1 = g\sigma, \qquad \lambda_1 = \lambda/g, \mu_1 = \mu/g, \nu_1 = \nu/g^2.$$

A rotation alone leaves ρ and μ invariant and does not change the absolute values of the other six simpler cases in Definition 5.8.1. However, $\gamma, \varepsilon, \alpha, \beta$, are more complicated. For example, under a boost by g and rotation by ϑ,

$$\begin{aligned}
\varepsilon_1 &= \tfrac{1}{2}\big[-<\nabla_{gk}(gk), \ell/g> + <\nabla_{gk}(e^{i\vartheta}m), e^{-i\vartheta}\bar{m}>\big] \\
&= \tfrac{1}{2}\big[-<\nabla_k(gk), \ell> + g<\nabla_k(m), \bar{m}> + ge^{-i\vartheta}k[e^{i\vartheta}]<m, \bar{m}>\big] \\
&= g\varepsilon + \tfrac{1}{2}k[g] + \tfrac{1}{2}igk[\vartheta].
\end{aligned}$$

Now we turn to Newman–Penrose treatment of curvature, in particular, its decomposition into Weyl and Ricci parts and its expression in terms of spin coefficients.

The Weyl curvature tensor C of a spacetime has a concise expression.

Definition 5.8.5 *Let* $k, \ell, m, \overline{m}$ *be a null tetrad field in a spacetime. The* Weyl scalars *of the tetrad are the complex-valued functions*

$$\psi_0 = C_{kmkm}, \ \psi_1 = C_{k\ell km}, \ \psi_2 = C_{km\overline{m}\ell}, \ \psi_3 = C_{\ell k\ell\overline{m}}, \ \psi_4 = C_{\ell\overline{m}\ell\overline{m}}$$

where C *is the Weyl tensor and, for example,* $C_{k\ell m\overline{m}} = C(k, \ell, m, \overline{m})$.

Note that tetrad reversal $k \leftrightarrow \ell, m \leftrightarrow \overline{m}$ produces the reversals

$$\psi_0 \leftrightarrow \psi_4, \ \psi_1 \leftrightarrow \psi_3, \ \text{with } \psi_2 \text{ unchanged.}$$

In later computations we need to know how the tetrad components of the Weyl tensor depend on these scalars.

Lemma 5.8.6 *The components* $C_{(xy)(vw)}$ *of the Weyl tensor relative to a null tetrad* $k, \ell, m, \overline{m}$, *are expressed in terms of the Weyl scalars as follows:*

	$k\ell$	km	$k\overline{m}$	ℓm	$\ell\overline{m}$	$m\overline{m}$
$k\ell$	$\psi_2 + \overline{\psi}_2$	ψ_1	$\overline{\psi}_1$	$-\overline{\psi}_3$	$-\psi_3$	$-\psi_2 + \overline{\psi}_2$
km		ψ_0	0	0	$-\psi_2$	$-\psi_1$
$k\overline{m}$			$\overline{\psi}_0$	$-\overline{\psi}_2$	0	$\overline{\psi}_1$
ℓm				$\overline{\psi}_4$	0	$-\overline{\psi}_3$
$\ell\overline{m}$					ψ_4	ψ_3
$m\overline{m}$						$\psi_2 + \overline{\psi}_2$

(Since $C_{(xy)(vw)}$ *is symmetrical by pairs, subdiagonal entries are omitted.)*

Proof. Because C is tracefree, for any vectors v, w from the tetrad,

$$0 = \Sigma g^{ab} C_{avbw} = -C_{kv\ell w} - C_{\ell vkw} + C_{mv\overline{m}w} + C_{\overline{m}vmw}.$$

Various choices of v and w lead to

1. $C_{kmk\overline{m}} = C_{\ell m\ell\overline{m}} = C_{km\ell m} = 0$.
2. $C_{k\ell k\ell} = C_{m\overline{m}m\overline{m}} = \psi_2 + \overline{\psi}_2$. (real)
3. $C_{km\overline{m}m} = \psi_1$, and $C_{\ell m\overline{m}m} = \psi_3$.

The cyclic symmetry $C_{k\ell m\bar{m}} + C_{km\bar{m}\ell} + C_{k\bar{m}\ell m} = 0$ gives

4. $C_{k\ell m\bar{m}} = -\psi_2 + \bar{\psi}_2$. (imaginary).

These relations together with complex conjugation suffice to determine all components of C.   □

The principal null directions are elegantly expressed in terms of Weyl scalars.

Lemma 5.8.7 *Let k be a principal null vector of multiplicity m. Then for one (hence every) tetrad field $k, \ell, m, \bar{m}$,*

$$m \geq 1 \Leftrightarrow \psi_0 = 0,$$
$$m \geq 2 \Leftrightarrow \psi_0 = \psi_1 = 0,$$
$$m \geq 3 \Leftrightarrow \psi_0 = \psi_1 = \psi_2 = 0,$$
$$m \geq 4 \Leftrightarrow \psi_0 = \psi_1 = \psi_2 = \psi_3 = 0$$

(Thus, for example, $m = 2 \Leftrightarrow \psi_0 = \psi_1 = 0$ and $\psi_2 \neq 0$.)

Proof. This result is a direct translation of Proposition 5.5.5.

(1) If $m \geq 1$, then by the proposition, $C_{kxkx} = C_{kyky} = C_{kxky} = 0$. But

$$\psi_0 = C_{kmkm} = C_{kxkx} - C_{kyky} + 2iC_{kxky},$$

hence $\psi_0 = 0$. Conversely, if $\psi_0 = 0$, then $C_{kxkx} - C_{kyky} = C_{kxky} = 0$. Because C is tracefree, $C_{kxkx} + C_{kyky} = 0$. Consequently, the function $C_{k \cdot k \cdot}$ is identically zero on $k^{\perp} = \text{span}\{k, x, y\}$, so by the proposition, $m \geq 1$.

(2) If $m \geq 2$, then by the proposition,

$$C_{kxkx} = C_{kyky} = C_{kxky} = C_{kxk\ell} = C_{kyk\ell} = 0.$$

But $\psi_1 = C_{k\ell km} = C_{k\ell kx} - iC_{k\ell ky}$, hence both ψ_0 and ψ_1 are zero.

Conversely, suppose $\psi_0 = \psi_1 = 0$. The latter equality gives $C_{k\ell kv} = 0$ for all $v \in k^{\perp}$, and $\psi_0 = 0$ gives $C_{kxkx} = C_{kyky} = C_{kxky} = 0$. Thus Proposition 5.5.5 yields $m \geq 2$.

The other cases are similar.   □

Remark 5.8.8 In the Newman–Penrose formalism the decomposition of curvature into Weyl and Ricci parts is gotten, as usual, by substitution into the formula near the end of Section 5.1. Recall that $g_{k\ell} = -1$ and $g_{m\overline{m}} = +1$; then, writing $R_{k\ell} = \mathrm{Ric}(k, \ell)$, we find, for example,

$$R_{kmkm} = C_{kmkm} = \psi_0, \qquad R_{k\ell km} = C_{k\ell km} + \tfrac{1}{2}R_{km} = \psi_1 + \tfrac{1}{2}R_{km},$$

$$R_{km\overline{m}\ell} = C_{km\overline{m}\ell} + \tfrac{1}{2}(R_{m\overline{m}} - R_{k\ell}) - S/6 = \psi_2 + S/12,$$

the latter since $S = \mathrm{CRic} = \Sigma g^{ab}R_{ab} = 2(R_{m\overline{m}} - R_{k\ell})$.

Analogous to the classical formulas for Riemannian curvature in terms of Christoffel symbols are tetrad expressions for curvature in terms of spin coefficients. In the Newman–Penrose formalism the emphasis is on derivatives of spin coefficients, and more than the minimum (20 real) equations are canonically supplied. However, we need only five of these curvature identities: the two that generalize the Sachs equations (Section 5.7), and the three that involve the Weyl scalar ψ_1. These are derived from the following formula for Riemannian curvature:

$$\begin{aligned}
R_{xyvw} &= -<R_{xy}v, w> = -<\nabla_x\nabla_y v, w> + <\nabla_y\nabla_x v, w> + <\nabla_{[x,y]}v, w> \\
&= -x<\nabla_y v, w> + <\nabla_y v, \nabla_x w> + y<\nabla_x v, w> \qquad (*) \\
&\quad - <\nabla_x v, \nabla_y w> + <\nabla_{[x,y]}v, w>.
\end{aligned}$$

Proposition 5.8.9 *For a tetrad field $k, \ell, m, \overline{m}$,*

(1) $k[\rho] - \overline{m}[\kappa] = \rho^2 + \sigma\overline{\sigma} + \rho(\varepsilon + \overline{\varepsilon}) - \overline{\kappa}\tau - \kappa(3\alpha + \overline{\beta} - \pi) + \tfrac{1}{2}\mathrm{Ric}(k, k).$

(2) $k[\sigma] - m[\kappa] = \sigma(\rho + \overline{\rho} + 3\varepsilon - \overline{\varepsilon}) - \kappa(\tau - \overline{\pi} + \overline{\alpha} + 3\beta) + \psi_0.$

(3) $k[\tau] - \ell[\kappa] = \rho(\tau + \overline{\pi}) + \sigma(\overline{\tau} + \pi) + \tau(\varepsilon - \overline{\varepsilon}) - \kappa(3\gamma + \overline{\gamma}) + \psi_1 + \tfrac{1}{2}\mathrm{Ric}(k, m).$

(4) $k[\beta] - m[\varepsilon] = \sigma(\alpha + \pi) + \beta(\overline{\rho} - \overline{\varepsilon}) - \kappa(\mu + \gamma) - \varepsilon(\overline{\alpha} - \overline{\pi}) + \psi_1.$

(5) $m[\rho] - \overline{m}[\sigma] = \rho(\overline{\alpha} + \beta) - \sigma(3\alpha - \overline{\beta}) + \tau(\rho - \overline{\rho}) + \kappa(\mu - \overline{\mu}) - \psi_1 + \tfrac{1}{2}\mathrm{Ric}(k, m).$

Proof. To generalize the Sachs equation (S1), we must expand $k[\rho] = -k<\nabla_{\overline{m}}k, m>$. Comparing this with the term $-x<\nabla_y v, w>$ in the general formula invites the substitutions $x = k$, $y = \overline{m}$, $v = k$, $w = m$. Then

$$\begin{aligned}
-k<\nabla_{\overline{m}}k, m> + \overline{m}<\nabla_k k, m> &= <\nabla_k k, \nabla_{\overline{m}}m> - <\nabla_{\overline{m}}k, \nabla_k m> \\
&\quad - <\nabla_{[k,\overline{m}]}k, m> + R_{k\overline{m}km}.
\end{aligned}$$

The left side here is $k[\rho] - \overline{m}[\kappa]$. On the right, Corollary 5.8.2 gives

$$\begin{aligned}
<\nabla_k k, \nabla_{\overline{m}}m> &= <(\varepsilon + \overline{\varepsilon})k - \overline{\kappa}m - \kappa\overline{m}, \overline{\mu}k - \rho\ell + (\alpha - \overline{\beta})m> \\
&= \rho(\varepsilon + \overline{\varepsilon}) - \kappa(\alpha - \overline{\beta}).
\end{aligned}$$

Similarly, $-\langle\nabla_{\overline{m}}k, \nabla_k m\rangle = -\kappa(\alpha + \bar{\beta}) + \rho(\varepsilon - \bar{\varepsilon})$. And since from before,

$$[k, \overline{m}] = (\pi - \alpha - \bar{\beta})k - \bar{\kappa}\ell + (\bar{\varepsilon} - \varepsilon + \rho)\overline{m} + \bar{\sigma}m,$$

we find $-\langle\nabla_{[k,\overline{m}]}k, m\rangle = (\pi - \alpha - \bar{\beta})\kappa - \bar{\kappa}\tau + (\bar{\varepsilon} - \varepsilon + \rho)\rho + \bar{\sigma}\sigma$. Adding these three terms gives $\rho^2 + \sigma\bar{\sigma} + \rho(\varepsilon + \bar{\varepsilon}) - \bar{\kappa}\tau - \kappa(3\alpha + \bar{\beta} - \pi)$. To complete the proof of (1) it remains only to recognize the curvature term $R_{k\overline{m}km}$ as $\frac{1}{2}R_{kk}$.

Similarly, $k[\sigma] = -k\langle\nabla_m k, m\rangle$ leads to equation (2), since, as noted in Remark 5.8.8, $R_{kmkm} = C_{kmkm} = \psi_0$.

Now we consider those curvature identities involving ψ_1. Since $\psi_1 = C_{k\ell km} = C_{km\overline{m}m}$, with Remark 5.8.8 in mind, we apply formula $(*)$ above to $R_{k\ell km}$ and $R_{km\overline{m}m}$ and, after using symmetry by pairs, to $R_{kmk\ell}$ and $R_{\overline{m}mkm}$.

The first of these has the form $R_{k\ell km} = k[\tau] - \ell[\kappa] + \ldots$ and continuing as above we get equation (3) since, as noted earlier, $R_{k\ell km} = \psi_1 + \frac{1}{2}\mathbf{R}_{km}$. Similarly, $R_{\overline{m}mkm}$ produces equation (5).

Because of the double definitions of the spin coefficients α, β, ε, the two remaining expansions have the form

$$R_{kmk\ell} = k[\bar{\alpha} + \beta] - m[\varepsilon + \bar{\varepsilon}] + \ldots$$
$$R_{km\overline{m}m} = -k[\bar{\alpha} - \beta] + m[\bar{\varepsilon} - \varepsilon] + \ldots$$

Hence $\frac{1}{2}[R_{kmk\ell} + R_{km\overline{m}m}] = k[\beta] - m[\varepsilon] + \ldots$, which leads to equation (4), since $R_{k\ell km} = \psi_1 + \frac{1}{2}\mathbf{R}_{km}$ and $R_{km\overline{m}m} = \psi_1 - \frac{1}{2}\mathbf{R}_{km}$.   □

Equations (1) and (2) in Proposition 5.8.9 generalize the Sachs equations (S1) and (S2) in the preceding section. In fact, if k is geodesic, then as mentioned earlier, $\kappa = 0$ and $\varepsilon + \bar{\varepsilon} = 0$. Thus equation (1) reduces to

$$k[\rho] = \rho^2 + \sigma\bar{\sigma} + \frac{1}{2}\mathrm{Ric}(k, k).$$

This is equation (S1) from the end of the preceding section, where $\rho = -\theta + i\omega$. Next, setting $\kappa = 0$ in equation (2) in Proposition 5.8.9 gives

$$k[\sigma] = \sigma(\rho + \bar{\rho} + 3\varepsilon - \bar{\varepsilon}) + \psi_0.$$

For equation (S2), it is assumed that the vector fields x and y—hence m and $\overline{m}$—are parallel in the k direction; hence $\varepsilon - \bar{\varepsilon} = 0$. Thus $\varepsilon = 0$, so the preceding equation becomes

$$k[\sigma] = \sigma(\rho + \bar{\rho}) + \psi_0 = -2\theta\sigma + \psi_0.$$

Here ψ_0 is exactly the curvature term in equation (S2), for we know that $R_{kmkm} = C_{kmkm}$, hence

$$\psi_0 = \tfrac{1}{2} R_{k(x-iy)k(x-iy)} = \tfrac{1}{2}[R_{kxkx} - R_{kyky} - i\,2R_{kxky}].$$

Thus equation (S2) shows that the Weyl scalar ψ_0 produces both expansion and shear, that is, $\psi_0 \neq 0$ implies $\theta \neq 0$ and $\sigma \neq 0$.

5.9   Bianchi Identities and Type D

To establish geometric properties of the principal null congruences in type D spacetimes such as Kerr's—and for other Petrov types as well—the Bianchi identities must be called on (see Section 1.3). These are the deepest of the general properties of curvature, and their full expression in explicit terms makes a formidable array in the Newman–Penrose formalism or any other notational system. Fortunately, we need only the cases in which the Weyl scalars ψ_0, ψ_1, ψ_2 are prominent.

Let $k, \ell, m, \overline{m}$ be a tetrad field and let v, w be arbitrary vector fields. Consider the product-rule tensor expansion

$$k[R(\ell, m, v, w)] = (\nabla_k R)(\ell, m, v, w) + R(\nabla_k \ell, m, v, w) + R(\ell, \nabla_k m, v, w)$$
$$+ R(\ell, m, \nabla_k v, w) + R(\ell, m, v, \nabla_k w). \qquad (*)$$

(We use k, ℓ, m for definiteness; these can be exchanged later on.) In view of the Bianchi identity

$$(\nabla_k R)_{\ell m} + (\nabla_\ell R)_{mk} + (\nabla_m R)_{k\ell} = 0,$$

the covariant derivatives of curvature in the equation $(*)$ can be eliminated by summing over the cyclic permutations of k, ℓ, m. This is the way the Bianchi identities are expressed in Newman–Penrose terms.

In the following derivation of two of the Bianchi identities, Ricci curvature terms are neglected since our applications are to Ricci-flat spacetimes.

Lemma 5.9.1 *For a tetrad field $k, \ell, m, \overline{m}$,*

(1) $k[\psi_1] - \overline{m}[\psi_0] = (\pi - 4\alpha)\psi_0 + 2(2\rho + \varepsilon)\psi_1 - 3\kappa\psi_2 +$ *(Ricci terms).*

(2) $m[\psi_1] - \ell[\psi_0] = (\mu - 4\gamma)\psi_0 + 2(2\tau + \beta)\psi_1 - 3\sigma\psi_2 +$ *(Ricci terms).*

Proof.   The definition of early Weyl scalars suggests setting $v = k$ and $w = m$ in equation ($*$) with the result:

$$k[R(\ell, m, k, m)] = (\nabla_k R)(\ell, m, k, m) + R(\nabla_k \ell, m, k, m) + R(\ell, \nabla_k m, k, m)$$
$$+ R(\ell, m, \nabla_k k, m) + R(\ell, m, k, \nabla_k m).$$

Since we are neglecting Ricci we can suppose Ric $= 0$, so R is replaced by C. Now add over the cyclic permutations of the first three vectors k, ℓ, m. We denote the resulting equation by $B(k\ell m)(km)$:

$$k[C(\ell, m, k, m)] + \ell[C(m, k, k, m)] + m[C(k, \ell, k, m)] =$$
$$C(\nabla_k \ell, m, k, m) + C(\ell, \nabla_k m, k, m) + C(\ell, m, \nabla_k k, m) + C(\ell, m, k, \nabla_k m)$$
$$+ C(\nabla_\ell m, k, k, m) + C(m, \nabla_\ell k, k, m) + C(m, k, \nabla_\ell k, m) + C(m, k, k, \nabla_\ell m)$$
$$+ C(\nabla_m k, \ell, k, m) + C(k, \nabla_m \ell, k, m) + C(k, \ell, \nabla_m k, m) + C(k, \ell, k, \nabla_m m).$$

On the left side (the first line) of this equation, $C(\ell, m, k, m) = 0$, leaving just $-\ell[\psi_0] + m[\psi_1]$, as in equation (2) of this lemma. Using the symmetries of curvature, the right side can be expressed as the sum of five terms

$$a = C(k, m, \nabla_k \ell, m) + 2C(k, m, m, \nabla_\ell k)$$
$$b = C(k, m, \ell, [k, m]) + C(\ell, m, k, \nabla_k m) + C(k, \ell, \nabla_m k, m)$$
$$c = C(\nabla_k k, m, \ell, m)$$
$$d = -2C(k, m, k, \nabla_\ell m) + C(k, m, k, \nabla_m \ell).$$
$$e = C(k, \ell, k, \nabla_m m).$$

Consider the term a. Using first the covariant derivatives in Corollary 5.8.2 and then the Weyl curvature expressions in Lemma 5.8.6 we find

$$a = C(k, m, \overline{m}, m)\overline{\pi} + 2C(k, m, m, k)(\gamma + \overline{\gamma}) + 2C(k, m, m, \overline{m})(-\tau)$$
$$= \psi_1 \overline{\pi} - 2\psi_0(\gamma + \overline{\gamma}) + 2\psi_1 \tau$$
$$= -2(\gamma + \overline{\gamma})\psi_0 + (2\tau + \overline{\pi})\psi_1,$$

Similarly,
$$b = (2\overline{\alpha} + 2\beta - \overline{\pi})\psi_1 - 2\sigma\psi_2 + \sigma\overline{\psi}_2 + \kappa\overline{\psi}_3,$$
$$c = -\kappa\overline{\psi}_3,$$
$$d = (-2\gamma + 2\overline{\gamma} + \mu)\psi_0 + (-\overline{\alpha} - \beta + 2\tau)\psi_1,$$
$$e = -\sigma\psi_2 - \sigma\overline{\psi}_2 + (-\overline{\alpha} + \beta)\psi_1.$$

Hence $a + b + c + d + e = (-4\gamma + \mu)\psi_0 + 2(2\tau + \beta)\psi_1 - 3\sigma\psi_2$, which proves the asserted equation (2).

An analogous computation based on $B(k\overline{m}m)(km)$ gives equation (1).   □

Proposition 5.9.2 *Let M be a Ricci-flat spacetime. If k is a repeated principal null vector field (i.e., one of multiplicity $m \geq 2$) then any tetrad field $k, \ell, m, \overline{m}$ has $\kappa \psi_2 = 0$ and $\sigma \psi_2 = 0$. Thus if k has multiplicity exactly 2, then k is pregeodesic and shearfree.*

Proof. By Lemma 5.8.7, $m \geq 2$ implies $\psi_0 = \psi_1 = 0$, and since Ric $= 0$, the equations in the preceding lemma collapse to $\kappa \psi_2 = 0$ and $\sigma \psi_2 = 0$.   □

This result is already enough to deal with type D spacetimes.

Proposition 5.9.3 *On a type D spacetime M the two principal null congruences are geodesic and shearfree. Furthermore, on M—indeed on any open set in M— these are the only geodesic shearfree congruences.*

Proof. Locally, let k and ℓ be tangent to the principal null directions. Each has multiplicity 2. Adjusting k or ℓ if necessary so that $<k, \ell> = -1$, we can construct a tetrad field $k, \ell, m, \overline{m}$. Lemma 5.8.7 applied to k gives $\psi_0 = \psi_1 = 0$, and tetrad reversal gives $\psi_3 = \psi_4 = 0$. Since M has type D everywhere the Weyl tensor cannot vanish, hence $\psi_2 \neq 0$. Then Proposition 5.9.2 shows that k is pregeodesic and shearfree. The same is true symmetrically for ℓ.

To prove the uniqueness assertion, suppose n is tangent to a geodesic null congruence. If it is shearfree, then the Sachs equation (S2), $k[\sigma] = -2\theta\sigma + \psi_0$, shows at once that $\psi_0 = 0$. Thus n is a principal null congruence. But k and ℓ already have total multiplicity 4, so by the classification in Proposition 5.5.3, n is collinear with either k or ℓ.   □

This result shows that it is no accident that in Kerr spacetime the ingoing and outgoing principal null congruences are geodesic and shearfree.

From the Newman–Penrose viewpoint, type D spacetimes are remarkably simple.

Corollary 5.9.4 *On any type D spacetime if k, ℓ, x, y is a (local) null tetrad field with k and ℓ in the principal null directions, then the spin coefficients $\kappa, \sigma, \nu, \lambda$ vanish as do all the Weyl scalars except ψ_2, which is the repeated eigenvalue of the Weyl tensor.*

(Since the spin coefficient vanishes, λ can be used unambiguously to denote this eigenvalue, which for Kerr spacetime we found to be $\lambda = -\text{M}(r - ia C)^{-3}$.)

Proof. Only the last assertion in the corollary requires verification. (This point is not quite trivial, because it links the eigenstructure view of Petrov type and the Newman–Penrose view.) For such a null tetrad, span$\{k, \ell\}$ is the principal plane Π.

Define $e_0 = (k + \ell)/\sqrt{2}$, $e_1 = (k - \ell)/\sqrt{2}$. These vector fields can be enlarged to an orthonormal frame field e_0, e_1, e_2, e_3. Since span$\{e_0, e_1\}$ is the principal plane, Lemma 5.6.3 asserts that $e_0 \wedge e_1$, and hence $e_2 \wedge e_3$, are eigenvectors of $\mathbb{C}$ belonging to its nonrepeated eigenvalue -2λ Thus $e_3 \wedge e_1$ and $e_1 \wedge e_2$ span the λ eigenspace. In short, e_0, e_1, e_2, e_3 is a canonical frame as defined in Section 5.6. Then Lemma 5.9.5 (below) shows that in the type D case, with $\lambda = \lambda_2 = \lambda_3$, we have $\psi_2 = \lambda$. $\square$

This simplicity made it possible for William Kinnersley (1968) to enlarge the scattered collection of known type D Ricci-flat spacetimes to a complete classification. There are ten families, given locally by explicit coordinate formulas for the metric tensor. The approach is to assume that k, ℓ, x, y is a null tetrad field on such a spacetime M and to introduce a coordinate system such that k is a coordinate vector field. Then with the aid of a variety of coordinate transformations and changes of tetrad, all the Newman–Penrose equations—first order differential equations—are solved, giving explicit coordinate formulas for the tetrad and the metric tensor.

Here is an indication of how type D spacetimes relate to the generic case, type I.

Lemma 5.9.5 *Let M be a spacetime whose Weyl operator $\mathbb{C} \neq 0$ is diagonaliz-able, with eigenvalues $\lambda, \lambda_2, \lambda_3$; so M has either type I $(\lambda_1, \lambda_2, \lambda_3 \neq)$ or type D $(\lambda_2 = \lambda_3)$. For the associated tetrad field k, ℓ, x, y of a canonical frame field,*

$$\psi_0 = \psi_4 = \frac{1}{2}(\lambda_3 - \lambda_2), \qquad \psi_2 = \frac{1}{2}(\lambda_2 + \lambda_3) = -\frac{1}{2}\lambda_1, \qquad \psi_1 = \psi_2 = 0.$$

Proof. The components of the Weyl tensor are given in Lemma 5.6.2 in terms of a canonical frame e_0, e_1, e_2, e_3. The associated null tetrad field is given by

$$\sqrt{2}k = e_0 + e_1, \quad \sqrt{2}\ell = e_0 - e_1, \quad \sqrt{2}m = e_2 - ie_3, \quad \sqrt{2}\overline{m} = e_2 + ie_3,$$

hence its Weyl scalars can be found by routine calculation. For example, with $\lambda_j = a_j + ib_j$,

$$\psi_2 = C_{km\overline{m}\ell} = (1/4)C(e_0 + e_1, e_2 - ie_3, e_2 + ie_3, e_0 - e_1)$$
$$= (1/4)[C_{0220} + C_{0330} - C_{1221} - C_{1331} - iC_{1320} + iC_{1230} + iC_{0321} - iC_{0231}]$$
$$= (1/4)[a_2 + a_3 + a_3 + a_2 + i(b_2 + b_3 + b_3 + b_2)] = \frac{1}{2}(\lambda_2 + \lambda_3).$$

Since $\Sigma\lambda_j = 0$ this can also be written as $\psi_2 = -\frac{1}{2}\lambda_1$. $\square$

Note that for type I, since ψ_0 and ψ_4 are nonvanishing, the null vectors k and ℓ of this tetrad are not principal. A less symmetrical alternative would be to take k and ℓ in two of the four principal null directions, thus getting $\psi_0 = \psi_4 = 0$.

5.10   Goldberg–Sachs Theorem

The occurrence of shearfree geodesic null congruences in Ricci-flat spacetimes is completely described by a fundamental result due to J. N. Goldberg and R. K. Sachs (1962).

Theorem 5.10.1 *In a Ricci-flat (but not flat) spacetime, a null congruence is geodesic and shearfree if and only if it is a repeated ($m \geq 2$) principal null congruence of the Weyl tensor.*

Thus in a Ricci-flat (vacuum) spacetime M such congruences exist if and only if M is algebraically degenerate (not of Petrov type I). The following proof derives from the paper in which Newman and Penrose presented their formalism (1962). We separate the theorem into its two halves.

PART A. *If k is a repeated ($m \geq 2$) principal null vector field, then k is geodesic and shearfree.*

The proof is a straightforward application of Bianchi identities. If the multiplicity m of k is exactly 2 we have already shown that k is geodesic and shearfree. So suppose $m \geq 3$. Then by Lemma 5.8.7, $\psi_0 = \psi_1 = \psi_2 = 0$ for any tetrad $k, \ell, m, \overline{m}$, and the array in Lemma 5.8.6 simplifies to

	$k\ell$	km	$k\overline{m}$	ℓm	$\ell\overline{m}$	$m\overline{m}$
$k\ell$	0	0	0	$-\overline{\psi}_3$	$-\psi_3$	0
km		0	0	0	0	0
$k\overline{m}$			0	0	0	0
ℓm				$\overline{\psi}_4$	0	$-\overline{\psi}_3$
$\ell\overline{m}$					ψ_4	ψ_3
$m\overline{m}$						0

To involve ψ_3 but not $\psi_4 = C_{\ell\bar{m}\ell\bar{m}}$, we should avoid the remaining Bianchi identities that contain three vectors from ℓ, m, ℓ, m. In the large equation $B(k\ell m)(km)$ in the preceding section, replacing the second m by $\bar{m}$ gives, as $B(k\ell m)(k\bar{m})$:

$$k[C(\ell, m, k, \bar{m})] + \ell[C(m, k, k, \bar{m})] + m[C(k, \ell, k, \bar{m})]$$
$$= C(\nabla_k \ell, m, k, \bar{m}) + C(\ell, \nabla_k m, k, \bar{m}) + C(\ell, m, \nabla_k k, \bar{m}) + C(\ell, m, k, \nabla_k \bar{m})$$
$$+ C(\nabla_\ell m, k, k, \bar{m}) + C(m, \nabla_\ell k, k, \bar{m}) + C(m, k, \nabla_\ell k, \bar{m}) + C(m, k, k, \nabla_\ell \bar{m})$$
$$+ C(\nabla_m k, \ell, k, \bar{m}) + C(k, \nabla_m \ell, k, \bar{m}) + C(k, \ell, \nabla_m k, \bar{m}) + C(k, \ell, k, \nabla_m \bar{m}).$$

The abundant zeros in the Weyl tensor shrink this array to just

$$C(\ell, m, \nabla_k k, \bar{m}) + C(\ell, m, k, \nabla_k \bar{m}) + C(k, \ell, \nabla_m k, \bar{m}) = 0.$$

Corollary 5.8.2 shows these three terms to be

$$C(\ell, m, \nabla_k k, \bar{m}) = C(\ell, m, (\varepsilon + \bar{\varepsilon})k - \bar{\kappa}m - \kappa\bar{m}, \bar{m})$$
$$= -\bar{\kappa}\, C(\ell, m, m, \bar{m}) = \bar{\kappa}\psi_3,$$
$$C(\ell, m, k, \nabla_k \bar{m}) = C(\ell, m, k, \pi k - \bar{\kappa}\ell - (\varepsilon - \bar{\varepsilon})\bar{m}) = -\bar{\kappa}\, C(\ell, m, k, \ell) = \bar{\kappa}\psi_3,$$
$$C(k, \ell, \nabla_m k, \bar{m}) = C(k, \ell, (\bar{\alpha} + \beta)k - \bar{\rho}m - \sigma\bar{m}, \bar{m}) = 0.$$

Thus $2\bar{\kappa}\psi_3 = 0$, proving $\kappa = 0$ in this case.

Similarly, replacing the first k in the $B(k\ell m)(k\bar{m})$ equation above by $\bar{m}$ leads to a corresponding proof that $\sigma = 0$.

In the final case, only ψ_4 is nonzero and the Bianchi identities $B(k\ell m)(\ell \bar{m})$ and $B(k\bar{m}m)(\ell m)$ quickly give $\kappa = \sigma = 0$, thus proving Part A.   □

PART B. *If k is a shearfree geodesic null vector field, then k is a repeated principal null vector field.*

Since the result is local, there is a null tetrad $k, \ell, m, \bar{m}$ containing k. For it, $\kappa = \sigma = 0$. By Lemma 5.8.7, we must show $\psi_0 = \psi_1 = 0$. The first of these is immediate from the Sachs equation $k[\sigma] = (\rho + \bar{\rho})\sigma + \psi_0$. However, $\psi_2 = 0$ is not as easy.

According to Remark 5.8.4 a suitable boost annihilates the spin coefficient ε without affecting the vanishing of κ and σ. This simplifies matters, as follows.

Corollary 5.10.2 *On a Ricci-flat spacetime, if $k, \ell, m, \bar{m}$ is a tetrad field with $\psi_0 = 0$ and $\kappa = \varepsilon = \sigma = 0$, then*

Curvature Identities	*Bianchi Identities*
(3') $k[\tau] = (\tau + \bar{\pi})\rho + \psi_1$	(1') $k[\psi_1] = 4\rho\psi_1$
(4') $k[\beta] = \bar{\rho}\beta + \psi_1$	(2') $m[\psi_1] = 2(2\tau + \beta)\psi_1$
(5') $m[\rho] = (\bar{\alpha} + \beta)\rho + (\rho - \bar{\rho})\tau - \psi_1$	

Proof. The curvature identities are from equations (3), (4), (5) in Proposition 5.8.9, and the Bianchi identities are from Lemma 5.9.1.   □

Continuing with the proof of Part B, we look for further simplification of the tetrad field. A first try might be to arrange for it to be parallel along the curves of the congruence, but Corollary 5.8.2 shows that this only produces $\pi = 0$. Inspection of the formulas in Corollary 5.10.2 shows that it is more valuable to eliminate τ. This can be accomplished (somewhat indirectly) by considering the following two cases.

Case 1. $\rho = 0$ on an open set $\mathcal{U}$. Then equation (5′) shows at once that $\psi_1 = 0$ on $\mathcal{U}$.

Case 2. ρ is nonvanishing on an open set $\mathcal{V}$. If we prove this case, then by continuity $\psi_1 = 0$ holds on the support of ψ_1, that is, the closure of the set $\{p \in M | \rho(p) \neq 0\}$. Since the other points of M (if any) are covered by Case 1, the theorem is true.

Thus we can suppose that ρ *is never zero* on the entire spacetime. Then according to Remark 5.8.4, a Class I change of tetrad—which leaves $\kappa, \rho, \sigma, \varepsilon$ unchanged— replaces the spin coefficient τ by

$$\tau_1 = |f|^2 \kappa + \tau + \bar{f}\sigma + f\rho.$$

Since $\kappa = \sigma = 0$ and ρ is nonvanishing, taking $f = -\tau/\rho$ gives a new tetrad for which τ *is identically zero*. Thus equation (3′) reduces to $\psi_1 = -\bar{\pi}\rho$.

A different computation of ψ_1 leads to $\psi_1 = 0$. First, rewrite the Bianchi equations above as

$$k[\ln \psi_1] = 4\rho, \qquad m[\ln \psi_1] = 2\beta.$$

Now apply the bracket $[k, m] = km - mk$ to $\ln \psi_1$. Using equations (4′) and (5′) yields

$$[k, m](\ln \psi_1) = 2k(\beta) - 4m(\rho) = 2\bar{\rho}\beta - 4(\bar{\alpha} + \beta)\rho + 6\psi_1. \qquad (*)$$

Next we recompute $[k, m](\ln \psi_1)$, using the Newman–Penrose formula for $[k, m]$ that follows Corollary 5.8.2. Since $\kappa = \sigma = \varepsilon = 0$, this reduces to $[k, m] = (\bar{\pi} - \bar{\alpha} - \beta)k + \bar{\rho}m$, and we find

$$\begin{aligned}
[k, m](\ln \psi_1) &= (\bar{\pi} - \bar{\alpha} - \beta)k[\ln \psi_1] + \bar{\rho}m[\ln \psi_1] \\
&= 4(\bar{\pi} - \bar{\alpha} - \beta)\rho + 2\bar{\rho}\beta.
\end{aligned} \qquad (**)$$

Equating the right sides of equations (∗) and (∗∗) gives

$$2\bar\rho\beta - 4(\bar\alpha + \beta)\rho + 6\psi_1 = 4(\bar\pi - \bar\alpha - \beta)\rho + 2\bar\rho\beta,$$

hence $\psi_1 = (2/3)\bar\pi\rho$. But above we found $\psi_1 = -\bar\pi\rho$. Since ρ is never zero, these two equations imply $\pi = 0$, hence $\psi_1 = 0$. This completes the proof of Part B and hence of Theorem 5.10.1.

The Goldberg-Sachs theorem shows that shearfree geodesic null congruences are highly exceptional in Ricci-flat spacetimes; none can exist, even locally, in the generic Petrov type, I. Although all the algebraically special types have such a congruence, only type D has two. The theorem has been generalized by replacing Ric = 0 by suitable Ricci curvature conditions; see the Kundt–Thompson theorem in Kundt, Thompson (1962) and Robinson, Schild (1962).

In this section we have given ad hoc proofs for the particular Bianchi identitities that were needed. This is not in the spirit of the Newman–Penrose formalism, where *all* the formulas are listed, having been proved once and for all. To attack a particular problem, one starts by scanning the full array to find the relevant formulas and then, if possible, simplifing them by suitable tetrad changes.

For example, consider following question: Since Petrov type D derives from type I when two eigenvalues of the Weyl operator $\mathbb{C}$ are equal, say $\lambda_2 = \lambda_3$, should not another equally distinctive type result when $\lambda_2 = -\lambda_3$? The following negative answer in the Ricci-flat case was given by Carl Brans (1975). (Recall that trace $\mathbb{C} = 0$, so $\lambda_2 = -\lambda_3$ is equivalent to $\lambda_1 = 0$.)

Proposition 5.10.3 *Suppose that the Weyl tensor $\mathbb{C}$ of a Ricci-flat spacetime M is diagonalizable. If an eigenvalue of $\mathbb{C}$ vanishes, then M is flat.*

Proof.   Since $\mathbb{C}$ is diagonalizable, at each point, M has type I, D, or O (we do not assume uniformity over M). Since M is Ricci-flat, type O (conformally flat) is just flat. Type D is ruled out since its eigenvalues are never zero. Thus it remains to eliminate type I. So assume M has type I at a point p, hence on some neighborhood $\mathcal{U}$ of p (since the eigenvalues remain distinct). Let λ_1 be the vanishing eigenvalue, and let f be the smooth nonvanishing function $f = \lambda_3 = -\lambda_2$ on $\mathcal{U}$. Then Lemma 5.9.5 shows that there is a null tetrad $k, \ell, m, \bar{m}$ near p whose only nonzero Weyl scalars are $\psi_0 = \psi_4 = \frac{1}{2}(\lambda_3 - \lambda_2) = f$.

Among the Newman–Penrose Bianchi identities in Chadrasekhar (1983) we find

$$-k[\psi_3] + \bar{m}[\psi_2] = 2\lambda\psi_1 - 3\pi\psi_2 + 2(\varepsilon - \rho)\psi_3 + \kappa\psi_4 + \text{(Ricci) and}$$
$$\ell[\psi_2] - m[\psi_3] = 2\nu\psi_1 - 3\mu\psi_2 + 2(\beta - \tau)\psi_3 + \sigma\psi_4 + \text{(Ricci).}$$

Under the conditions above, these reduce to simply $\kappa f = \sigma f = 0$. Since f is nonzero on $\mathcal{U}$, $\kappa = \sigma = 0$. But the generalized Sachs equation (2) of Proposition 5.8.9 then implies $\psi_0 = 0$, that is, $f = 0$, a contradiction. $\square$

As illustrated in Section 5.9, the Newman–Penrose formalism has been particularly successful in the discovery and classification of locally defined spacetimes satisfying various natural conditions. In the collaborative work by Kramer et al., *Exact Solutions of Einstein's Field Equations* (1980), a vast collection of spacetimes is coherently organized on the basis of Petrov type and isometry groups, usually by means of special choices of tetrad fields.

In the opposite direction lies the detailed study of the physics of the very small number of spacetimes presently known to have physical significance. Here the Newman–Penrose formalism must compete with tensor calculus, the Cartan calculus, and a variety of other approaches. An outstanding case where the Newman–Penrose approach is taken as fundamental is S. Chandrasekhar's book, *The Mathematical Theory of Black Holes* (1983), which stresses the

remarkable fact that the black-hole solutions of general relativity are all of Petrov type D [which], therefore, enables their analysis in a null tetrad-frame in which the spin coefficients κ, σ, λ, ν and all the Weyl scalars, except ψ_2, vanish.

Finally, we note that because of the extremely formal character of the Newman–Penrose approach its calculations can often be carried out, or at least checked, by computer.

Appendix A
UNITS

Relativity is conveniently expressed in terms of a simple system of physical units, called *geometric units*, in which the speed c of light and the gravitational constant G are both set equal to the (unitless) number 1. In any conventional system of units, its various units can be reduced to powers of a single freely-choosen *base unit U*. A familiar instance is time in years and then distance in *(light) years*.

In geometric units, distance, time, and mass are always measured in the same units since $1 = c =$ distance/time, and Newton's laws give $|a| = |F|/m = GM/r^2$, hence distance/time^2 = mass/distance^2.

The conversion factors between a particular set of conventional units and geometric units derive from the value of c and G in those units. For example, in *cgs* units (using approximate values),

$$c_{cgs} = 3 \times 10^{10} \text{ cm/sec}, \quad \text{and} \quad G_{cgs} = 6.67 \times 10^{-8} \text{ cm}^3/(g \text{ sec}^2).$$

Taking centimeters as the basic unit gives 1 sec = 3×10^{10}cm, and then

$$1g = 6.67 \times 10^{-8} \text{ cm}^3/\text{sec}^2 = 7.4 \times 10^{-29} \text{ cm}.$$

For instance, the mass of our sun, about $2 \times 10^{33}g$, becomes 14.8×10^4cm, roughly 1.5 km. Thus our sun is not very relativistic. A black hole of one solar mass would have horizon radii $r_\pm < 2M = 3$ km; so the sun's bulk easily eliminates horizons.

Tensor fields can also have consistent units. It is easy to invent ones that do not, but naturally occuring tensors often do. Aside from their obvious physical significance, units are valuable as a check against gross computational errors (see below).

For Kerr spacetime we take the base unit U to be an (unspecified) distance $\sim$ mass $\sim$ time unit. Thinking of the Boyer–Lindquist coordinate vector field ∂_t as the operator d/dt shows that ∂_t has units U^{-1}. Newtonian formulas show that the Kerr parameter a (angular momentum per unit mass) has $a \sim U$. Angles and trigonometric functions are pure numbers since they are given by $U/U = U^0$. Hence $\Delta = r^2 - 2Mr + a^2$ and $\rho^2 = r^2 + a^2C^2$ both have units U^2.

Some examples of units are as follows:

U^4 $Q \sim \mathcal{K}$ (Carter constant)
U^2 Δ, ρ^2, g (metric), $q = ds^2, g_{\vartheta\vartheta}, g_{\varphi\varphi}, L$
U^1 $M, a, r, t, g_{t\varphi}, E, V$ (canonical vector field), τ (proper time)
U^0 $\vartheta, \varphi, \partial_\vartheta, \partial_\varphi, g_{rr}, g_{tt}, W, \Omega^i{}_j, C = \cos\vartheta, S = \sin\vartheta$
U^{-1} ∂_r, ∂_t
U^{-2} sectional (and Gaussian) curvature

Thus, for instance, we can see at once that the metric identity (m3) of Lemma 2.1, is *not* $ag_{\varphi\varphi} + (r^2 + a^2)g_{t\varphi} = \Delta S^2$.

Units can be altered by simplifying choices of parameters. The principal case occurs in Chapter 4, where material particles ($q = -m^2$) are studied in terms of test particles ($q = -1$). Orbits are not changed by this reparametrization but the units of first-integrals are. Energy E, for example, becomes energy per unit mass (specific energy), a unitless number. Specifically, $E \sim U^0$, $L \sim U^1$, $Q \sim \mathcal{K} \sim U^2$.

A more radical simplification formally sets the mass M of the black hole equal to 1, thus taking M as the geometric unit. The effect of the two simplifications, $m = M = 1$, can be found by considering the function

$$R(r) = (E^2 - m^2)r^4 + 2Mm^2r^3 + [a^2(E^2 - m^2) - L^2 - Q]r^2$$
$$+ 2M[Q + (L - aE)^2]r - a^2Q \qquad (A.1)$$

from Proposition 4.37. Dividing by m^2M^4 yields

$$\boldsymbol{R}(r) = (\boldsymbol{E}^2 - 1)\boldsymbol{r}^4 + 2\boldsymbol{r}^3 + [\boldsymbol{a}^2(\boldsymbol{E}^2 - 1) - \boldsymbol{L}^2 - \boldsymbol{Q}]\boldsymbol{r}^2$$
$$+ 2[\boldsymbol{Q} + (\boldsymbol{L} - \boldsymbol{aE})^2]\boldsymbol{r} - \boldsymbol{a}^2\boldsymbol{Q} \qquad (A.2)$$

provided we define the following (unitless) quantities:

$$\boldsymbol{a} = a/M, \quad \boldsymbol{r} = r/M, \quad \boldsymbol{E} = E/m, \quad \boldsymbol{L} = L/(mM), \quad \boldsymbol{Q} = Q/(m^2M^2).$$

In a lengthy computation equation (A.2) can be significantly easier to use than that of (A.1). However, to avoid ambiguity (or additional notation) we do not adopt $M = 1$ in the text proper, although it is a natural choice in numerical examples.

Appendix B
DIFFERENTIAL FORMS

Classically, a *differential form* θ of *degree* p on a smooth manifold is an anti-symmetric covariant tensor field of rank p. Equivalently, θ can be described as a cross section of a suitable vector bundle. A full treatment of differential forms is necessarily rather extensive, but our use of forms—though crucial—is limited to those of degree $p \le 2$, so a brief account should suffice. For further details, see, for example, Abraham, Marsden, Ratzu (1983) or Wald (1984). We always assume that forms are smooth, that is, C^∞.

- A zero-form is just a (smooth) real-valued function f on M;
- A one-form is, as defined in Section 1.1, a "covector field" on M.

In particular, the one-forms constitute a module $\mathfrak{X}^*(M)$ over the ring $\mathfrak{F}(M)$ of smooth functions. There are two main operations on forms: *wedge product* $\wedge$, and *exterior derivative d*.

- The wedge product of a zero-form f and a form ω is just $f\omega$, as usual.
- The wedge product of two one-forms is given by

$$(\theta \wedge \phi)(v, w) = \begin{vmatrix} \theta(v) & \theta(w) \\ \phi(v) & \phi(w) \end{vmatrix} \qquad \text{for all tangent vectors } v, w.$$

(There are two commonly used definitions of wedge product; the only evidence of our choice is the absence of a factor $1/2$ in the formula above). It follows that for *one-forms*, $\theta \wedge \phi = -\phi \wedge \theta$, and hence $\theta \wedge \theta = 0$.

- The *exterior derivative* of a zero-form f is just its differential df.
- The *exterior derivative* of a one-form ω is the two-form defined by

$$(d\omega)(X, Y) = X(\omega Y) - Y(\omega X) - \omega([X, Y]) \quad \text{for all vector fields } X, Y.$$

The differential has the familiar properties

$$d(\text{const}) = 0, \quad d(f + g) = df + dg, \quad d(fg) = g\,df + f\,dg,$$
$$d(F(f)) = F'(f)\,df.$$

The essential properties of the d on one-forms are easily proved:

$$d(\omega + \theta) = d\omega + d\theta, \quad d(f\omega) = df \wedge \omega + f\,d\omega, \quad d(df) = 0.$$

As shown in Section 1.1, a one-form ω can be written in terms of a coordinate system $x^1, \ldots, x^n$ as $\omega = \Sigma f_i\,dx^i$, where $f_i = \omega(\partial_i)$ for $i = 1, \ldots, n$. Then, since $d(dx^i) = 0$,

$$d\omega = \Sigma\, d(f_j dx^j) = \Sigma\, df_j \wedge dx^j = \Sigma\, (\partial f_j / \partial x^i)\, dx^i \wedge dx^j$$

and since differentials anticommute,

$$d\omega = \sum_{i<j} (\partial f_j / \partial x^i - \partial f_i / \partial x^j)\, dx^i \wedge dx^j.$$

The generalization of these results to forms of higher degree is simple in principle, but permutations complicate matters.

At one point in Chapter 2 we do need a form of degree 3, as follows: A nonvanishing vector field V on a semi-Riemannian manifold is *hypersurface-orthogonal* (a long but descriptive term) if the $(n - 1)$-dimensional distribution $V^\perp$ is integrable. There is a simple criterion for this, which is expressed in terms of the *metric dual* θ of V. (Invariantly, θ is the one-form such that $\theta(y) = <V, y>$ for all tangent vectors y; classically, θ_i results from lowering the contravariant index V^i of V.)

Lemma. *A vector field V is hypersurface-orthogonal if and only if $\theta \wedge d\theta = 0$, where θ is the metric dual of V.*

The proof derives from Frobenius' theorem (Section 1.7).

Appendix C
CARTER CONSTANT

In Section 4.2 the existence of the Carter constant is proved in the special case of a polar plane. We now apply the same scheme to the general case. In terms of Boyer–Lindquist coordinates the Lagrangian form of the ϑ geodesic equation for a Kerr geodesic is $(\rho^2 \vartheta')' = \mathrm{I} + \mathrm{II}$, where

$$\mathrm{I} = \tfrac{1}{2}\left[\partial_\vartheta g_{rr} r'^2 + \partial_\vartheta g_{\vartheta\vartheta} \vartheta'^2\right] \quad \text{and}$$
$$\mathrm{II} = \tfrac{1}{2}\partial_\vartheta g_{tt} t'^2 + \partial_\vartheta g_{t\varphi} t'\varphi' + \tfrac{1}{2}\partial_\vartheta g_{\varphi\varphi}\varphi'^2$$

The first summand I is the same as for the polar plane, namely, $-a^2 SC(r'^2/\Delta + \vartheta'^2)$. As before we can eliminate r' using the first integral $q = ds^2$, expressed now by Lemma 4.2.1. This yields

$$\mathrm{I} = -\rho^{-4} a^2 SC[q\rho^2 + \mathbb{P}^2/\Delta - \mathbb{D}^2/S^2].$$

The remaining terms II (nonexistent in the polar plane case) are not as tractable. However, the derivatives φ' and t' can be converted to r and ϑ using Proposition 4.1.5. Write $\mathrm{II} = \mathcal{A}t' + \mathcal{B}\varphi'$, where

$$\mathcal{A} = \tfrac{1}{2}[\partial_\vartheta g_{tt} t' + \partial_\vartheta g_{t\varphi}\varphi'], \quad \mathcal{B} = \tfrac{1}{2}[\partial_\vartheta g_{t\varphi} t' + \partial_\vartheta g_{\varphi\varphi}\varphi'].$$

Rather than compute all four of these partial derivative we differentiate the metric identities (m4) and (m3) in Lemma 2.1.1 to get

$$(r^2 + a^2)\partial_\vartheta g_{tt} + a\partial_\vartheta g_{t\varphi} = 0, \tag{C.1}$$

$$(r^2 + a^2)\partial_\vartheta g_{t\varphi} + a\partial_\vartheta g_{\varphi\varphi} = 2a\Delta SC. \tag{C.2}$$

Using equation (C.1) in $\mathcal{A}$ and equation (C.2) in $\mathcal{B}$ eliminates one partial derivative from each, giving

$$\mathcal{A} = (1/2a)\partial_\vartheta g_{tt}[at' - (r^2 + a^2)\varphi'],$$
$$\mathcal{B} = (1/2a)\partial_\vartheta g_{t\varphi}[at' - (r^2 + a^2)\varphi'] + \Delta SC\varphi'.$$

The common term $[at' - (r^2 + a^2)\varphi']$ is shown by Lemma 4.1.4 to be $-\mathbb{D}/S^2$, so

$$\text{II} = \mathcal{A}t' + \mathcal{B}\varphi' = -\mathbb{D}/(2aS^2)[\partial_\vartheta g_{tt}t' + \partial_\vartheta g_{t\varphi}\varphi'] + \Delta SC\varphi'^2.$$

Now we can now substitute for t' and φ' from Proposition 4.1.5. The term T within square brackets above then simplifies by the use of equation (C.1) to

$$T = \mathbb{D}/(S^2\rho^2)[aS^2\partial_\vartheta g_{tt} + \partial_\vartheta g_{t\varphi}].$$

Differentiating the metric identity (m2) gives

$$aS^2\partial_\vartheta g_{tt} + \partial_\vartheta g_{t\varphi} + 2aSCg_{tt} = -2aSC.$$

Hence

$$T = -\mathbb{D}/(S^2\rho^2)2aSC(1 + g_{tt}) = -\mathbb{D}/(S^2\rho^2)2aSC(2Mr/\rho^2)$$
$$= -4MarSC\mathbb{D}/(S^2\rho^4).$$

Inserting this into the expression above for II and substituting for φ' gives

$$\text{II} = \mathcal{A}t' + \mathcal{B}\varphi' = -\mathbb{D}/(2aS^2)\left[-4MarSC\mathbb{D}/(S^2\rho^4)\right]$$
$$+ \Delta SC[\Delta^{-1}a\mathbb{P}/\Delta + S^{-2}\mathbb{D}]^2/\rho^4$$
$$= 2Mr\rho^{-4}S^{-3}C\mathbb{D}^2 + \rho^{-4}SC[\Delta^{-1}a^2\mathbb{P}^2/\Delta + 2aS^{-2}\mathbb{P}\mathbb{D} + \Delta S^{-4}\mathbb{D}^2].$$

When this is added to $\text{I} = -a^2\rho^{-4}SC[q\rho^2 + \mathbb{P}^2/\Delta - \mathbb{D}^2/S^2]$, the $\mathbb{P}^2$ terms cancel leaving

$$\text{I} + \text{II} = \mathbb{D}^2\rho^{-4}S^{-3}C[2Mr + \Delta + a^2S^2] + 2\mathbb{P}\mathbb{D}aS^{-1}C\rho^{-4} - qa^2\rho^{-2}SC$$
$$= \mathbb{D}\rho^{-4}S^{-3}C[\mathbb{D}(r^2 + a^2 + a^2S^2) + 2aS^2\mathbb{P}] - qa^2\rho^{-2}SC.$$

The definitions of $\mathbb{P}$ and $\mathbb{D}$ show at once that $\mathbb{P} + a\mathbb{D} = E\rho^2$ so the square-bracketed term above is

$$\mathbb{D}(r^2 + a^2 + a^2 S^2) + 2aS^2[-a\mathbb{D} + E\rho^2] = \rho^2[\mathbb{D} + 2aS^2 E],$$

Hence, multiplication by ρ^2 gives

$$\rho^2(\rho^2\vartheta')' = \rho^2(\mathrm{I} + \mathrm{II}) = \mathbb{D}S^{-3}C[\mathbb{D} + 2aS^2 E] - qa^2 SC.$$

Since its right side depends only on ϑ, *this equation is integrable*, as in the polar case. One way to integrate it is to use $d\mathbb{D}/d\vartheta = -2aESC$ to get

$$\rho^2(\rho^2\vartheta')' = S^{-4}[\mathbb{D}^2 SC - S^2\mathbb{D}\, d\mathbb{D}/d\vartheta] - qa^2 SC.$$

Then comparison with $d(\mathbb{D}^2/S^2)/d\vartheta = 2S^{-4}[S^2\mathbb{D}\, d\mathbb{D}/d\vartheta - \mathbb{D}^2 SC]$ gives

$$2(\rho^2\vartheta')(\rho^2\vartheta')' = -(\mathbb{D}/S^2)' - 2qa^2 SC\vartheta'.$$

Hence

$$\rho^4\vartheta'^2 = -\mathbb{D}/S^2 + qa^2 C^2 + \mathcal{K}.$$

This completes the proof of Theorem 4.2.2 since, as noted in Section 4.2, each of its two equations implies the other.

Several other derivations are known, for example, Carter's Hamiltonian approach (Carter 1968) and the Newman–Penrose method (Walker, Penrose 1970).

Appendix D
EXTERIOR PRODUCTS

Let V be an n-dimensional vector space, over the real field, though this is not essential. For each integer p, $0 \le p \le n$, there is a vector space $\Lambda^p(V)$ called the pth *exterior product* of V. Intuitively, $\Lambda^p(V)$ consists of all linear combinations of p-tuple products $v_1 \wedge v_2 \wedge \ldots, \wedge v_p$ of vectors of V, where the *wedge product* $\wedge$ is bilinear, associative, and *alternating* on vectors of V (i.e., $w \wedge v = -v \wedge w$). Exceptionally, $\Lambda^0 V = \mathbf{R}$, and $\Lambda^1 V = V$. We record here two properties that uniquely characterize $\Lambda^p(V)$ and deduce some consequences needed in Chapter 5. For more details, see, for example, Abraham, Marsden, Ratiu (1983); Bishop, Goldberg (1980), and Warner (1983).

Universal Mapping Property. *Given any antisymmetric multilinear mapping A from $V \times \ldots, \times V$ (p factors) to a vector space W, there exists a unique linear transformation $\ell \colon \Lambda^p(V) \to W$ such that $\ell(v_1 \wedge v_2 \wedge \ldots, \wedge v_p) = A(v_1, v_2, \ldots, v_p)$ for all $v_1, v_2, \ldots, v_p \in V$.*

In general, elements of $\Lambda^p(V)$ are called *p-vectors*, but 2-vectors are *bivectors*, and since $\Lambda^1(V) = V$, 1-vectors are just *vectors*.

The alternation property of $\wedge$ implies $v \wedge v = 0$, and it follows as with one-forms that "repeats kill," that is, if a product $v_1 \wedge v_2 \wedge \ldots, \wedge v_p$ contains two identical vectors, then it is zero.

Canonical Basis Property. *If $e_1, \ldots, e_n$ is a basis for V, then the set of all p-vectors $e_{i_1} \wedge \ldots, e_{i_p}$ with $1 \leq i_1 < \ldots, < i_p \leq n$ is a basis for $\Lambda^p(V)$, said to be* canonical.

Thus, the dimension of $\Lambda^p(V)$ is the binomial coefficient $C(n, p)$. In particular, $\Lambda^n(V)$ is 1-dimensional, with basis $e_1 \wedge \ldots \wedge e_n$. Further properties of exterior products can be derived from the two above.

Property 1. *Vectors $v_1, v_2, \ldots, v_p$ in V are linearly independent if and only if $v_1 \wedge v_2 \wedge \ldots, \wedge v_p \neq 0$.*

Proof. If $\{v_1, v_2, \ldots, v_p\}$ is independent, then it is part of a basis for V. Thus, $v_1 \wedge v_2 \wedge \ldots, \wedge v_p$ (possibly reordered) is part of a canonical basis for $\Lambda^p(V)$, hence is nonzero. On the other hand, if $\{v_1, v_2, \ldots, v_p\}$ is dependent, then, say, $v_p = \Sigma_{i<p} c_i e_i$. Substituting this in $v_1 \wedge v_2 \wedge \ldots, \wedge v_p$ gives a linear combination of products all containing repeats, hence $v_1 \wedge v_2 \wedge \ldots, \wedge v_p = 0$.   □

Property 2. *If $0 \neq v \in V$ and α is a p-vector such that $v \wedge \alpha = 0$, then there is a $(p-1)$-vector γ such that $\alpha = v \wedge \gamma$.*

Proof. Express α in terms of the canonical basis derived from a basis $v, e_2, \ldots, e_n$ for V. Since $v \wedge \alpha = 0$, every basis p-vector in that basis must contain v as one factor. Extracting v from these p-vectors leaves the required α.   □

A *p-vector* that is a product $v_1 \wedge v_2 \wedge \ldots, \wedge v_p$ of vectors of V is said to be *decomposable.* We can picture it, roughly, as a parallelopiped in V whose edges are the vectors $v_1, v_2, \ldots, v_p$.

The wedge product can be extended in a natural way from vectors to p-vectors (use the universal property twice). The alternating property of the wedge on vectors produces *anticommutativity:* $\alpha \wedge \beta = (-1)^{pq} \beta \wedge \alpha$, where $\alpha \in \Lambda^p(V)$ and $\beta \in \Lambda^q(V)$. For example, bivectors commute with all p-vectors.

Property 3. *A bivector β is decomposable if and only if $\beta \wedge \beta = 0$.*

Proof. The wedge condition is certainly implied by decomposabilility. We prove the converse only for $n = \dim V \leq 4$. For $n \leq 3$, every bivector α has $\alpha \wedge \alpha = 0$ and is decomposable. This is trivial for $n < 3$. For $n = 3$, expressing α in terms of a canonical basis gives $\alpha \wedge \alpha = 0$. The map $x \to x \wedge \alpha$ from V to $\Lambda^3 V \approx \mathbf{R}$ is linear so there is an x such that $x \wedge \alpha = 0$. Thus, by Property 2, α is decomposable.

Now suppose $n = 4$, and let e_0, e_1, e_2, e_3 be a basis for V. Write $\beta = e_0 \wedge x + \alpha$, where α does not involve e_0. Then α is in $\Lambda^2 W$, where $W = \text{span}\{e_1, e_2, e_3\}$, so

α is decomposable, say $\alpha = v \wedge w$. Now $0 = \beta \wedge \beta = e_0 \wedge x \wedge v \wedge w$, so these vector are dependent. Thus β is in some $\Lambda^2 U$, where dim $U \leq 3$, hence is decomposable. □

Property 4. *A decomposable bivector $\delta = 0$ determines a unique 2-plane $P(\delta)$, namely span$\{v, w\}$, where $\delta = v \wedge w$.*

Proof. We must show that if $x \wedge y = v \wedge w$ then span$\{x, y\} = $ span$\{v, w\}$. Since $v \wedge w$ is nonzero there is a basis for V of the form $v, w, e_3, \ldots, e_n$. If say, x is not in span$\{v, w\}$, then its expression in terms of this basis involves a vector e_j, $j \geq 3$. But then $x \wedge v \wedge w \neq 0$, contradicting $x \wedge x \wedge y = 0$. □

It follows at once that nonzero decomposable bivectors β and δ are dependent (that is, $\beta = c\delta$) if and only if $P(\beta) = P(\delta)$.

Also, if x is an nonzero vector in $P(\delta)$ there is a vector $y \in P(\delta)$ such that $\delta = v \wedge w$. In fact, if $\delta = v \wedge w$, then $x = av + bw$. One of these coefficients must be nonzero, say a, then setting $y = w/a$ gives $x \wedge y = v \wedge w$.

Property 5.   *Decomposable bivectors β and δ such that $\beta \wedge \delta = 0$ share a vector; that is, there exist vectors u, v, w such that $\beta = u \wedge v$ and $\delta = u \wedge w$.*

Proof. Write $\beta = v \wedge w$ and $\delta = x \wedge y$. If $P(\beta) \cap P(\delta) = 0$, then the vectors v, w, x, y are independent, so $\beta \wedge \delta \neq 0$, which is impossible. If $P(\beta) = P(\delta)$, then x and y are linear combinations of v and w. This implies $x \wedge y = c(v \wedge w)$, so $\delta = v \wedge (cw)$. Finally, suppose $P(\beta) \cap P(\delta)$ is 1-dimensional, say $\mathbf{R}u$. Then, as noted earlier, there are vectors $v \in P(\beta)$ and $w \in P(\delta)$ such that $\beta = u \wedge v$ and $\delta = u \wedge w$. □

The alternation property of the wedge product produces exactly the signs needed in the classical formula for the determinant of a square matrix.

Property 6.   *If $v_i = \Sigma A_{ij} w_j$ for $1 \leq i, j \leq q$, then*

$$v_1 \wedge \ldots, \wedge v_q = (\det A) w_1 \wedge \ldots, \wedge w_q.$$

Thus, for example, a decomposable bivector δ determines not only a 2-dimensional subspace P (as in Property 4) but also an orientation of P, expressed, say, as the set of all bases $\{v, w\}$ for P such that $v \wedge w$ is a *positive* scalar multiple of δ. Evidently this property generalizes to arbitrary dimensions $p \leq $ dim V.

Considering four-dimensional manifolds for definiteness, we have defined M to be orientable if it possesses a continuous field ω of 4-vectors. In this case,

M is oriented by the choice of such a field, and at each point $p \in M$ those bases $e_0, \ldots, e_3$ such that $e_0 \wedge \ldots \wedge e_3 = \omega_p$ are positively oriented. In view of Property 6, a diffeomorphism of (oriented) spacetimes preserves [reverses] orientation provided its Jacobian matrix at a single point has positive [negative] determinant. Since spacetimes are connected and the determinant cannot vanish, the same sign is maintained at every point.

Index of Notations

Chapter number and section number are given for the first appearance. Generic notations (such as M for manifold) and ad hoc local notations are not included.

The abbreviations $S = \sin \vartheta$ and $C = \cos \vartheta$ are used throughout. Several lower case Greek letters are used exceptionally in the Newman–Penrose formalism (Sections 5.7 through 5.10 only).

Bibliography

Abraham, R., J. E. Marsden, and T. Ratiu, 1983. *Manifolds, Tensor Analysis, and Applications*, Addison-Wesley, Reading, MA.

Ashtekar, A., and R. O. Hansen, 1978. A Unified Treatment of Null and Spatial Infinity in General Relativity I, J. Math. Phys. 19, 1542–1566.

Bardeen, J. M., 1975. Timelike and null geodesics in the Kerr metric, in *Black Holes*, ed. by C. DeWitt and B. S. DeWitt, Gordon and Breach, New York.

Beem, J. K., and P. E. Ehrlich, 1981. *Global Lorentzian Geometry*, Dekker, New York.

Bishop, R. L., and S. Goldberg, 1980. *Tensor Analysis on Manifolds*, Dover, New York.

Boyer, R. H., and R. W. Lindquist, 1967. Maximal analytic extension of the Kerr metric, J. Math. Phys. 8, 265–281.

Brans, C. H., 1975. Some restrictions on algebraically general vacuum metrics, J. Math. Phys. 16, 1008–1010.

Carter, B., 1968. Global structure of the Kerr family of gravitational fields, Phys. Rev. 174, 1559–1571.

Carter, B., 1973. Properties of the Kerr metric, in *Black Holes*, ed. by C. DeWitt and B. S. DeWitt, Gordon and Breach, New York.

Chandrasekhar, S., 1983. *The Mathematical Theory of Black Holes*, Oxford Univ. Press, Oxford, England.

Choquet-Bruhat, Y., C. DeWitt-Morette, and M. Dillard-Bleick, 1977. *Analysis, Manifolds, and Physics*, North Holland, Amsterdam.

Debever, R., 1958. La super-énergie en relativité général, Bull. Soc. Math. Belg. 10, 112–147.

de Felice, F., 1968. Equatorial Geodesic Motion in the Gravitational Field of a Rotating Source, Nuovo Cim. B 57, 351–388.

de Felice, F., 1972. Orbital and Vortical Motion in the Kerr Metric, Nuovo Cim. 10 B, 447–458.

Dodson, C. T. J., and T. Poston, 1991. *Tensor Geometry*, 2nd ed., Springer, New York.

Eisenhart, L. P., 1926. *Riemannian Geometry*, Princeton Univ. Press, Princeton, NJ.

Ellis, G. F. R., and B. G. Schmidt, 1977. Singular Spacetimes, Gen. Rel. Grav. 8, 915–953.

Geroch, R. P., 1970. Singularities, in *Relativity*, ed. by S. Fickler, M. Carmeli, and L. Witten, Plenum Press, New York.

Geroch, R. P., A. Held, and R. Penrose, 1973. A space-time calculus based on pairs of null directions, J. Math. Phys. 14, 874–881.

Greenberg, M. J., and J.R. Harper, 1981. *Algebraic Topology*, Benjamin/Cummings, Menlo Park, CA.

Goldberg, J. N., and R. K. Sachs, 1962. A theorem on Petrov types, Acta. Phys. Polonica 22 (Supp. 13), 13–23.

Harpaz, A., 1992, *Concepts of the Theory of Relativity*, Jones and Bartlett, Boston.

Hawking, S. W., and G. F. R. Ellis, 1973. *The Large Scale Structure of Space-Time*, Cambridge Univ. Press, Cambridge, England.

Helgason, S., 1978. *Differential Geometry, Lie Groups, and Symmetric Spaces*, Academic Press, New York.

Helliwell, T. M., and A. J. Mallinckrodt, 1975. Null geodesics in the extended Kerr manifold, Phys. Rev. D 12, 2993–3003.

Kerr, R. P., 1963. Gravitational field of a spinning mass as an example of algebraically special metrics, Phys. Rev. Lett. II, 237–238.

Kinnersley, W., 1968. Type D vacuum metrics, J. Math. Phys. 10, 1195–1203.

Kobayashi, S., and K. Nomizu, 1963, 1969. *Foundations of Differential Geometry*, vol. 1, vol. 2, Wiley, New York.

Kramer, D., H. Stephani, E. Herlt, and M. A. H. MacCallum, 1980. *Exact Solutions of Einstein's Field Equations*, Cambridge Univ. Press, Cambridge, England.

Krivenko, O. P., K. A. Pyragas, and I. T. Zhuk, 1976. On the Second Integrals of Geodesics in the Kerr Field, J. Astrophysics and Space Science 40, 39–61.

Kruskal, M. D., 1960. Maximal extension of Schwarzschild metric, Phys. Rev. 119, 1743–1745.

Kundt, W., and A. Thompson, 1962. Le tenseur de Weyl et une congruence associée de géodésiques isotropes sans distorsion, C. R. Acad. Sci. (Paris) 254, 4257.

Mal'cev, A. I., 1963. *Foundations of Linear Algebra*, Freeman, San Francisco.

Massey, W. S., 1987. *Algebraic Topology: An Introduction*, Springer, New York.

Misner, C. W., K. S. Thorne, and J. A. Wheeler, 1973. *Gravitation*, Freeman, San Francisco.

Newman, E. T., et al, 1965. Metric of rotating, charged mass, J. Math. Phys. 6, 918–919.

Newman, E. T., and R. Penrose, 1962. An Approach to Gravitational Radiation by a Method of Spin Coefficients, J. Math. Phys. 3, 566–578; errata, 4, 998.

O'Neill, B., 1966. *Elementary Differential Geometry*, Academic Press, New York.

O'Neill, B., 1983. *Semi-Riemannian Geometry*, Academic Press, New York.

Pais, A., 1982. *Subtle is the Lord: The Science and Life of Albert Einstein*, Oxford Univ. Press, Oxford, England.

Penrose, R., 1969. Gravitational collapse: The role of general relativity, Nuovo Cim. 1, 252–271.

Petrov, A. Z., 1954. Sci. Trans. Kazan State Univ. 114, book 8, 55–69. This paper is summarized in Thorpe, 1969.

Pirani, F. A. E., 1962. Gravitational Radiation, in *Gravitation: An Introduction to Current Research*, ed. by L. Witten, Wiley, New York.

Redheffer, R., 1991. *Differential Equations: Theory and Applications*, Jones and Bartlett, Boston.

Robinson, D. C., 1977. A Simple Proof of the Generalization of Israel's Theorem, Gen. Rel. Grav. 8, 695–698.

Robinson, I, and A. Schild, 1962. Generalization of a theorem by Goldberg and Sachs, J. Math. Phys. 4, 484.

Sachs, R. K., 1962. Gravitational Waves in General Relativity VI: The Outgoing Radiation Condition, Proc. Roy. Soc. London A 264, 309–337 (1961), ibid. 270, 103–126.

Sachs, R. K., 1964. Gravitational Radiation, in *Relativity, Groups, and Topology*, ed. by DeWitt and DeWitt, Gordon and Breach, New York.

Sachs, R. K., and H. Wu, 1977. *General Relativity for Mathematicians*, Springer, New York.

Sharp, N. A., 1979. Geodesics in black hole space-times, Gen. Rel. Grav. 10, 659–670.

Stewart, J. M., and M. Walker, 1973. *Black Holes: The Outside Story*, Springer Tracts in Modern Physics, vol. 69, Springer, Berlin.

Steenrod, N., 1951. *Topology of Fibre Bundles*, Princeton Univ. Press, Princeton, NJ.

Stoghianidis, E. and D. Tsoubelis, 1987. Polar Orbits in the Kerr Space-Time, Gen. Rel. Grav. 19, 1235–1249.

Thorpe, J. A., Curvature and the Petrov Canonical Forms, J. Math. Phys. 10, 1–7.

Tipler, F. J., C. J. S. Clark, and G. F. R. Ellis, 1980. Singularities and Horizons, In *General Relativity and Gravitation: one hundred years after the birth of Albert Einstein*, vol. 2, ed. by A. Held, Plenum Press.

Vishveshvara, C. V., 1968. Generalization of the Schwarzschild Surface to arbitrary static and stationary Metrics, J. Math. Phys. 9, 1319–1322.

Wald, R., 1984. *General Relativity*, Univ. Chicago Press, Chicago, IL.

Walker, M. and R. Penrose, 1970. On Quadratic First Integrals of the Geodesic Equuations for Type [22] Spacetimes, Commun. Math. Phys. 18, 265–274.

Warner, F., 1983. *Foundations of Differentiable Manifolds and Lie Groups*, Springer, New York.

Wilkins, D. C., 1972. Bound Geodesics in the Kerr Metric, Phys. Rev. D 5, 814–822.

Index